# Graphs & Digraphs

*Graphs & Digraphs, Seventh Edition* masterfully employs student-friendly exposition, clear proofs, abundant examples, and numerous exercises to provide an essential understanding of the concepts, theorems, history, and applications of graph theory.

This classic text, widely popular among students and instructors alike for decades, is thoroughly streamlined in this new, seventh edition, to present a text consistent with contemporary expectations.

Changes and updates to this edition include:
- A rewrite of four chapters from the ground up.
- Streamlining by over a third for efficient, comprehensive coverage of graph theory.
- Flexible structure with foundational Chapters 1–6 and customizable topics in Chapters 7–11.
- Incorporation of the latest developments in fundamental graph theory.
- Statements of recent groundbreaking discoveries, even if proofs are beyond scope.
- Completely reorganized chapters on traversability, connectivity, coloring, and extremal graph theory to reflect recent developments.

The text remains the consummate choice for an advanced undergraduate level or introductory graduate-level course exploring the subject's fascinating history, while covering a host of interesting problems and diverse applications.

Our major objective is to introduce and treat graph theory as the beautiful area of mathematics we have always found it to be. We have striven to produce a reader-friendly, carefully written book that emphasizes the mathematical theory of graphs, in all their forms. While a certain amount of mathematical maturity, including a solid understanding of proof, is required to appreciate the material, with a small number of exceptions this is the only pre-requisite.

In addition, owing to the exhilarating pace of progress in the field, there have been countless developments in fundamental graph theory ever since the previous edition, and many of these discoveries have been incorporated into the book. Of course, some of the proofs of these results are beyond the scope of the book, in which cases we have only included their statements. In other cases, however, these new results have led us to completely reorganize our presentation. Two examples are the chapters on coloring and extremal graph theory.

# Textbooks in Mathematics

Series editors:
Al Boggess, Kenneth H. Rosen

**Modeling Change and Uncertainty**
Machine Learning and Other Techniques
*William P. Fox and Robert E. Burks*

**Abstract Algebra**
A First Course, Second Edition
*Stephen Lovett*

**Multiplicative Differential Calculus**
*Svetlin Georgiev, Khaled Zennir*

**Applied Differential Equations**
The Primary Course
*Vladimir A. Dobrushkin*

**Introduction to Computational Mathematics: An Outline**
*William C. Bauldry*

**Mathematical Modeling the Life Sciences**
Numerical Recipes in Python and MATLAB™
*N. G. Cogan*

**Classical Analysis**
An Approach through Problems
*Hongwei Chen*

**Classical Vector Algebra**
*Vladimir Lepetic*

**Introduction to Number Theory**
*Mark Hunacek*

**Probability and Statistics for Engineering and the Sciences with Modeling using R**
*William P. Fox and Rodney X. Sturdivant*

**Computational Optimization: Success in Practice**
*Vladislav Bukshtynov*

**Computational Linear Algebra: with Applications and MATLAB® Computations**
*Robert E. White*

**Linear Algebra With Machine Learning and Data**
*Crista Arangala*

**Discrete Mathematics with Coding**
*Hugo D. Junghenn*

**Applied Mathematics for Scientists and Engineers**
*Youssef N. Raffoul*

**Graphs & Digraphs, Seventh Edition**
*Gary Chartrand, Heather Jordon, Vincent Vatter, and Ping Zhang*

https://www.routledge.com/Textbooks-in-Mathematics/book-series/CANDHTEXBOOMTH

# Graphs & Digraphs
## Seventh Edition

Gary Chartrand, Heather Jordon,
Vincent Vatter, and Ping Zhang

## CRC Press
Taylor & Francis Group
Boca Raton  London  New York

CRC Press is an imprint of the
Taylor & Francis Group, an **informa** business

A CHAPMAN & HALL BOOK

Seventh edition published 2024
by CRC Press
2385 NW Executive Center Drive, Suite 320, Boca Raton FL 33431

and by CRC Press
4 Park Square, Milton Park, Abingdon, Oxon, OX14 4RN

*CRC Press is an imprint of Taylor & Francis Group, LLC*

First edition published by Prindle, Weber and Schmidt 1979
Second edition published by Taylor and Francis 1986
Third edition published by Taylor and Francis 1996
Fourth edition published by Taylor and Francis 2004
Fifth edition published by Taylor and Francis 2010
Sixth edition published by Taylor and Francis 2015

**Library of Congress Cataloging-in-Publication Data**

Names: Chartrand, Gary, author. | Jordon, Heather, author. | Vatter,
    Vincent, author. | Zhang, Ping, 1957- author.
Title: Graphs and digraphs / Gary Chartrand, Heather Jordon, Vincent
    Vatter, Ping Zhang.
Description: Seventh edition. | Boca Raton : Taylor and Francis, 2024. |
    Includes bibliographical references and index.
Identifiers: LCCN 2023026342 | ISBN 9781032133409 (hardback) | ISBN
    9781032606989 (paperback) | ISBN 9781003461289 (ebk)
Subjects: LCSH: Graph theory. | Directed graphs.
Classification: LCC QA166 .C4525 2024 | DDC 551/.5--dc23/eng/20231010
LC record available at https://lccn.loc.gov/2023026342

ISBN: 978-1-032-13340-9 (hbk)
ISBN: 978-1-032-60698-9 (pbk)
ISBN: 978-1-003-46128-9 (ebk)

DOI: 10.1201/9781003461289

Typeset in Latin Modern font
by KnowledgeWorks Global Ltd.

*Publisher's note:* This book has been prepared from camera-ready copy provided by the authors.

# Contents

# Preface to the seventh edition

The field of graph theory has seen tremendous advancement since the first edition of *Graphs & Digraphs* was published in 1979. This progress is reflected in the vast range of books available on the subject, from introductory texts for beginners to advanced works focusing on specialized topics.

Our primary aim with *Graphs & Digraphs* has always been to introduce and develop graph theory as the captivating area of mathematics that we find it to be. This seventh edition remains accessible to students with a solid understanding of mathematical proofs, but its structure will reveal additional depth for those with a more advanced mathematical background. We have carefully crafted it to be suitable for courses taught at both undergraduate and graduate levels. Recognizing the diversity of our readers, this edition has been designed for a one- or two-semester course that could be paced according to the course's needs.

The seventh edition has been significantly streamlined, reducing its size by over one-third for a comprehensive yet efficient coverage of graph theory. This consolidation brings the book closer to the concise forms of the third and fourth editions. The transformation also marks our pleasure in welcoming two new authors, Heather Jordon and Vincent Vatter, who have brought fresh perspectives to the text, while Gary Chartrand and Ping Zhang have begun to step back from textbook writing.

The first six chapters serve as a foundation, with the subsequent chapters offering customizable topics. The substantial reduction from 21 chapters in the sixth edition to 11 in this edition reflects our aim to compile related material for improved cohesion and efficiency. In this light, no chapter from the sixth edition remains unchanged. For example, the chapter on embeddings has been reintroduced as Chapter 10, and the chapter on graphs and algebra (now Chapter 11) has been expanded.

In terms of potential syllabi, the book's core is encapsulated in Chapters 1 through 6, covering traversability, connectivity, planarity, and coloring, after a thorough introduction to graphs and then digraphs. Beyond the core, the order in which Chapters 7 through 11 are taught can be customized according to the course's requirements. While it would take a full-year course to cover the entire book, we recommend Chapters 1 through 6 for a single semester course. Depending on the pace of the course, instructors might find time to cover one or more of the additional chapters. Instructors who wish for a faster trip through the core of the book may omit Sections 2.4, 3.4, 3.5, 5.3, and 5.4 without significant consequence.

Many new findings since the sixth edition of this text was published have been incorporated into this edition, providing statements of these novel theories even when their proofs might exceed the scope of this text. Such recent advancements led to a complete reorganization of our presentation, most notably in the chapters on coloring and extremal graph theory. Visual representation of concepts has been improved as well. Nearly all of the figures previously drawn in PicTeX have been redrawn in Ti$k$Z for enhanced clarity.

In conclusion, we believe that this thoroughly revised seventh edition will serve as an effective introduction to graph theory, while also catering to those seeking a more advanced exploration of the subject. We hope that the extensive changes and new organization of this edition will help facilitate your understanding and appreciation of this vibrant area of mathematics.

We continue to be grateful to Bob Ross, senior editor of CRC Press, for his support and assistance throughout the writing process. We also thank Michael Albert, Brian Alspach, Teegan Bailey, Robert Brignall, Brian Cai, Brandon Duebel, Saad El-Zanati, Michael Engen, Zachary Hamaker, Madeline Hastie, Michael Jones, Sean Mandrick, Zachary Melson, Jay Pantone, Andrew Penton, Andrew Vince, and Gjergji Zaimi for their numerous and diverse contributions to this edition of the text.

— Gary Chartrand, Heather Jordon, Vincent Vatter, and Ping Zhang

# About the authors

**Gary Chartrand** is professor emeritus at Western Michigan University. He was awarded all three of his university degrees in mathematics from Michigan State University. His main interest in mathematics has always been graph theory.

He has authored or co-authored over 300 research articles in graph theory. He served as the first managing editor of the *Journal of Graph Theory* (for seven years) and was a member of the editorial boards of the *Journal of Graph Theory* and *Discrete Mathematics*. He served as a vice president of the Institute of Combinatorics and Its Applications. He directed the dissertations of 22 doctoral students at Western Michigan University.

He is the recipient of the University Distinguished Faculty Scholar Award and the Alumni Association Teaching Award from Western Michigan University and the Distinguished Faculty Award from the State of Michigan. He also received an award as managing editor of the best new journal (*Journal of Graph Theory*) by the Association of American Publishers in the scientific, medical, and technical category.

**Heather Jordon** earned her PhD in mathematics from Western Michigan University in 1996 under the direction of Gary Chartrand. She is currently an Associate Editor for *Mathematical Reviews*, produced by the American Mathematical Society.

**Vincent Vatter** earned his PhD in mathematics from Rutgers University in 2006, studying under Doron Zeilberger. Prior to that, he received his bachelor's degree in mathematics from Michigan State University in 2001. Currently, he is a professor of mathematics at the University of Florida, where he resides with his two daughters, Madison and Vienna. He has authored or co-authored over 60 research articles in enumerative combinatorics, graph theory, order theory, and theoretical computer science, and has directed the dissertations of five doctoral students.

**Ping Zhang** earned her PhD in mathematics from Michigan State University. After spending a year at the University of Texas at El Paso, she joined Western Michigan University, where she currently serves as a professor. In 2017, she was named a Distinguished University Faculty Scholar. Her primary research interests are in algebraic combinatorics and graph theory.

Dr. Zhang has co-authored six textbooks, notably *Graphs & Digraphs and Chromatic Graph Theory*, and is a co-editor of *The Handbook of Graph Theory*,

*Second Edition*, all published by CRC Press. She has also authored or co-authored over 340 research articles and given more than 80 talks at various universities and conferences. At Western Michigan University, she has directed the dissertations of 26 doctoral students.

# Graphs

We begin our study of graphs and digraphs by introducing many of the basic concepts that we shall encounter throughout this book.

## 1.1    Fundamentals

A **graph** $G$ is a finite nonempty set $V$ of objects called **vertices** (the singular is **vertex**) together with a possibly empty set $E$ of 2-element subsets of $V$ called **edges**. To indicate that a graph $G$ has **vertex set** $V$ and **edge set** $E$, we write $G = (V, E)$. To emphasize that $V$ and $E$ are the vertex set and edge set of a graph $G$, we often write $V$ as $V(G)$ and $E$ as $E(G)$. Each edge $\{u, v\}$ of $G$ is usually denoted by $uv$ or $vu$. If $e = uv$ is an edge of $G$, then $e$ is said to **join** $u$ and $v$.

    A graph $G$ can be represented by a diagram, where each vertex is represented by a point or small circle and an edge joining two vertices is represented by a line segment or curve joining the corresponding points in the diagram. It is customary to refer to such a diagram as the graph $G$ itself. In addition, the points in the diagram are referred to as the vertices of $G$ and the line segments are referred to as the edges of $G$. For example, the graph $G$ with vertex set $V(G) = \{u, v, w, x, y\}$ and edge set $E(G) = \{uv, uy, vx, vy, wy, xy\}$ is shown in Figure 1.1. Even though the edges $vx$ and $wy$ cross in Figure 1.1, their point of intersection is not a vertex of $G$. If $e = uv$ is an edge of a graph $G$, then $u$ and $v$ are **adjacent vertices** while $u$ and $e$ are **incident**, as are $v$ and $e$. If $uv$ and $vw$ are distinct edges, then $uv$ and $vw$ are **adjacent edges**.

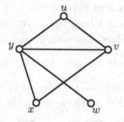

**Figure 1.1.** A graph

Two adjacent vertices are referred to as **neighbors** of each other. The set of neighbors of a vertex $v$ is called the **open neighborhood** of $v$ (or simply the **neighborhood** of $v$) and is denoted by $N_G(v)$, or $N(v)$ if the graph $G$ is understood. The set $N[v] = N(v) \cup \{v\}$ is called the **closed neighborhood** of $v$.

For the graph $G$ of Figure 1.1, the vertices $u$ and $v$ are adjacent while the vertices $u$ and $x$ are not adjacent; the edges $uv$ and $vx$ are adjacent while the edges $vx$ and $wy$ are not adjacent. The vertex $v$ is incident with the edge $uv$ but is not incident with the edge $wy$. The neighborhood of $v$ is $N(v) = \{u, x, y\}$.

The number of vertices in a graph $G$ is called the **order** of $G$ and the number of edges is called the **size** of $G$. The order of the graph $G$ of Figure 1.1 is 5 and its size is 6. We typically use $n$ and $m$ for the order and size, respectively, of a graph. The **degree of a vertex $v$** in a graph $G$ is the number of vertices that are adjacent to $v$. Equivalently, the degree of $v$ is the number of edges incident with $v$. The degree of a vertex $v$ is denoted by $\deg_G v$ or, more simply, by $\deg v$ if the graph $G$ under discussion is clear. A vertex of degree 0 is referred to as an **isolated vertex** and a vertex of degree 1 is an **end-vertex**. The largest degree among the vertices of $G$ is called the **maximum degree** of $G$ and is denoted by $\Delta(G)$. The **minimum degree** of $G$ is denoted by $\delta(G)$. Thus, if $v$ is a vertex of a graph $G$ of order $n$, then

$$0 \leq \delta(G) \leq \deg v \leq \Delta(G) \leq n - 1.$$

For the graph $G$ of Figure 1.1,

$$\deg w = 1, \deg u = \deg x = 2, \deg v = 3, \text{and} \deg y = 4.$$

Thus, $\delta(G) = 1$ and $\Delta(G) = 4$. Note that the sum of the degrees of the graph $G$ of Figure 1.1 is 12, which is exactly twice its size. That this sum always equals twice the size is a useful observation, often referred to as the first theorem of graph theory, with interesting consequences as we shall see.

**Theorem 1.1 (First theorem of graph theory).** *If $G$ is a graph of size $m$, then*

$$\sum_{v \in V(G)} \deg v = 2m.$$

*Proof.* When summing the degrees of the vertices of $G$, each edge is counted twice, once for each of its two incident vertices. ■

A vertex in a graph $G$ is **even** or **odd**, according to whether its degree is even or odd. The graph $G$ of Figure 1.1 has three even vertices and two odd vertices. It turns out that every graph must have an even number of odd vertices.

**Corollary 1.2.** *Every graph has an even number of odd vertices.*

*Proof.* Suppose that $G$ is a graph of size $m$. Let $W$ be the set of odd vertices of $G$ and let $U$ be the set of even vertices of $G$. By Theorem 1.1,

$$\sum_{v \in V(G)} \deg v = \sum_{v \in W} \deg v + \sum_{v \in U} \deg v = 2m.$$

Certainly $\sum_{v \in U} \deg v$ is even and since $\sum_{v \in V(G)} \deg v$ is even as well, it follows that $\sum_{v \in W} \deg v$ must also be even. Hence, $|W|$ must be even and thus $G$ has an even number of odd vertices. ■

# 1.2 Isomorphism

Two graphs often have the same structure, differing only in the way their vertices and edges are labeled or in the way they are drawn. For example, there is only one graph of order 1, two graphs of order 2, four graphs of order 3 and 11 graphs of order 4. All 18 of these graphs are shown in Figure 1.2. To make this idea more precise, a mathematical formulation for when two graphs are the same, called isomorphic, is required.

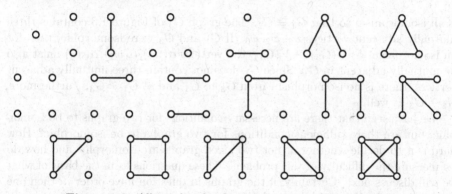

**Figure 1.2.** All graphs of order 4 or less, up to isomorphism

Two graphs $G$ and $H$ are **isomorphic** if there is a bijection $\phi\colon V(G) \to V(H)$ such that vertices $u$ and $v$ are adjacent in $G$ if and only if $\phi(u)$ and $\phi(v)$ are adjacent in $H$. The function $\phi$ is called an **isomorphism** from $G$ to $H$, and we write $G \cong H$. If there is no such function $\phi$ as described above, then $G$ and $H$ are **nonisomorphic graphs** and we write $G \ncong H$. It is easy to see that 'is isomorphic to' is an equivalence relation on the set of all graphs (see Exercise 1.8). Hence, this relation divides the set of all graphs into equivalence classes, where two graphs are nonisomorphic if they belong to different equivalence classes.

Suppose that two graphs $G$ and $H$ are isomorphic. Then it must be the case that $G$ and $H$ have the same order and size, and the degrees of the vertices of $G$ are exactly the degrees of the vertices of $H$ (see Exercise 1.9). However, these conditions are not sufficient: two graphs can have the same order, size, and degrees and yet fail to be isomorphic. The next example illustrates this.

Consider the graphs $G_1$, $G_2$, and $G_3$ of Figure 1.3. Each graph has order 6, size 9, and every vertex has degree 3, yet $G_1$ and $G_3$ are not isomorphic while $G_1$ and $G_2$ are.

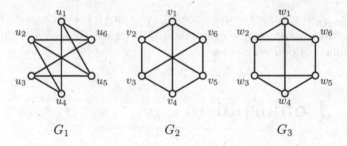

**Figure 1.3.** Isomorphic and nonisomorphic graphs

The function $\phi\colon V(G_1) \to V(G_2)$ defined by

$$\phi(u_1) = v_1, \phi(u_2) = v_3, \phi(u_3) = v_5,$$
$$\phi(u_4) = v_2, \phi(u_5) = v_4, \phi(u_6) = v_6$$

is an isomorphism so that $G_1 \cong G_2$. The graph $G_3$ of Figure 1.3 contains three mutually adjacent vertices $w_1, w_2, w_6$. If $G_1$ and $G_3$ were isomorphic, then for an isomorphism $\alpha\colon V(G_3) \to V(G_1)$, the vertices $\alpha(w_1), \alpha(w_2), \alpha(w_6)$ must also be mutually adjacent in $G_1$. Since $G_1$ does not contain three mutually adjacent vertices, there is no isomorphism from $G_3$ to $G_1$ and so $G_1 \not\cong G_3$. Furthermore, $G_2 \not\cong G_3$ as well.

We have seen that there are necessary conditions for two graphs to be isomorphic, but are there sufficient conditions for two graphs to be isomorphic? How hard is it to decide whether or not two given graphs are isomorphic and how do we measure the difficulty of this problem? These questions form the basis of what we will discuss next. Certainly, if the graphs in question have order $n$, then one approach might be to check all $n!$ bijections between their vertex sets to see if one (or more) is an isomorphism. If some bijection is an isomorphism, we would answer *yes*, while if none of the $n!$ bijections is an isomorphism, we would answer *no*. This does not seem like a very efficient approach as $n!$ grows rapidly as $n$ grows, and it may be the case that we have to check *all* $n!$ bijections if the graphs in question are in fact nonisomorphic.

To decide if an algorithm is efficient or not, we determine its complexity. The **complexity** of an algorithm refers to the number of basic computational steps (ordinarily the arithmetic operations and comparisons) required to execute the algorithm in the worst case. This number usually depends on the nature of the input as well as the size of input. For a graph of order $n$ and size $m$, the complexity typically depends on $n$ and/or $m$. So, the algorithm described above of enumerating all the bijections between two vertex sets with $n$ vertices already has complexity at least $n!$, and that is not including the computational steps required for checking whether a given bijection is indeed an isomorphism. *Efficient* algorithms are those whose complexity is polynomial in terms of the input size and are called **polynomial-time** algorithms. For example, if the input is a graph of order $n$ and the complexity of an algorithm is at most $cn^p$

for some constant $c$ and positive integer $p$, then the algorithm is a polynomial-time algorithm and we denote the complexity by $O(n^p)$. Thus, the approach, or algorithm, of enumerating all $n!$ bijections and checking whether any one of them is an isomorphism is not an efficient algorithm.

A **decision problem** is a question that can be answered *yes* or *no*. The decision problem of deciding whether or not two given graphs are isomorphic is called the **graph isomorphism problem**. The class of all decision problems with polynomial-time algorithms solving every instance of the problem is denoted by **P**. At this time, it is not known whether the graph isomorphism problem belongs to P.

The class of all decision problems for which an instance of the problem can be verified in polynomial time is called **NP**, which stands for **nondeterministic polynomial-time**. One way to think about the problems in NP is that they are difficult to solve in general, but it is easy to determine whether a given instance is a solution to the problem. For example, we can easily verify, in polynomial time, whether a given bijection between the vertex sets of two graphs is in fact an isomorphism, but in general it is very difficult to decide if two graphs are isomorphic. Thus, the graph isomorphism problem belongs to the class NP.

The problems in NP and the problems in P have one property in common: given a solution to a problem in either case, the solution can be verified in polynomial time. Thus, P $\subseteq$ NP. One of the best known open problems in mathematics asks whether every problem in NP is also in P, that is, if P = NP. This problem is one of the seven Millennium Prize Problems selected by the Clay Mathematics Institute, each of which carries a \$1,000,000 prize for the first correct solution, and is considered by many to be *the* most important problem in theoretical computer science.

In order to answer this question of whether P = NP, it is not necessary to find a polynomial-time algorithm for *every* problem in NP but to find a polynomial-time algorithm for *one* of the problems in a subset of problems in NP, called NP-complete; that is, a problem in NP is **NP-complete** if a polynomial-time algorithm for a solution to the problem would result in polynomial-time solutions for all problems in NP. The NP-complete problems are among the most difficult in the set NP and can be reduced from and to all other NP-complete problems in polynomial time. The concept of NP-completeness was initiated in 1971 by Cook [55] who gave an example of the first NP-complete problem. The following year Karp [139] described some 20 diverse NP-complete problems, one of which we will study in Chapter 3.

The graph isomorphism problem is one of a small number of natural algorithmic problems with unsettled complexity status: it is not expected to be NP-complete, as there are many classes of graphs for which deciding whether two graphs from a given class are isomorphic can be done in polynomial time, but it is not yet known to be solvable in polynomial time in general. In 2016, Babai [14] (see also [15]) showed that the graph isomorphism problem can be decided in quasipolynomial time. Quasipolynomial time is slower than polynomial time but not as slow as exponential time; that is, for an input of size $n$, quasipolynomial time means that the worst case running time is $2^{O((\log n)^c)}$. Babai used the fact that the graph isomorphism problem reduces to the string isomorphism problem, pointed out by Luks [160] by encoding each graph as a sequence of 0s and 1s depending on which edges are present, and the Classification of Finite Simple Groups to show that the string isomorphism problem is decidable in quasipolynomial time, and hence so is the graph isomorphism problem. As a result of Babai's work, many people now believe that the graph isomorphism problem is closer to being in P than being NP-complete.

## Subgraphs

A graph $H$ is a **subgraph** of a graph $G$ if $V(H) \subseteq V(G)$ and $E(H) \subseteq E(G)$, in which case we write $H \subseteq G$. If $V(H) = V(G)$, then $H$ is a **spanning subgraph** of $G$. If $H$ is a subgraph of a graph $G$ with either $V(H)$ a proper subset of $V(G)$ or $E(H)$ a proper subset of $E(G)$, then $H$ is a **proper subgraph** of $G$.

Figure 1.4 shows six graphs, namely $G$ and the graphs $G_i$ for $i = 1, 2, \ldots, 5$. All six of these graphs are proper subgraphs of $G$, except $G$ itself and $G_1$. Although $G$ is a subgraph of itself, it is not a proper subgraph of $G$. The graph $G_1$ contains the edge $uz$, which is not an edge of $G$ and so $G_1$ is not a subgraph of $G$. The graph $G_3$ is a spanning subgraph of $G$ since $V(G_3) = V(G)$.

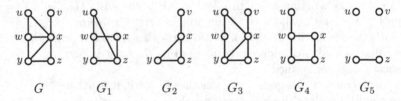

**Figure 1.4.** Graphs and subgraphs

For a nonempty subset $S$ of $V(G)$, the **subgraph $G[S]$ of $G$ induced by $S$** has $S$ as its vertex set and two vertices $u$ and $v$ are adjacent in $G[S]$ if and only if $u$ and $v$ are adjacent in $G$. A subgraph $H$ of a graph $G$ is called an **induced subgraph** if there is a nonempty subset $S$ of $V(G)$ such that $H = G[S]$. Thus $G[V(G)] = G$. For a nonempty set $X$ of $E(G)$, the **subgraph $G[X]$ induced by $X$** has $X$ as its edge set, and a vertex $v$ belongs to $G[X]$ if $v$ is incident with at least one edge in $X$. A subgraph $H$ of $G$ is **edge-induced** if there is a nonempty subset $X$ of $E(G)$ such that $H = G[X]$. Thus, $G[E(G)] = G$ if and only if $G$ has no isolated vertices.

Once again, consider the graphs shown in Figure 1.4. Since $xy \in E(G)$ but $xy \notin E(G_4)$, the subgraph $G_4$ is not an induced subgraph of $G$. On the other hand, the subgraphs $G_2$ and $G_5$ are both induced subgraphs of $G$. Indeed, for $S_1 = \{v, x, y, z\}$ and $S_2 = \{u, v, y, z\}$, we have $G_2 = G[S_1]$ and $G_5 = G[S_2]$. The subgraph $G_4$ of $G$ is edge-induced; in fact, if $X = \{uw, wx, wy, xz, yz\}$, then $G_4 = G[X]$.

For a vertex $v$ and an edge $e$ in a nonempty graph $G = (V, E)$, the subgraph $G - v$, obtained by deleting $v$ from $G$, is the induced subgraph $G[V - \{v\}]$ of $G$ and the subgraph $G - e$, obtained by deleting $e$ from $G$, is the spanning subgraph of $G$ with edge set $E - \{e\}$. More generally, for a proper subset $U$ of $V$, the graph $G - U$ is the induced subgraph $G[V - U]$ of $G$. For a subset $X$ of $E$, the graph $G - X$ is the spanning subgraph of $G$ with edge set $E - X$. If $u$ and $v$ are distinct nonadjacent vertices of $G$, then $G + uv$ is the graph with $V(G + uv) = V(G)$ and $E(G + uv) = E(G) \cup \{uv\}$. Thus, $G$ is a spanning subgraph of $G + uv$. For the graph $G$ of Figure 1.5, the set $U = \{t, x\}$ of vertices and the set $X = \{tw, ux, vx\}$ of edges, the subgraphs $G - u$, $G - wx$, $G - U$, and $G - X$ of $G$ are also shown in Figure 1.5, as is the graph $G + uv$.

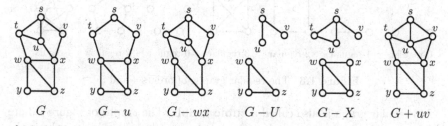

**Figure 1.5.** Deleting vertices and edges from and adding edges to a graph

## 1.3  Families of graphs

There are certain classes of graphs that occur so often that they deserve special mention and, in some cases, special notation. We describe some of the most prominent of these now.

A graph of order 1 is called a **trivial graph**. A **nontrivial graph** therefore has two or more vertices. A graph of size 0 is called an **empty graph**. A **nonempty graph** then has one or more edges. In an empty graph, no two vertices are adjacent. At the other extreme is a **complete graph** in which every two distinct vertices are adjacent. The size of a complete graph of order $n$ is $\binom{n}{2} = n(n-1)/2$. Therefore, if $G$ is a graph of order $n$ and size $m$, it follows that $0 \le m \le \binom{n}{2}$. The complete graph of order $n$ is denoted by $K_n$.

Two other classes of graphs that are often encountered are the paths and cycles. For an integer $n \ge 1$, the **path** $P_n$ is a graph of order $n$ and size $n - 1$ whose vertices can be labeled by $v_1, v_2, \ldots, v_n$ and whose edges are $v_i v_{i+1}$ for $i = 1, 2, \ldots, n - 1$. For an integer $n \ge 3$, the **cycle** $C_n$ is a graph of order $n$ and size $n$ whose vertices can be labeled by $v_1, v_2, \ldots, v_n$ and whose edges are $v_1 v_n$ and $v_i v_{i+1}$ for $i = 1, 2, \ldots, n - 1$. The cycle $C_n$ is also referred to as an

$n$-cycle and the 3-cycle is also called a **triangle**. Observe that $P_1 = K_1$, $P_2 = K_2$ and $C_3 = K_3$.

A graph $G$ is **regular** if the vertices of $G$ have the same degree and is **regular of degree $r$** if this degree is $r$. Such graphs are also called **$r$-regular**. The complete graph of order $n$ is therefore a regular graph of degree $n-1$ and every cycle is 2-regular. In Figure 1.6, all (nonisomorphic) regular graphs of orders 4 and 5, including the cycles $C_4$ and $C_5$ and the complete graphs $K_4$ and $K_5$, are shown. Since no graph has an odd number of odd vertices, there is no 1-regular or 3-regular graph of order 5. Indeed, the pairs $r, n$ of integers for which there exist $r$-regular graphs of order $n$ are predictable (see Exercise 1.15).

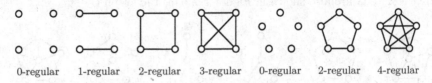

0-regular    1-regular    2-regular    3-regular    0-regular    2-regular    4-regular

**Figure 1.6.** The regular graphs of orders 4 and 5

A 3-regular graph is also called a **cubic graph**. The graphs of Figure 1.3 are cubic as is the complete graph $K_4$. One of the best known cubic graphs is the **Petersen graph**, named for Julius Petersen whose work [190] is often credited as the beginning of the theoretical study of graphs. Three different drawings of the Petersen graph are shown in Figure 1.7. We will have many occasions to encounter this graph.

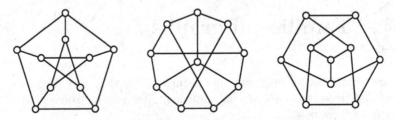

**Figure 1.7.** Three drawings of the Petersen graph

Another class of graphs that we often encounter are the bipartite graphs. A graph $G$ is **bipartite** if $V(G)$ can be partitioned into two sets $U$ and $W$ (called **partite sets**) so that every edge of $G$ joins a vertex of $U$ and a vertex of $W$. The graph shown on the left in Figure 1.8 is bipartite with partite sets $\{v_1, v_3, v_5, v_7\}$ and $\{v_2, v_4, v_6\}$. This graph is redrawn on the right in Figure 1.8 to show more clearly that it is bipartite. If $G$ is an $r$-regular bipartite graph, $r \geq 1$, with partite sets $U$ and $W$, then $|U| = |W|$. This follows since the size of $G$ is $r|U| = r|W|$.

A graph $G$ is a **complete bipartite graph** if $V(G)$ can be partitioned into two sets $U$ and $W$ (called **partite sets** again) so that $uw$ is an edge of $G$ if and only if $u \in U$ and $w \in W$. If $|U| = s$ and $|W| = t$, then this complete bipartite graph has order $s + t$ and size $st$ and is denoted by $K_{s,t}$ (or $K_{t,s}$). The complete bipartite

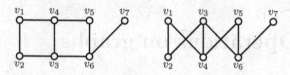

**Figure 1.8.** A bipartite graph

graph $K_{1,t}$ is called a **star**. The complete bipartite graphs $K_{1,3}$, $K_{2,2}$, $K_{2,3}$, and $K_{3,3}$ are shown in Figure 1.9. Observe that $K_{2,2} = C_4$. The star $K_{1,3}$ is sometimes referred to as a **claw**.

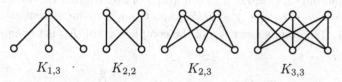

| $K_{1,3}$ | $K_{2,2}$ | $K_{2,3}$ | $K_{3,3}$ |

**Figure 1.9.** Complete bipartite graphs

Since the size of the complete bipartite graph $K_{\lfloor \frac{n}{2} \rfloor, \lceil \frac{n}{2} \rceil}$ is $\lfloor n/2 \rfloor \cdot \lceil n/2 \rceil = \lfloor n^2/4 \rfloor$, there are bipartite graphs of order $n$ and size $\lfloor n^2/4 \rfloor$. No bipartite graph of order $n$ has more edges, however (see Exercise 1.19).

**Theorem 1.3.** *Every bipartite graph of order $n$ has at most $\lfloor n^2/4 \rfloor$ edges.*

Bipartite graphs belong to a more general class of graphs. For an integer $k \geq 1$, a graph $G$ is a **$k$-partite graph** if $V(G)$ can be partitioned into $k$ subsets $V_1, V_2, \ldots, V_k$ (again called **partite sets**) such that every edge of $G$ joins vertices in two different partite sets. A 1-partite graph is then an empty graph and a 2-partite graph is bipartite. A **complete $k$-partite graph** $G$ is a $k$-partite graph with the property that two vertices are adjacent in $G$ if and only if the vertices belong to different partite sets. If $|V_i| = n_i$ for $1 \leq i \leq k$, then $G$ is denoted by $K_{n_1, n_2, \ldots, n_k}$ (the order in which the numbers $n_1, n_2, \ldots, n_k$ are written is not important). If $n_i = 1$ for all $i$ $(1 \leq i \leq k)$, then $G$ is the complete graph $K_k$. A **complete multipartite graph** is a complete $k$-partite graph for some integer $k \geq 2$. Some complete multipartite graphs are shown in Figure 1.10.

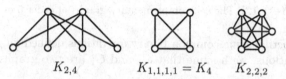

| $K_{2,4}$ | $K_{1,1,1} = K_4$ | $K_{2,2,2}$ |

**Figure 1.10.** Some complete multipartite graphs

# 1.4    Operations on graphs

There are many ways of producing a new graph from one or more given graphs. The most common of these is the complement of a graph. The **complement** $\overline{G}$ of a graph $G$ is the graph with vertex set $V(G)$ in which two vertices are adjacent in $\overline{G}$ if and only if these vertices are not adjacent in $G$. Any isomorphism from a graph $G$ to a graph $H$ is also an isomorphism from $\overline{G}$ to $\overline{H}$. Consequently, $\overline{G} \cong \overline{H}$ if and only if $G \cong H$. If $G$ is a graph of order $n$ and size $m$, then $\overline{G}$ is a graph of order $n$ and size $\binom{n}{2} - m$. A graph $G$ and its complement are shown in Figure 1.11. The complement $\overline{K}_n$ of the complete graph $K_n$ is the empty graph of order $n$.

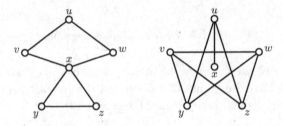

**Figure 1.11.** A graph (left) and its complement (right)

A graph $G$ is **self-complementary** if $G$ is isomorphic to $\overline{G}$. Certainly, if $G$ is a self-complementary graph of order $n$, then its size is $m = \binom{n}{2}/2 = n(n-1)/4$. Since only one of $n$ and $n-1$ is even, either $4 \mid n$ or $4 \mid (n-1)$; that is, if $G$ is a self-complementary graph of order $n$, then either $n \equiv 0 \pmod{4}$ or $n \equiv 1 \pmod{4}$. In fact, the converse is also true (see Exercise 1.29). The self-complementary graphs of order 5 or less are shown in Figure 1.12.

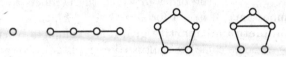

**Figure 1.12.** The self-complementary graphs of order 5 or less

We next describe some common binary operations defined on graphs. In the following definitions, we assume that $G_1$ and $G_2$ are two graphs with disjoint vertex sets.

The **union** $G = G_1 \cup G_2$ of $G_1$ and $G_2$ has vertex set $V(G) = V(G_1) \cup V(G_2)$ and edge set $E(G) = E(G_1) \cup E(G_2)$. The union $G \cup G$ of two disjoint copies of $G$ is denoted by $2G$. Indeed, if a graph $G$ consists of $k$ $(\geq 2)$ disjoint copies of a graph $H$, then we write $G = kH$. The graph $2K_1 \cup 3K_2 \cup K_{1,3}$ is shown on the left in Figure 1.13.

The **join** $G = G_1 \vee G_2$ of $G_1$ and $G_2$ has vertex set $V(G) = V(G_1) \cup V(G_2)$ and edge set

$$E(G) = E(G_1) \cup E(G_2) \cup \{uv : u \in V(G_1), v \in V(G_2)\}.$$

Using the join operation, we see that $\overline{K}_s \vee \overline{K}_t = K_{s,t}$. Another illustration is given on the right in Figure 1.13.

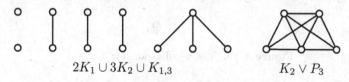

$$2K_1 \cup 3K_2 \cup K_{1,3} \qquad\qquad K_2 \vee P_3$$

**Figure 1.13.** The union and join of graphs

The **cartesian product** $G$ of two graphs $G_1$ and $G_2$, commonly denoted by $G_1 \square G_2$ or $G_1 \times G_2$, has vertex set

$$V(G) = V(G_1) \times V(G_2)$$

in which two distinct vertices $(u, v)$ and $(x, y)$ of $G_1 \square G_2$ are adjacent if either

(1) $u = x$ and $vy \in E(G_2)$ or (2) $v = y$ and $ux \in E(G_1)$.

A convenient way of drawing $G_1 \square G_2$ is to first place a copy of $G_2$ at each vertex of $G_1$ (see Figure 1.14) and then join corresponding vertices of $G_2$ in those copies of $G_2$ placed at adjacent vertices of $G_1$ (again see Figure 1.14). Equivalently, $G_1 \square G_2$ can be constructed by placing a copy of $G_1$ at each vertex of $G_2$ and adding the appropriate edges. As expected, $G_1 \square G_2 \cong G_2 \square G_1$ for all graphs $G_1$ and $G_2$.

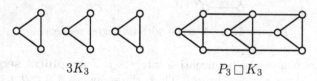

$$3K_3 \qquad\qquad P_3 \square K_3$$

**Figure 1.14.** The cartesian product of two graphs

The **$n$-cube** $Q_n$ is an important class of graphs that can be defined recursively in terms of cartesian products: $Q_1 = K_2$ and for $n \geq 2$, $Q_n = Q_{n-1} \square K_2$, the cartesian product of $Q_{n-1}$ and $K_2$. The $n$-cube can also be defined as that graph whose vertex set is the set of ordered $n$-tuples $(a_1, a_2, \ldots, a_n)$ or $a_1 a_2 \cdots a_n$ where $a_i$ is 0 or 1 for $1 \leq i \leq n$ (commonly called **$n$-bit strings**) in which two vertices are adjacent if and only if the corresponding ordered $n$-tuples differ at precisely one coordinate. The graph $Q_n$ is an $n$-regular graph of order $2^n$. The $n$-cubes for $n = 1, 2, 3, 4$ are shown in Figure 1.15, where (for $n \leq 3$) their vertices are labeled by $n$-bit strings. The graphs $Q_n$ are also called **hypercubes**.

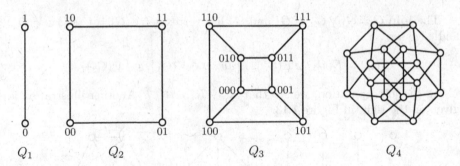

**Figure 1.15.** The $n$-cubes for $n = 1, 2, 3, 4$

 **1.5**   **Degree sequences**

We saw in the first theorem of graph theory (Theorem 1.1) that the sum of the degrees of the vertices of a graph $G$ is twice the size of $G$ and in Corollary 1.2 that $G$ must have an even number of odd vertices. We now consider the degrees of the vertices of a graph in more detail.

The **degree sequence** of a graph $G$ of order $n$ whose vertices are labeled $v_1, v_2, \ldots, v_n$ is $\deg v_1, \deg v_2, \ldots, \deg v_n$. For example, the degree sequence of the graph $G$ of Figure 1.16 is $4, 3, 2, 2, 1$ (or $1, 2, 2, 3, 4$ or $2, 1, 4, 2, 3$, etc.). We commonly write the degree sequence of a graph as a nonincreasing sequence.

**Figure 1.16.**  A graph with its degree sequence

A finite sequence $s$ of nonnegative integers is a **graphical sequence** if $s$ is a degree sequence of some graph. Thus, the sequence $4, 3, 2, 2, 1$ is graphical. There are some obvious necessary conditions for a sequence $s: d_1, d_2, \ldots, d_n$ of $n$ nonnegative integers to be graphical. While the conditions that $d_i \leq n - 1$ for $i = 1, 2 \ldots, n$ and $\deg v_1 + \deg v_2 + \cdots + \deg v_n$ is even are necessary for $s$ to be graphical, they are not sufficient. For example, the sequence $3, 3, 3, 1$ satisfies both conditions, but it is not graphical, for if three vertices of a graph of order 4 have degree 3 then the remaining vertex must have degree 3 as well.

It is not all that unusual for a graphical sequence to be the degree sequence of more than one graph. For example, the graphical sequence $3, 2, 2, 2, 1$ is the degree sequence of the two (nonisomorphic) graphs in Figure 1.17. On the other hand, each of the 18 graphs in Figure 1.2 has a degree sequence possessed by no other graph.

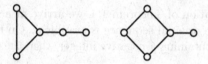

**Figure 1.17.** Two graphs with the same degree sequence

There are necessary and sufficient conditions for a finite sequence of nonnegative integers to be graphical. One of these is due to Havel [122] and Hakimi [117], and is often referred to as the Havel–Hakimi theorem, despite the fact that Havel and Hakimi gave independent proofs and wrote separate papers containing this theorem.

**Theorem 1.4 (Havel–Hakimi theorem).** *A sequence* $s\colon d_1, d_2, \ldots, d_n$ *of nonnegative integers with* $\Delta = d_1 \geq d_2 \geq \cdots \geq d_n$ *and* $\Delta \geq 1$ *is graphical if and only if the sequence*

$$s_1\colon d_2 - 1, d_3 - 1, \ldots, d_{\Delta+1} - 1, d_{\Delta+2}, \ldots, d_n$$

*is graphical.*

*Proof.* First, assume that $s_1$ is graphical. Then there exists a graph $G_1$ of order $n-1$ such that $s_1$ is a degree sequence of $G_1$. Thus, the vertices of $G_1$ can be labeled as $v_2, v_3, \ldots, v_n$ so that

$$\deg_{G_1} v_i = \begin{cases} d_i - 1 & \text{if } 2 \leq i \leq \Delta + 1 \\ d_i & \text{if } \Delta + 2 \leq i \leq n. \end{cases}$$

A new graph $G$ can now be constructed by adding a new vertex $v_1$ to $G_1$ together with the $\Delta$ edges $v_1 v_i$ for $2 \leq i \leq \Delta + 1$. Since $\deg_G v_i = d_i$ for $1 \leq i \leq n$, it follows that $s\colon \Delta = d_1, d_2, \ldots, d_n$ is a degree sequence of $G$ and so $s$ is graphical.

Conversely, let $s$ be a graphical sequence. Thus there is at least one graph of order $n$ with degree sequence $s$. Among all such graphs, let $G$ be the one in which $V(G) = \{v_1, v_2, \ldots, v_n\}$ with $\deg v_i = d_i$ for $i = 1, 2, \ldots, n$ and with the sum of the degrees of the vertices adjacent to $v_1$ a maximum. We wish to show that $v_1$ is adjacent to vertices having degrees $d_2, d_3, \ldots, d_{d_1+1}$.

Suppose, to the contrary, that $v_1$ is not adjacent to vertices with degrees $d_2, d_3, \ldots, d_{d_1+1}$. Then there exist vertices $v_j$ and $v_k$ with $d_j > d_k$ such that $v_1$ is adjacent to $v_k$ but not to $v_j$. Since the degree of $v_j$ exceeds the degree of $v_k$, there exists a vertex $v_\ell$ such that $v_\ell$ is adjacent to $v_j$ but not to $v_k$ Removing the edges $v_1 v_k$ and $v_j v_\ell$ and adding the edges $v_1 v_j$ and $v_k v_\ell$ results in a graph $G'$ with the same degree sequence as $G$ yet the sum of the degrees of the vertices adjacent to $v_1$ in $G'$ is larger than in $G$, contradicting the choice of $G$.

Thus, $v_1$ is adjacent to vertices of degrees $d_2, d_3, \ldots, d_{d_1+1}$ and hence the graph $G - v_1$ has degree sequence $s_1$ so that $s_1$ is graphical. ∎

Theorem 1.4 actually provides us with a polynomial-time algorithm for determining whether a given finite sequence of nonnegative integers is graphical. If,

upon repeated application of Theorem 1.4, we arrive at a sequence every term
of which is 0, then the original sequence is graphical. On the other hand, if we
arrive at a sequence containing a negative integer, then the given sequence is not
graphical.

We now illustrate Theorem 1.4 with the sequence

$$s: 5, 3, 3, 3, 3, 2, 2, 2, 1, 1, 1.$$

After one application of Theorem 1.4 (deleting 5 from $s$ and subtracting 1 from
the next five terms), we obtain

$$s_1': 2, 2, 2, 2, 1, 2, 2, 1, 1, 1.$$

Reordering this sequence, we have

$$s_1: 2, 2, 2, 2, 2, 2, 1, 1, 1, 1.$$

Continuing in this manner, we get

$$s_2': 1, 1, 2, 2, 2, 1, 1, 1, 1$$
$$s_2: 2, 2, 2, 1, 1, 1, 1, 1, 1$$
$$s_3' = s_3: 1, 1, 1, 1, 1, 1, 1, 1$$
$$s_4': 0, 1, 1, 1, 1, 1, 1$$
$$s_4: 1, 1, 1, 1, 1, 1, 0$$
$$s_5': 0, 1, 1, 1, 1, 0$$
$$s_5: 1, 1, 1, 1, 0, 0$$
$$s_6': 0, 1, 1, 0, 0$$
$$s_6: 1, 1, 0, 0, 0$$
$$s_7' = s_7: 0, 0, 0, 0.$$

Therefore, $s$ is graphical. Of course, if we observe that some sequence prior
to $s_7$ is graphical, then we can conclude by Theorem 1.4 that $s$ is graphical. For
example, the sequence $s_3$ is clearly graphical since it is the degree sequence of the
graph $G_3 = 4K_2$ in Figure 1.18. By Theorem 1.4, each of the sequences $s_2$, $s_1$,
and $s$ is also graphical. To construct a graph with degree sequence $s_2$, we proceed
in reverse from $s_3' = s_3$ to $s_2$, observing that a vertex should be added to $G_3$
so that it is adjacent to two vertices of degree 1. We thus obtain a graph $G_2$
with degree sequence $s_2$ (or $s_2'$). Proceeding from $s_2'$ to $s_1$, we again add a new
vertex joining it to two vertices of degree 1 in $G_2$. This gives a graph $G_1$ with
degree sequence $s_1$ (or $s_1'$). Finally, we obtain a graph $G$ with degree sequence $s$
by considering $s_1'$; that is, we add a new vertex to $G_1$ joining it to vertices of
degrees $2, 2, 2, 2, 1$. This procedure is illustrated in Figure 1.18.

It should be pointed out that the graph $G$ in Figure 1.18 is not the only graph
with degree sequence $s$. However, there are graphs that cannot be produced
by the method used to construct the graph $G$ in Figure 1.18. For example, the
graph $2P_3$ shown in Figure 1.19 is such a graph.

Suppose that $s: d_1, d_2, \ldots, d_n$ is a graphical sequence with $d_1 \geq d_2 \geq \cdots \geq d_n$.
Then there exists a graph $G$ of order $n$ with $V(G) = \{v_1, v_2, \ldots, v_n\}$ such that

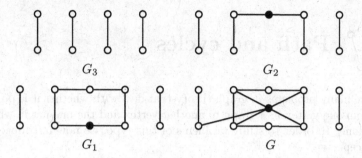

**Figure 1.18.** Construction of a graph $G$ with a given degree sequence

**Figure 1.19.** The graph $2P_3$ cannot be constructed by the method following Theorem 1.4

$\deg v_i = d_i$ for all $i$ $(1 \leq i \leq n)$. Of course, $\deg v_1 + \deg v_2 + \cdots + \deg v_n$ is even. Let $k$ be an integer with $1 \leq k \leq n - 1$. Suppose that $V_1 = \{v_1, v_2, \ldots, v_k\}$ and $V_2 = \{v_{k+1}, v_{k+2}, \ldots, v_n\}$. Now the sum $\deg v_1 + \deg v_2 + \cdots + \deg v_k$ counts every edge in $G[V_1]$ twice and counts each edge from a vertex of $V_1$ to a vertex of $V_2$ once. The size of $G[V_1]$ is at most $\binom{k}{2} = k(k-1)/2$, while for each $i$ $(k+1 \leq i \leq n)$ the number of edges joining $v_i$ and $V_1$ is at most $\min\{k, d_i\}$. Thus,

$$\sum_{i=1}^{k} d_i \leq 2\binom{k}{2} + \sum_{i=k+1}^{n} \min\{k, d_i\} = k(k-1) + \sum_{i=k+1}^{n} \min\{k, d_i\}. \qquad (1.1)$$

Hence, for every graphical sequence $s$, we must have that $\deg v_1 + \deg v_2 + \cdots + \deg v_n$ is even and that (1.1) is satisfied for every integer $k$ with $1 \leq k \leq n - 1$. These conditions are not only necessary for a nonincreasing sequence of nonnegative integers to be graphical, they are sufficient as well, as established by Erdős and Gallai [81]. We omit the proof.

**Theorem 1.5 (Erdős–Gallai theorem).** *A sequence $s$: $d_1, d_2, \ldots, d_n$ $(n \geq 2)$ of nonnegative integers with $d_1 \geq d_2 \geq \cdots \geq d_n$ is graphical if and only if $\deg v_1 + \deg v_2 + \cdots + \deg v_n$ is even and for each integer $k$ with $1 \leq k \leq n - 1$,*

$$\sum_{i=1}^{k} d_i \leq k(k-1) + \sum_{i=k+1}^{n} \min\{k, d_i\}.$$

# 1.6 Path and cycles

There are many problems in graph theory that deal with whether it is possible to travel from one vertex in a graph to another vertex and the manner in which this can be done. In order to study problems of this type, we now introduce several new concepts.

For two (not necessarily distinct) vertices $u$ and $v$ in a graph $G$, a **$u$-$v$ walk** $W$ is a sequence of vertices, beginning with $u$ and ending at $v$ such that consecutive vertices in $W$ are adjacent. Such a walk $W$ can be expressed as

$$W : u = v_0, v_1, \ldots, v_k = v,$$

where $v_i v_{i+1} \in E(G)$ for $0 \le i \le k - 1$. Note that nonconsecutive vertices in a walk need not be distinct. The walk $W$ can therefore be thought of as beginning at the vertex $u = v_0$, proceeding along the edge $v_0 v_1$ to the vertex $v_1$, then along the edge $v_1 v_2$ to the vertex $v_2$, and so forth, until finally arriving at the vertex $v = v_k$. The walk $W$ is said to **contain** each vertex $v_i$ for $0 \le i \le k$ and each edge $v_i v_{i+1}$ for $0 \le i \le k - 1$. The number of edges contained in $W$ (including multiplicities) is the **length** of $W$. Hence, the length of the walk $W$ is $k$.

A walk whose initial and terminal vertices are distinct is an **open walk**; otherwise, it is a **closed walk**. A walk is allowed to consist of a single vertex, in which case it is a **trivial walk**, and trivial walks are closed.

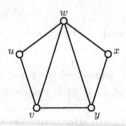

**Figure 1.20.** Walks in a graph

For example, in the graph $G$ of Figure 1.20,

$$W : x, w, y, w, v, u, w$$

is an $x$-$w$ walk of length 6. This is an open walk, and it encounters the vertex $w$ three times and the edge $wy$ twice.

A walk in which no edge is repeated is called a **trail**. For example, in the graph $G$ of Figure 1.20, $T : u, v, y, w, v$ is a $u$-$v$ trail of length 4 because although it repeats the vertex $v$, it does not repeat an edge. On the other hand, a walk in which no vertex is repeated is called a **path**. Every nontrivial path is necessarily an open walk. Thus, $P: u, v, w, y$ is $u$-$y$ path of length 3 in the graph $G$ of Figure 1.20. Many proofs in graph theory make use of $u$-$v$ walks or $u$-$v$ paths of minimum length (or of maximum length) for some pair $u, v$ of vertices of a graph. The proof of the following theorem illustrates this.

**Theorem 1.6.** *Let $u$ and $v$ be distinct vertices of a graph $G$. For every $u$-$v$ walk $W$ in $G$, there exists a $u$-$v$ path $P$ such that every edge of $P$ belongs to $W$.*

*Proof.* Let $W$ be a $u$-$v$ walk. Among all $u$-$v$ walks in $G$, every edge of which belongs to $W$, let

$$P: u = u_0, u_1, \ldots, u_k = v$$

be one of minimum length. Thus, the length of $P$ is $k$. We claim that $P$ is a $u$-$v$ path. Assume, to the contrary, that this is not the case. Then some vertex of $G$ must be repeated in $P$, say $u_i = u_j$ for some $i$ and $j$ with $0 \le i < j \le k$. If we then delete the vertices $u_{i+1}, u_{i+2}, \ldots, u_j$ from $P$, we arrive at the $u$-$v$ walk

$$W': u = u_0, u_1, \ldots, u_{i-1}, u_i = u_j, u_{j+1}, \ldots, u_k = v$$

whose length is less than $k$ and such that every edge of $W'$ belongs to $W$, producing a contradiction. Hence $P$ is the required $u$-$v$ path. ∎

A closed trail is called a **circuit**. Thus a circuit is a walk without repeated edges that ends at the same vertex from which it begins. Note that we consider the edgeless walk to be a trivial circuit. An example of a nontrivial circuit is given by the walk

$$C: u, w, x, y, w, v, u$$

in the graph $G$ of Figure 1.20. In addition to the necessary repetition of $u$ in this circuit, $w$ is repeated as well. This is acceptable since no edge is repeated in $C$. A circuit

$$C: v = v_0, v_1, \ldots, v_k = v,$$

with $k \ge 2$ and for which the vertices $v_i$, $0 \le i \le k - 1$, are distinct is called a **cycle**. Thus

$$u, v, y, x, w, u$$

is a cycle of length 5 in the graph $G$ of Figure 1.20. As with the class of graphs $C_k$, $k \ge 3$, called cycles earlier in this chapter, the cycle $C$ above is called a **$k$-cycle**. Once again, a 3-cycle is referred to as a **triangle**. A cycle of even length is an **even cycle**, while a cycle of odd length is an **odd cycle**.

The subgraph induced by the edges in a path $v_1, v_2, \ldots, v_k$ or, for $k \ge 3$, a cycle $v_1, v_2, \ldots, v_k, v_1$ is itself called a *path* or *cycle*, respectively. Consequently, paths and cycles have more than one interpretation: a means to proceed between vertices in a graph, as subgraphs in a graph and as a class of graphs. This is also the case with trails and circuits. The length of a smallest cycle in a graph $G$ (containing cycles) is the **girth** of $G$, denoted by $g(G)$, and the length of a longest cycle is the **circumference**, denoted by $c(G)$. Thus, $g(K_n) = 3$ for $n \ge 3$ and $c(K_n) = n$; while $g(K_{s,t}) = 4$ for $2 \le s \le t$ and $c(K_{s,t}) = 2s$. Since the Petersen

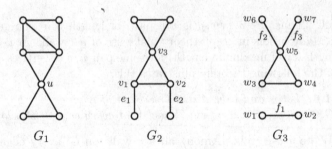

$$G_1 \qquad\qquad G_2 \qquad\qquad G_3$$

**Figure 1.21.** Two connected graphs and a disconnected graph

graph contains 5-cycles but no 3- or 4-cycles, its girth is 5. As we will see in Chapter 3, the Petersen graph does not have a 10-cycle; however, since it does have 9-cycles, its circumference is 9.

# 1.7   Connected graphs and distance

Two vertices $u$ and $v$ in a graph $G$ are **connected** if $G$ contains a $u$-$v$ walk. The graph $G$ itself is **connected** if every two vertices of $G$ are connected. By Theorem 1.6, a graph $G$ is connected if and only if $G$ contains a $u$-$v$ *path* for every two vertices $u$ and $v$, and we frequently take this to be the definition instead. A graph $G$ that is not connected is a **disconnected graph**. The graph $F$ of Figure 1.21 is connected since $F$ contains a $u$-$v$ path for every two vertices $u$ and $v$ in $F$. On the other hand, the graph $H$ is disconnected since, for example, $H$ contains no $y_4$-$y_5$ path.

A maximal connected subgraph of a graph $G$ is called a **component** of $G$. Thus, every component of $G$ is an induced subgraph of $G$, and every vertex of $G$ lies in a unique component. The graphs $G_1$ and $G_2$ of Figure 1.21 each have only one component because they are connected, while the graph $G_3$ from this figure has precisely two components, the induced subgraphs $G_3[\{w_1, w_2\}]$ and $G_3[\{w_3, w_4, w_5, w_6, w_7\}]$.

## Cut-vertices and bridges

A vertex $v$ is a **cut-vertex** of a graph $G$ if $G - v$ has more components than $G$. Returning to the graphs of Figure 1.21, we see that $G_1$ has a single cut-vertex $u$; the cut-vertices of $G_2$ are $v_1$, $v_2$, and $v_3$; and the cut-vertices of $G_3$ are $w_1$, $w_2$, and $w_5$. Not all graphs have cut-vertices, of course; for example, the complete graph $K_n$ has no cut-vertices for all $n \geq 1$, and the cycle $C_n$ has no cut-vertices for all $n \geq 3$. While these graphs have no cut-vertices because they cannot be disconnected by removing a single vertex, $\overline{K}_n$ has no cut-vertices for $n \geq 1$ for a different reason: removing a vertex of $\overline{K}_n$ cannot create any additional components. At the other extreme, for $n \geq 2$, only two vertices of the path $P_n$ are *not* cut-vertices, namely its two end-vertices. That this is the other extreme is verified in the following result.

**Theorem 1.7.** *Every connected graph of order $n \geq 2$ contains at least two vertices that are not cut-vertices.*

*Proof.* Let $G$ be a nontrivial connected graph and let $P$ be a longest path in $G$. Label the end-vertices of $P$ with $u$ and $v$. We show that $u$ and $v$ are not cut-vertices. Assume, to the contrary, that $u$ is a cut-vertex of $G$. Then $G - u$ is disconnected and so contains two or more components. Let $w$ be the vertex on $P$ that is adjacent to $u$ and let $P'$ be the $w$-$v$ subpath of $P$. Necessarily, $P'$ belongs to a component, say $G_1$ of $G - u$. Let $G_2$ be another component of $G - u$. Then $G_2$ contains some vertex $x$ that is adjacent to $u$. This produces an $x$-$v$ path $P'$ that contains $P$. However, $P'$ is longer than $P$, which is impossible. Similarly, $v$ is not a cut-vertex of $G$. ∎

The edge analogues of cut-vertices are bridges, so an edge $e$ is a **bridge** of a graph $G$ if $G - e$ has more components than $G$. Returning to the graphs of Figure 1.21, we see that $G_1$ has no bridges; the bridges of $G_2$ are $e_1$ and $e_2$; and the bridges of $G_3$ are $f_1$, $f_2$, and $f_3$.

If $e = uv$ is a bridge of a graph $G$, then $u$ is a cut-vertex unless $\deg u = 1$. This implies that $K_2$ is the only connected graph with a bridge but without a cut-vertex. Beyond $K_2$, we see that for $n \geq 3$, the complete graph $K_n$ and the cycle have no bridges, while every edge of the path $P_n$ is a bridge for all $n \geq 2$. These facts also follow from the following useful characterization of bridges.

**Theorem 1.8.** *An edge $e$ in a graph $G$ is a bridge of $G$ if and only if $e$ lies on no cycle in $G$.*

*Proof.* We may assume that $G$ is connected, for otherwise we can consider a component of $G$ containing $e$. First, suppose that $e = uv$ is an edge of $G$ that is not a bridge. Since $G - e$ is connected, there is a $u$-$v$ path $P$ in $G - e$. Then $P$ together with $e$ produce a cycle in $G$ containing $e$.

For the converse, assume that $e = uv$ is an edge of $G$ belonging to a cycle of $G$. Since $e$ lies on a cycle of $G$, there is a $u$-$v$ path $P'$ in $G$ not containing $e$. We show that $G - e$ is connected and, consequently, that $e$ is not a bridge. Let $x$ and $y$ be two vertices of $G$. Since $G$ is connected, $G$ contains an $x$-$y$ path $Q$. If $e$ does not lie on $Q$, then $Q$ is an $x$-$y$ path in $G - e$ as well. If, on the other hand, $e$ lies on $Q$, then replacing $e$ in $Q$ by the $u$-$v$ path $P'$ produces an $x$-$y$ walk in $G - e$. Thus $G - e$ is connected. ∎

# Distance

The **distance** $d_G(u, v)$ from a vertex $u$ to a vertex $v$ in a connected graph $G$ is
the minimum length of a $u$-$v$ path. If the graph $G$ being considered is understood,
then this distance is written simply as $d(u, v)$. In the graph $G$ of Figure 1.22, the
path $P$: $v_1, v_5, v_6, v_{10}$ is a shortest $v_1$-$v_{10}$ path; thus, $d(v_1, v_{10}) = 3$. In addition,

$$d(v_1, v_1) = 0, d(v_1, v_2) = 1, d(v_1, v_6) = 2, d(v_1, v_7) = 3, \text{and } d(v_1, v_8) = 4.$$

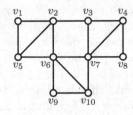

**Figure 1.22.** Distances in a graph

The distance $d$ defined above satisfies the following properties in a connected
graph $G$:
  (1) $d(u, v) \geq 0$ for every two vertices $u$ and $v$ of $G$;
  (2) $d(u, v) = 0$ if and only if $u = v$;
  (3) $d(u, v) = d(v, u)$ for all $u, v \in V(G)$ (the **symmetric property**);
  (4) $d(u, w) \leq d(u, v) + d(v, w)$ for all $u, v, w \in V(G)$ (the **triangle inequality**).
Since $d$ satisfies the four properties (1)–(4), $d$ is a **metric** on $V(G)$ and $(V(G), d)$
is a **metric space**. Because $d$ satisfies the symmetric property, we can speak of
the distance *between* two vertices $u$ and $v$ rather than the distance *from $u$ to $v$*.

The **diameter** of a connected graph $G$, denoted by $\text{diam}(G)$, is defined to be
the maximum distance between any pair of vertices of $G$. The following result
will be used in our upcoming characterization of bipartite graphs.

**Theorem 1.9.** *If a graph contains a closed walk of odd length, then it contains
an odd cycle.*

*Proof.* We prove the result by induction on the length of the walk. Suppose that
a graph contains a closed walk $W$ : $v_1, v_2, \ldots, v_k, v_1$ of odd length, so $k$ is odd. In
the case $k = 3$, it is clear that the three vertices $v_1$, $v_2$, and $v_3$ must be distinct,
and so the graph contains a triangle. Now suppose that $k \geq 5$ and that the claim
is true for all shorter closed walks of odd length. If $v_i \neq v_j$ for all $1 \leq i < j \leq k$,
then $W$ is itself a cycle and we are done. Otherwise, we have $v_i = v_j$ for some pair
of indices satisfying $1 \leq i < j \leq k$. We can then view $W$ as the concatenation of
two closed walks: $v_i, v_{i+1}, \ldots, v_j = v_i$ and $v_i = v_j, v_{j+1}, \ldots, v_k, v_1, \ldots, v_i$. As $W$
has odd length, precisely one of these two walks must have odd length, and thus
the graph must contain an odd cycle by induction. ∎

We are now ready to characterize the bipartite graphs.

**Theorem 1.10.** *A graph is bipartite if and only if it contains no odd cycles.*

*Proof.* Suppose first that $G$ is bipartite. Then $V(G)$ can be partitioned into partite sets $U$ and $W$ (and so every edge of $G$ joins a vertex of $U$ and a vertex of $W$). Let $C: v_1, v_2, \ldots, v_k, v_1$ be a $k$-cycle of $G$. We may assume that $v_1 \in U$. Thus, $v_2 \in W$, $v_3 \in U$ and so forth. In particular, $v_i \in U$ for every odd integer $i$ with $1 \leq i \leq k$ and $v_j \in W$ for every even integer $j$ with $2 \leq j \leq k$. Since $v_1 \in U$, it follows that $v_k \in W$ and so $k$ is even.

It suffices to prove the converse for connected graphs. Suppose that $G$ is a connected graph containing no odd cycles, and define

$$U = \{x \in V(G) : d(u, x) \text{ is even}\} \text{ and } W = \{x \in V(G) : d(u, x) \text{ is odd}\}.$$

We claim that $G$ is bipartite with partite sets $U$ and $W$. Suppose to the contrary that two vertices of $U$ were adjacent, say $u_1$ and $u_2$. By the definition of $U$, there is a $u$-$u_1$ path $P_1$ of even length and a $u$-$u_2$ path $P_2$ of even length. Thus $P_1$ followed by the edge $u_1u_2$ and then the reverse of $P_2$ would be a closed walk of odd length, and thus would contain an odd cycle by Theorem 1.9, a contradiction. If two vertices of $W$ were adjacent, say $w_1$ and $w_2$, then we would have a $u$-$w_1$ path $P_1$ of odd length and a $u$-$w_2$ path $P_2$ of odd length, and thus the same construction would give us a walk of odd length and another contradiction, proving the result. ∎

 ## 1.8   Trees and forests

An **acyclic graph** is a graph without cycles, and we call such a graph a **forest**. Thus, by Theorem 1.8, every edge of a forest is a bridge. A **tree** is a connected acyclic graph. Thus, in a forest, every connected component is a tree. Trees are important in the understanding of the structure of graphs and are used to systematically visit their vertices. Outside of graph theory, trees are widely used in computer science as a means to organize data, in physics to model currents in electrical networks, and in chemistry to count certain types of chemical compounds. Figure 1.23 shows the six trees of order 6.

The vertices of degree 1 in a tree are called **leaves** and play an important role in the study of trees. In fact, every tree with at least two vertices has at least two leaves.

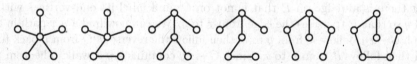

**Figure 1.23.** The six trees of order 6

**Lemma 1.11.** *Every tree with at least two vertices has at least two leaves.*

*Proof.* Let $T$ be a tree with at least two vertices. Consider a longest path $P : v_1, v_2, \ldots, v_k$ in $T$. Since $T$ is connected, it must be the case that $k \geq 2$, so $v_1 \neq v_k$. Since $P$ is a longest path, it follows that $\deg v_1 = \deg v_k = 1$. ∎

There are several well-known families of trees. For example, the paths $P_n$ and stars $K_{1,n-1}$ are trees of order $n \geq 2$. For $t \geq 2$, only one vertex of the star $K_{1,t}$ is not a leaf. A tree containing exactly two vertices that are not leaves (which are necessarily adjacent) is called a **double star**.

Reading the graphs of Figure 1.23 from left to right, the first is the star $K_{1,5}$, the second and third are double stars, and the fifth tree is the path $P_6$. In each of the trees of Figure 1.23, the size of the tree is one less than its order. This is not only true of these trees, it is true of all trees.

**Theorem 1.12.** *A tree of order $n$ has $n - 1$ edges.*

*Proof.* We proceed by induction on the order of a tree. There is only one tree of order 1, namely $K_1$, and it has no edges. Thus, the basis step of the induction is established. Now assume that every tree of order $n - 1 \geq 1$ has $n - 2$ edges, and let $T$ be a tree of order $n$. By Lemma 1.11, $T$ has at least two leaves. Let $v$ be a leaf of $T$. As we have observed, $T - v$ is a tree of order $n - 1$. By the induction hypothesis, $T - v$ has $n - 2$ edges. Thus, $T$ has $(n - 2) + 1 = n - 1$ edges. ∎

Since every component of a forest is a tree, the following result is a corollary of Theorem 1.12 (see Exercise 1.79).

**Corollary 1.13.** *A forest of order $n$ with $k$ components has $n - k$ edges.*

Since every tree is connected, every two vertices are connected by a path. In fact, even more can be said.

**Theorem 1.14.** *A graph $G$ is a tree if and only if every two vertices are connected by a unique path.*

*Proof.* First suppose that $G$ is a graph in which every two vertices are connected by a unique path. Certainly, $G$ is connected, so we need only show that it is acyclic. If $G$ were to contain a cycle $C : v_1, v_2, \ldots, v_k, v_1$, then $G$ would have two $v_1$-$v_k$ paths, namely, the path of length one $v_1, v_k$, and the path $v_1, v_2, \ldots, v_k$. Therefore, $G$ must also be acyclic, which proves that it is a tree.

Now suppose that $G$ is a tree and that $u$ and $v$ are two vertices of $G$. Since $G$ is connected, it contains some $u$-$v$ path, say $P$. Suppose, to the contrary, that $G$ contains a different $u$-$v$ path, say $Q$. Because $P \neq Q$, there must be some edge that lies on one of the paths and not the other. Suppose without loss of generality that there is an edge on $P$ that is not on $Q$, and label its end-vertices with $x$ and $y$ so that $P$ traverses the edge from $x$ to $y$. We can now find a $y$-$x$ path in the graph $G - xy$: follow $P$ from $y$ to $v$, then follow the reverse of $Q$ from $v$ back to $u$, and then follow $P$ from $u$ to $x$. Since $G - xy$ contains a $y$-$x$ walk, Theorem 1.6 shows that $G - xy$ contains a $y$-$x$ path. However, this implies that $G$ contains a cycle, which is a contradiction. ∎

Suppose $G$ is a graph of order $n$ and size $m$. By Theorem 1.12, if $G$ is a tree, then $m = n - 1$. It is easy to see that the converse is not true. However, there are several sufficient conditions for a graph to be a tree in addition to the characterization given in Theorem 1.14. These other characterizations of trees, any one of which could be used as the definition, are presented next.

**Theorem 1.15.** *For a graph $G$ with $n$ vertices, the following are equivalent:*
  *(a)  $G$ is a tree;*
  *(b)  $G$ is acyclic with precisely $n - 1$ edges; and*
  *(c)  $G$ is connected with precisely $n - 1$ edges.*

*Proof.* We have seen in Theorem 1.12 that a tree with $n$ vertices has $n - 1$ edges. Since trees are acyclic by definition, this shows that (a) implies (b). Now suppose that (b) holds, so $G$ is a forest with $n$ vertices. By Corollary 1.13, $G$ has $n - k$ edges, where $k$ denotes the number of components of $G$. Since our hypotheses state that $G$ has $n - 1$ edges, it must consist of a single component, proving that (b) implies (c).

It remains only to prove that (c) implies (a). Suppose that $G$ is connected and has precisely $n - 1$ edges. We want to show that $G$ is acyclic. Suppose, to the contrary, that $G$ were to contain one or more cycles. By Theorem 1.8, edges that lie on cycles are not bridges. Therefore, by successively deleting edges from cycles, the resulting subgraph will be connected and acyclic, and thus a tree, but will contain fewer than $n - 1$ edges. This contradicts Theorem 1.12, and hence $G$ is acyclic, proving that (c) implies (a). ■

The following result is essentially a restatement of Theorem 1.15.

**Corollary 1.16.** *Let $G$ be a graph of order $n$. Then any two of the following properties implies the third:*

  *(1) $G$ is acyclic,    (2) $G$ is connected,    (3) $G$ has $n - 1$ edges.*

There are several useful ways of constructing a new tree from a given tree. For example, if $v$ is a leaf in a tree $T$, then $T - v$ is also a tree. If a new vertex is added to $T$ and joined to any vertex of $T$, then another tree results. Similar conclusions can be drawn when adding or deleting edges from a tree.

**Corollary 1.17.** *Let $T$ be a tree.*
  • *$T - e$ is disconnected for every edge $e$ of $T$.*
  • *$T + f$ contains a cycle for every edge $f \in E(\overline{T})$.*

Another useful fact is that every connected graph contains a spanning tree.

**Corollary 1.18.** *Every connected graph has a spanning subgraph isomorphic to a tree.*

*Proof.* Let $G$ be a connected graph. If $G$ is a tree, then such a subgraph is the graph $G$ itself. Otherwise, $G$ has at least one cycle. By deleting an edge from each cycle in each resulting subgraph, a tree of order $n$ is obtained. ■

If $s: d_1, d_2, \ldots, d_n$ is the degree sequence of a tree of order $n \geq 2$, then, necessarily, $d_1 + d_2 + \cdots + d_n = 2n - 2$. This necessary condition is also sufficient for a sequence of positive integers to be the degree sequence of a tree (see Exercise 1.83).

**Theorem 1.19.** *A sequence* $s: d_1, d_2, \ldots, d_n$ *of* $n \geq 2$ *integers is the degree sequence of a tree with* $n$ *vertices if and only if* $d_i \geq 1$ *for all* $1 \leq i \leq n$ *and*

$$\sum_{i=1}^{n} d_i = 2n - 2.$$

As one would anticipate, graphs often contain many subgraphs that are trees. In fact, for any fixed tree, every graph with vertices of high enough degree has a subgraph isomorphic to that tree.

**Theorem 1.20.** *Let* $T$ *be a tree of order* $k$. *Every graph of minimum degree at least* $k - 1$ *contains a subgraph that is isomorphic to* $T$.

*Proof.* We proceed by induction on $k$. The result is immediate for $k = 1$ and $k = 2$.

Assume for every tree $T'$ of order $k - 1$ with $k \geq 3$ and for every graph $G'$ with $\delta(G') \geq k - 2$, that $G'$ contains a subgraph that is isomorphic to $T'$. Now, let $T$ be a tree of order $k$ and let $G$ be a graph with $\delta(G) \geq k - 1$. We show that $G$ contains a subgraph that is isomorphic to $T$.

Let $v$ be an end-vertex of $T$ and let $u$ be the vertex of $T$ that is adjacent to $v$. Then $T - v$ is a tree of order $k - 1$. Since $\delta(G) \geq k - 1 > k - 2$, it follows by the induction hypothesis that $G$ contains a subgraph $T'$ that is isomorphic to $T - v$. Let $u'$ denote the vertex of $T'$ corresponding to $u$ in $T - v$. Since $\deg_G u' \geq k - 1$ and the order of $T'$ is $k - 1$, the vertex $u'$ is adjacent to a vertex $v'$ that does not belong to $T'$. The tree obtained by adding $v'$ to $T'$ and joining it to $u'$ is isomorphic to $T$, completing the proof. ∎

 **1.9** # Multigraphs and pseudographs

In a graph $G$, there is at most one edge connecting any two distinct vertices. A **multigraph** is a set of vertices that can be connected by a finite number of edges. A multigraph can be represented as $H = (V, E)$, where $E$ is a multiset (a set that allows repeated elements) of 2-element subsets of $V$. If there are two or more edges connecting the same pair of distinct vertices in a multigraph, they are called **parallel edges**. The **underlying graph** of a multigraph $H$ is a graph $G$ with the same vertices as $H$ and an edge connecting two vertices if there is at least one edge connecting them in $H$.

An edge connecting a vertex to itself is called a **loop**. Structures that allow both parallel edges and loops (including parallel loops) are sometimes called **pseudographs**. In contrast, a graph only allows one edge between each pair of vertices and does not allow loops. In a multigraph, it is possible to have more than one edge between a pair of distinct vertices, but loops are not permitted. Pseudographs allow more than one edge between pairs of vertices and allow loops. Note that some authors refer to multigraphs or pseudographs as simply "graphs", and use the term "simple graph" for graphs as we have defined them. Thus when reading about graph theory, it is essential to be clear on how the term "graph" is being used. To reiterate, using the terminology of this book, every graph is a multigraph, and every multigraph is a pseudograph.

In Figure 1.24, $H_1$ and $H_4$ are multigraphs while $H_2$ and $H_3$ are pseudographs. Of course, $H_1$ and $H_4$ are also pseudographs while $H_4$ is the only graph in Figure 1.24. For a vertex $v$ in a multigraph $G$, the **degree** of $v$, denoted by $\deg v$, is the number of edges of $G$ incident with $v$. In a pseudograph, each loop at a vertex contributes 2 to its degree. Thus for the pseudograph $H_3$ of Figure 1.24, $\deg u = 5$ and $\deg v = 2$.

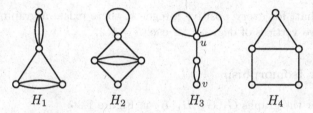

**Figure 1.24.** Multigraphs and pseudographs

When describing walks in multigraphs or in pseudographs, it is often necessary to list edges in the sequence as well as vertices in order to specify the edges being used in the walk. For example,

$$W : u, e_1, u, e_4, v, e_6, w, e_6, v, e_7, w$$

is a $u$-$w$ walk in the pseudograph $G$ of Figure 1.25.

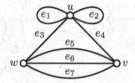

**Figure 1.25.** A pseudograph

# Exercises for Chapter 1

### Section 1.1. Fundamentals

**1.1.** A graph $G = (V, E)$ of order 8 has the power set of the set $S = \{1, 2, 3\}$ as its vertex set, that is, $V$ is the set of all subsets of $S$. Two vertices $A$ and $B$ of $V$ are adjacent if $A \cap B = \varnothing$. Draw the graph $G$, determine the degree of each vertex of $G$ and determine the size of $G$.

**1.2.** A graph $G$ of order 26 and size 58 has five vertices of degree 4, six vertices of degree 5, and seven vertices of degree 6. The remaining vertices of $G$ all have the same degree. What is this degree?

**1.3.** A graph $G$ has order $n = 3k + 3$ for some positive integer $k$. Every vertex of $G$ has degree $k + 1$, $k + 2$ or $k + 3$. Prove that $G$ has at least $k + 3$ vertices of degree $k + 1$ or at least $k + 1$ vertices of degree $k + 2$ or at least $k + 2$ vertices of degree $k + 3$.

**1.4.** Show that, for every positive integer $k$, there exists a graph of order $2k$ containing two vertices of degree $i$ for each $i = 1, 2, \ldots, k$.

### Section 1.2. Isomorphism

**1.5.** Consider the graphs $G_1, G_2, H_1, H_2$ in Figure 1.26.
  (a) Determine whether $G_1 \cong G_2$.
  (b) Determine whether $H_1 \cong H_2$.

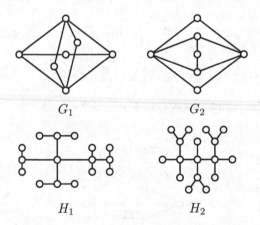

**Figure 1.26.** The graphs $G_1, G_2, H_1, H_2$ in Exercise 1.5

**1.6.** Determine all nonisomorphic graphs of order 5.

**1.7.** Determine the minimum size of a graph $G$ of order 5 such that every graph of order 5 and size 5 is isomorphic to some subgraph of $G$.

**1.8.** Show that "is isomorphic to" is an equivalence relation on the set of all graphs.

**1.9.** Prove that if two graphs $G$ and $H$ are isomorphic, then they have the same order and same size, and the degrees of the vertices of $G$ are the same as the degrees of the vertices of $H$.

## Section 1.3. Families of graphs

**1.10.** Show that if $G$ is a nonregular graph of order $n$ and size $rn/2$ for some integer $r$ with $1 \le r \le n - 2$, then $\Delta(G) - \delta(G) \ge 2$.

**1.11.** For each integer $k \ge 2$, give an example of $k$ nonisomorphic regular graphs, all of the same order and same size.

**1.12.** Give an example of a nonregular graph $G$ containing an edge $e$ and a vertex $u$ such that $G - e$ and $G - u$ are both regular.

**1.13.** Give an example of two nonisomorphic regular graphs $G_1$ and $G_2$ of the same order and same size such that (1) for every two vertices $v_1 \in V(G_1)$ and $v_2 \in V(G_2)$, $G_1 - v_1 \not\cong G_2 - v_2$ and (2) there exist 2-element subsets $S_1 \subseteq V(G_1)$ and $S_2 \subseteq V(G_2)$ such that $G_1 - S_1 \cong G_2 - S_2$.

**1.14.** Give an example of two nonisomorphic regular graphs $H_1$ and $H_2$ of the same order and same size such that (1) for every 2-element subsets $S_1 \subseteq V(H_1)$ and $S_2 \subseteq V(H_2)$, $H_1 - S_1 \not\cong H_2 - S_2$ and there exist 3-element subsets $S_1' \subseteq V(H_1)$ and $S_2' \subseteq V(H_2)$ such that $H_1 - S_1' \cong H_2 - S_2'$.

**1.15.** For integers $r$ and $n$, show that there exists an $r$-regular graph of order $n$ if and only if $0 \le r \le n - 1$ and $r$ and $n$ are not both odd.

**1.16.** Prove that for every graph $G$ and every integer $r \ge \Delta(G)$ there exists an $r$-regular graph containing $G$ as an induced subgraph.

**1.17.** For positive integers $k$ and $n$ with $n > 2k$, the graph $G_{n,k}$ is that graph whose vertices are the $k$-element subsets of an $n$-element set $S = \{1, 2, \ldots, n\}$ and where two vertices ($k$-element subsets) $A$ and $B$ are adjacent if $A$ and $B$ are disjoint. The graph $G_{n,k}$ is called the **Kneser graph**.
  (a) Determine the graphs $G_{6,1}$ and $G_{5,2}$. What familiar graph is $G_{5,2}$?
  (b) Show that $G_{n,k}$ is an $r$-regular graph for some integer $r$.

**1.18.** A bipartite graph $G$ of order $n$ has partite sets $U$ and $W$ where $|U| = 10$. Every vertex of $U$ has degree 6. In $W$, there are four vertices of degree 2 and three vertices of degree 4. All other vertices of $G$ have degree 8. What is $n$?

**1.19.** Prove Theorem 1.3: every bipartite graph of order $n$ has at most $\lfloor n^2/4 \rfloor$ edges.

**1.20.** Show for each integer $n \geq 2$ that there is exactly one bipartite graph of order $n$ having size $\lfloor n^2/4 \rfloor$.

**1.21.** Let $G$ be a 3-partite graph of order $n = 3k$ and size $m$. Prove that $m \leq 3k^2$.

**1.22.** Let $G$ be a nonempty graph with the property that whenever $uv \notin E(G)$ and $vw \notin E(G)$, then $uw \notin E(G)$. Prove that $G$ is a complete multipartite graph.

## Section 1.4.  Operations on graphs

**1.23.** Determine all bipartite graphs $G$ such that $\overline{G}$ is bipartite.

**1.24.** Let $G$ be a graph of odd order $n = 2k + 1 \geq 3$ for some positive integer $k$. Prove that if the vertices of $G$ have exactly the same degrees as the vertices of $\overline{G}$, then $G$ has an odd number of vertices of degree $k$.

**1.25.** Let $G$ be a graph.
  (a) Show that there are exactly two 4-regular graphs $G$ of order 7.
  (b) How many 6-regular graphs of order 9 are there?

**1.26.** Prove that there is no regular self-complementary graph of even order.

**1.27.** We have seen that $C_5$ is a self-complementary graph. Therefore, there is a regular self-complementary graph of order 5. Show that there is a regular self-complementary graph of order $5^n$ for every positive integer $n$.

**1.28.** Let $G_1$ and $G_2$ be self-complementary graphs, where $G_2$ has even order $n$. Let $G$ be the graph obtained from $G_1$ and $G_2$ by joining each vertex of $G_2$ whose degree is less than $n/2$ to every vertex of $G_1$. Show that $G$ is self-complementary.

**1.29.** Prove that there exists a self-complementary graph of order $n$ for every positive integer $n$ with $n \equiv 0 \pmod{4}$ or $n \equiv 1 \pmod{4}$.

**1.30.** Give an example of a graph $G$ of order 6 and size 7 such that $G$ is isomorphic to a subgraph $H$ of $\overline{G}$.

**1.31.** Give an example of a graph $G$ of order 7 and size 10 such that $G$ is isomorphic to a subgraph $H$ of $\overline{G}$.

**1.32.** Prove that for every integer $n \geq 3$ there exists a graph $G$ of order $n$ and size $\lfloor \binom{n}{2}/2 \rfloor$ that is isomorphic to a subgraph $H$ of $\overline{G}$.

**1.33.** For $i = 1, 2$, let $u_i$ be a vertex in a graph $G_i$ of order $n_i$ and size $m_i$.
  (a) Determine the degree of $u_1$ in $G_1 \cup G_2$.
  (b) Determine the degree of $u_1$ in $G_1 \vee G_2$.
  (c) Determine the degree of $(u_1, u_2)$ in $G_1 \,\square\, G_2$.

**1.34.** Determine the order and size of each of the graphs $P_3 \vee 2P_3$, $P_3 \square 2P_3$ and $Q_1 \cup Q_2 \cup Q_3$.

**1.35.** Let $n$ be a given positive integer and let $r$ and $s$ be nonnegative integers such that $r + s = n$ and $s$ is even.
  (a) Give an example of a graph containing $r$ even vertices and $s$ odd vertices.
  (b) Determine the minimum size of a graph $G$ satisfying the properties in (a).
  (c) Determine the maximum size of a graph $G$ satisfying the properties in (a).

## Section 1.5. Degree sequences

**1.36.** Let $s: d_1, d_2, \ldots, d_n$ be a graphical sequence with $\Delta = d_1 \geq d_2 \geq \cdots \geq d_n$. Show, for each integer $k$ with $1 \leq k \leq n$, that there exists a graph $G$ with $V(G) = \{v_1, v_2, \ldots, v_n\}$ where $\deg v_i = d_i$ for $1 \leq i \leq n$ having the property that $v_k$ is adjacent to either (1) the vertices of $\{v_1, v_2, \ldots, v_{d_k}\}$ if $k > d_k$ or (2) the vertices of $\{v_1, v_2, \ldots, v_{d_k+1}\} - \{v_k\}$ if $1 \leq k \leq d_k$.

**1.37.** Determine whether the following sequences are graphical. If so, construct a graph with the appropriate degree sequence.
  (a) $4, 4, 3, 2, 1$
  (b) $3, 3, 2, 2, 2, 2, 1, 1$
  (c) $7, 7, 6, 5, 4, 4, 3, 2$
  (d) $7, 6, 6, 5, 4, 3, 2, 1$
  (e) $7, 4, 3, 3, 2, 2, 2, 1, 1, 1$.

**1.38.** Prove that a sequence $d_1, d_2, \ldots, d_n$ is graphical if and only if the sequence $n - d_1 - 1, n - d_2 - 1, \ldots, n - d_n - 1$ is graphical.

**1.39.** Prove that for every integer $x$ with $0 \leq x \leq 5$, the sequence $5, 5, 3, 2, 1, x$ is not graphical.

**1.40.** For which integers $x$, $0 \leq x \leq 7$, if any, is the sequence $7, 6, 5, 4, 3, 2, 1, x$ graphical?

**1.41.** Use Theorem 1.5 to determine whether the sequence $s: 6, 6, 5, 4, 3, 2, 2$ is graphical.

**1.42.** Show that for every finite set $S$ of positive integers, there exists a positive integer $k$ such that the sequence obtained by listing each element of $S$ a total of $k$ times is graphical. Find the minimum such $k$ for $S = \{2, 6, 7\}$.

**1.43.** Two finite sequences $s_1$ and $s_2$ of nonnegative integers are called **bigraphical** if there exists a bipartite graph $G$ with partite sets $V_1$ and $V_2$ such that $s_i$ lists the degrees of the vertices of $G$ in $V_i$ for $i = 1, 2$. Prove that the sequences $s_1: a_1, a_2, \ldots, a_r$ and $s_2: b_1, b_2, \ldots, b_t$ of nonnegative integers with $r \geq 2$, $a_1 \geq a_2 \geq \cdots \geq a_r$, $b_1 \geq b_2 \geq \cdots \geq b_t$, $0 < a_1 \leq t$ and $0 < b_1 \leq r$ are bigraphical if and only if the sequences $s_1': a_2, a_3, \cdots, a_r$ and $s_2': b_1 - 1, b_2 - 1, \ldots, b_{a_1} - 1, b_{a_1+1}, \ldots, b_t$ are bigraphical.

**1.44.** A nontrivial graph $G$ is an **irregular graph** if $\deg u \neq \deg v$ for every pair of distinct vertices of $G$. Prove that no graph is irregular.

## Section 1.6.  Path and cycles

**1.45.** Give an example of a graph $G$ and two vertices $u$ and $v$ of $G$ such that there is a $u$-$v$ trail containing all vertices of $G$ but no $u$-$v$ path containing all vertices of $G$.

**1.46.** Give an example of a graph $G$ with three vertices $u$, $v$ and $w$ such that (1) every $u$-$v$ path avoids $w$, (2) every $u$-$w$ path avoids $v$ and no $v$-$w$ path avoids $u$.

**1.47.** Prove that every nontrivial circuit in a graph contains a cycle.

**1.48.** Let $G$ be a graph with $\delta(G) \geq 2$.
  (a) Prove that the circumference $c(G)$ of $G$ satisfies $c(G) \geq \delta(G) + 1$.
  (b) Show that $G$ has a path of length $\delta(G)$.

**1.49.** Prove that "is connected to" is an equivalence relation on the vertex set of a graph.

**1.50.** Show, for every two vertices $u$ and $v$ in a connected graph $G$, that there exists a $u$-$v$ walk containing all vertices of $G$.

**1.51.** Give an example of a cubic graph of order 10 containing a $k$-cycle for each integer $k$ with $3 \leq k \leq 10$.

**1.52.** Give an example of a cubic graph $G$ containing no $k$-cycle for some integer $k$ with $g(G) < k < c(G)$.

**1.53.** Determine $g(G)$ and $c(G)$ for $G = K_{k,2k,4k}$ for every positive integer $k$.

## Section 1.7.  Connected graphs and distance

**1.54.** Characterize the graphs having the property that all of their induced subgraphs are connected.

**1.55.** Show that, for every integer $k \geq 2$, there is a graph containing a cut-vertex of degree $k$.

**1.56.** Let $G$ be a connected graph with $u \in V(G)$. Prove that if $v$ is a vertex that is farthest away from $u$ in $G$, then $v$ is not a cut-vertex of $G$.

**1.57.** Prove or disprove: If $v$ is a cut-vertex of a connected graph $G$, and $H$ is a proper connected subgraph with at least three vertices including $v$, then $v$ is a cut-vertex of $H$.

**1.58.** Prove or disprove: If $G$ is a connected graph and every proper connected induced subgraph of $G$ with at least three vertices contains a cut-vertex, then $G$ also contains a cut-vertex.

**1.59.** Let $G$ be a graph of order $\dot{n}$. Prove each of the following.
(a) If $\deg u + \deg v \geq n - 1$ for every two nonadjacent vertices $u$ and $v$ of $G$, then $G$ is connected.
(b) If $\delta(G) \geq (n-1)/2$, then $G$ is connected.

**1.60.** Prove that if $G$ is a graph of order $n \geq 2$ and at least $\binom{n-1}{2} + 1$ edges, then $G$ is connected.

**1.61.** For each positive integer $k$, show that there exists a graph $G$ of order $2k+1$ such that every vertex of $G$ lies on one or more triangles but on no larger cycles.

**1.62.** Let $G$ be a connected graph of order $n$ and let $k$ be an integer such that $2 \leq k \leq n - 1$. Show that if $\deg u + \deg v \geq k$ for every pair of nonadjacent vertices $u$ and $v$, then $G$ contains a path of length $k$.

**1.63.** Let $G$ be a disconnected graph of order $n \geq 6$ having three components. Prove that $\Delta(\overline{G}) \geq 2n/3$.

**1.64.** Let $G$ be a connected graph.
(a) Show that if $G$ has the property that the degree of every vertex is one of three distinct numbers and each of these three numbers is the degree of at least one vertex of $G$, then there is a path in $G$ containing three vertices whose degrees are distinct.
(b) Is the statement in (a) true if "three" is replaced by "four"?

**1.65.** Let $G$ be a nontrivial connected graph that is not bipartite. Show that $G$ contains two adjacent vertices $u$ and $v$ such that $\deg u + \deg v$ is even.

**1.66.** Prove that if $G$ is a connected graph of order $n \geq 2$, then the vertices of $G$ can be listed as $v_1, v_2, \ldots, v_n$ such that each vertex $v_i$ $(2 \leq i \leq n)$ is adjacent to some vertex in the set $\{v_1, v_2, \ldots, v_{i-1}\}$.

**1.67.** Prove that a graph $G$ is connected if and only if for every partition $\{V_1, V_2\}$ of $V(G)$, there exists an edge of $G$ joining a vertex of $V_1$ and a vertex of $V_2$.

**1.68.** Suppose that the vertices of a graph $G$ of order $n \geq 2$ can be listed as $v_1, v_2, \ldots, v_n$ such that each vertex $v_i$ $(2 \leq i \leq n)$ is adjacent to some vertex in the set $\{v_1, v_2, \ldots, v_{i-1}\}$. Prove that $G$ is connected.

**1.69.** Let $u, v$ and $w$ be three vertices in a connected graph $G$. Prove that $d(u, v) + d(u, w) + d(v, w) \geq 2d(u, w)$.

**1.70.** Prove that a nontrivial graph $G$ is bipartite if and only if $G$ contains no *induced* odd cycle.

**1.71.** Let $G$ be a connected graph such that the length of a longest path in $G$ is $\ell$.
(a) Prove that no two paths of length $\ell$ in $G$ are vertex-disjoint.
(b) By (a), two paths of length $\ell$ cannot be vertex-disjoint. Prove that if $P$ and $Q$ are two paths of length $\ell$ that meet in a single vertex, then $\ell$ is even.

**1.72.** Prove that if $G$ is a disconnected graph, then $\overline{G}$ is connected and, in fact, $\operatorname{diam}(\overline{G}) \le 2$.

**1.73.** Prove that if $v$ is a cut-vertex of a connected graph $G$, then $v$ is *not* a cut-vertex of $\overline{G}$.

### Section 1.8. Trees and forests

**1.74.** Prove that a graph with $n$ vertices in which every vertex has degree 2 must contain a cycle.

**1.75.** Prove that a graph with $n$ vertices and at least $n$ edges must contain a cycle.

**1.76.** A **caterpillar** is a tree $T$ of order 3 or more, the removal of whose leaves produce a path (which is called the **spine** of $T$).
  (a) Which trees in Figure 1.23 are caterpillars?
  (b) Give an example of a tree of order 8 that is *not* a caterpillar.

**1.77.** Determine the average degree of a tree $T$ of order $n$ in terms of $n$.

**1.78.** Draw all forests of order 6.

**1.79.** Prove Corollary 1.13: A forest of order $n$ with $k$ components has $n - k$ edges.

**1.80.** Prove that a graph $G$ is a forest if and only if every induced subgraph of $G$ contains a vertex of degree at most 1.

**1.81.** Characterize those graphs with the property that every connected subgraph is an induced subgraph.

**1.82.** A graph $G$ of order 8 has the degree sequence $s: 3, 3, 3, 1, 1, 1, 1, 1$. Prove or disprove: $G$ is a tree.

**1.83.** Prove Theorem 1.19: a sequence $s: d_1, d_2, \ldots, d_n$ of $n \ge 2$ integers is the degree sequence of a tree with $n$ vertices if and only if $d_i \ge 1$ for all $1 \le i \le n$ and

$$\sum_{i=1}^{n} d_i = 2n - 2.$$

**1.84.** Determine all trees $T$ such that $\overline{T}$ is also a tree.

**1.85.** Let $T$ be a tree of order $n$ with degree sequence $d_1, d_2, \ldots, d_n$ such that $d_1 \ge d_2 \ge \cdots \ge d_n$. Prove that $d_i \le \left\lceil \frac{n-1}{i} \right\rceil$ for each integer $i$ with $1 \le i \le n$.

**1.86.** Let $T$ be a tree of order $n$. Prove that $T$ is isomorphic to a subgraph of $\overline{C}_{n+2}$.

**1.87.** Prove that if $T$ is a tree of order $n \geq 2$ that is not a star, then $T$ is isomorphic to a subgraph of $\overline{T}$.

**1.88.** Find all those graphs $G$ of order $n \geq 4$ such that the subgraph induced by every three vertices of $G$ is a tree, or show that no such graph exists.

**1.89.** Show that every tree with maximum degree $k$ has at least $k$ leaves.

**1.90.** Let $G$ be a connected graph of order $n \geq 4$ and size $m$.
(a) Show that if $m = n$, then for every two distinct vertices $u$ and $v$ of $G$, the graph $G$ contains at most two distinct $u$-$v$ paths.
(b) Is the statement in (a) true if $m = n + 1$?

**1.91.** Prove that a sequence $s_n : d_1, d_2, \ldots, d_n$ $(n \geq 3)$ of integers with $1 \leq d_i \leq n - 1$ for $1 \leq i \leq n$ is a degree sequence of a unicyclic graph of order $n$ if and only if at most $n - 3$ terms of $s_n$ are 1 and $\sum_{i=1}^{n} d_i = 2n$.

**1.92.** Prove that an edge $e$ of a connected graph is a bridge if and only if $e$ belongs to every spanning tree of $G$.

## Section 1.9. Multigraphs and pseudographs

**1.93.** We have seen that no graph is irregular (see Exercise 1.44). Give an example of an irregular multigraph (if such a multigraph exists) having degree sequence

(a)   $5, 4, 3, 2, 1$
(b)   $6, 5, 4, 3, 2, 1$
(c)   $7, 6, 5, 4, 3, 2, 1$.

**1.94.** Determine which of the following sequences are the degree sequences of a multigraph.

(a)   $s_1 : 3, 2, 1$        (b)   $s_2 : 5, 2, 1$
(c)   $s_3 : 6, 4, 2$        (d)   $s_4 : 3, 2, 2$
(e)   $s_5 : 4, 4, 2, 2$     (f)   $s_6 : 5, 3, 2, 1$
(g)   $s_7 : 4, 4, 4, 4$     (h)   $s_8 : 7, 5, 3, 1$.

**1.95.** Prove that a sequence $s : d_1, d_2, \ldots, d_n$ $(n \geq 1)$ of nonnegative integers with $d_1 \geq d_2 \geq \cdots \geq d_n$ is the degree sequence of a multigraph if and only if $\sum_{i=1}^{n} d_i$ is even and $d_1 \leq \frac{1}{2} \sum_{i=1}^{n} d_i$.

**1.96.** Let $G$ be a connected graph of order $n$ where the vertices of $G$ are labeled as $v_1, v_2, \ldots, v_n$ in some way. A multigraph $H$ of size $m$ with $V(H) = V(G)$ is obtained by replacing each edge $v_i v_j$ of $G$ by $\min\{i, j\}$ parallel edges.
(a) Find $m$ if $G = K_5$.
(b) Find sharp upper and lower bounds for $m$ if $G = C_5$.
(c) Find the minimum value of $m$ if $G$ is bipartite.

# Digraphs

**2**

In certain situations, a graph may not be the most suitable way to model a particular scenario. For example, the symmetric nature of graphs may not provide the desired structure to represent a given situation. In such cases, we can turn to the concept of directed graphs, also known as digraphs. This chapter will focus on studying digraphs, as well as strongly connected digraphs and tournaments.

## 2.1 Fundamentals

A **directed graph** or **digraph** for short $D$ is a finite nonempty set of objects called **vertices** together with a (possibly empty) set of ordered pairs of distinct vertices of $D$ called **arcs** or **directed edges**. As with graphs, the vertex set of $D$ is denoted by $V(D)$ or simply $V$ and the arc set (or directed edge set) of $D$ is denoted by $E(D)$ or $E$. A digraph $D$ with vertex set $V = \{u, v, w, x\}$ and arc set $E = \{(u, v), (v, u), (u, w), (w, v), (w, x)\}$ is shown in Figure 2.1. When a digraph is described by means of a diagram, the "direction" of each arc is indicated by an arrowhead. Observe that in a digraph, it is possible for two arcs to join the same pair of vertices if the arcs are directed oppositely.

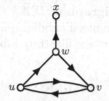

**Figure 2.1.** A digraph

Much of the terminology used for digraphs is quite similar to that used for graphs. The cardinality of the vertex set of a digraph $D$ is called the **order** of $D$ and is ordinarily denoted by $n$, while the cardinality of its arc set is the **size** of $D$ and is ordinarily denoted by $m$. If $a = (u, v)$ is an arc of a digraph $D$, then $u$ is said to be **adjacent to** $v$ and $v$ is **adjacent from** $u$. For a vertex $v$ in a digraph $D$, the **outdegree** od $v$ of $v$ is the number of vertices of $D$ to which $v$

is adjacent, while the **indegree** id $v$ of $v$ is the number of vertices of $D$ from which $v$ is adjacent. The **degree** deg $v$ of a vertex $v$ is defined by

$$\deg v = \operatorname{od} v + \operatorname{id} v.$$

For the vertex $v$ in the digraph of Figure 2.1, od $v = 1$, id $v = 2$ and deg $v = 3$. The directed graph version of Theorem 1.1 is stated next.

**Theorem 2.1 (First theorem of digraph theory).** *If $D$ is a digraph of size $m$, then*

$$\sum_{v \in V(D)} \operatorname{od} v = \sum_{v \in V(D)} \operatorname{id} v = m.$$

*Proof.* When the outdegrees of the vertices are summed, each arc is counted once. Similarly, when the indegrees of the vertices are summed, each arc is counted just once. ∎

A digraph $D_1$ is **isomorphic** to a digraph $D_2$, written $D_1 \cong D_2$, if there exists a bijection $\phi \colon V(D_1) \to V(D_2)$ such that $(u, v) \in E(D_1)$ if and only if $(\phi(u), \phi(v)) \in E(D_2)$. The function $\phi$ is called an **isomorphism** from $D_1$ to $D_2$.

There is only one digraph of order 1, namely the **trivial digraph**. Also, there is only one digraph of order 2 and size $m$ for each $m$ with $0 \le m \le 2$. There are four digraphs of order 3 and size 3, all of which are shown in Figure 2.2.

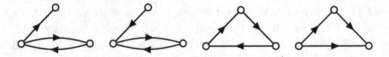

**Figure 2.2.** The digraphs of order 3 and size 3

A digraph $D_1$ is a **subdigraph** of a digraph $D$ if $V(D_1) \subseteq V(D)$ and $E(D_1) \subseteq E(D)$. We use $D_1 \subseteq D$ to indicate that $D_1$ is a subdigraph of $D$. A subdigraph $D_1$ of $D$ is a **spanning subdigraph** of $D$ if $V(D_1) = V(D)$. Vertex-deleted, arc-deleted, induced and arc-induced subdigraphs are defined in the expected manner. These last two concepts are illustrated for the digraph $D$ of Figure 2.3, where

$$V(D) = \{v_1, v_2, v_3, v_4\}, \ U = \{v_1, v_2, v_3\} \text{ and } X = \{(v_1, v_2), (v_2, v_4)\}.$$

We now consider certain types of digraphs that occur periodically. A digraph is **symmetric** if whenever $(u, v)$ is an arc of $D$, then $(v, u)$ is an arc of $D$ as well. There is a natural one-to-one correspondence between symmetric digraphs and graphs. The **complete symmetric digraph** $K_n^*$ of order $n$ has both arcs $(u, v)$ and $(v, u)$ for every two distinct vertices $u$ and $v$. A digraph is called an **oriented graph** if whenever $(u, v)$ is an arc of $D$, then $(v, u)$ is *not* an arc of $D$. Thus, an oriented graph $D$ can be obtained from a *graph* $G$ by assigning a direction to (or by "orienting") each edge of $G$, thereby transforming each edge of a graph $G$ into an arc and transforming $G$ itself into an oriented graph. The digraph $D$ is also

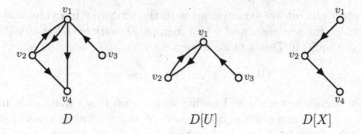

**Figure 2.3.** Induced and arc-induced subdigraphs

called an **orientation** of $G$. Consider the digraphs shown in Figure 2.4. The first of these (reading left to right) is a symmetric digraph, the second is an oriented graph, and the first is neither. The **underlying graph** of a digraph $D$ is the graph obtained by replacing each arc $(u, v)$ or symmetric pair $(u, v)$, $(v, u)$ of arcs by the edge $uv$. All three digraphs shown in Figure 2.4 have the same underlying graph, shown on the right in the figure.

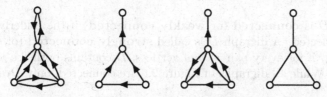

**Figure 2.4.** Digraphs with the same underlying graph

An orientation of a complete graph is called a **tournament** and will be studied in some detail later in this chapter. A digraph $D$ is **regular of degree $r$**, or **$r$-regular**, if od $v = $ id $v = r$ for every vertex $v$ of $D$. A 1-regular digraph $D_1$ and a 2-regular digraph $D_2$ are shown in Figure 2.5. The digraph $D_2$ is a tournament.

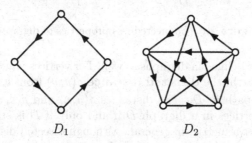

**Figure 2.5.** Regular digraphs

The terms walk, open and closed walk, trail, path, circuit and cycle for graphs have natural counterparts in digraph theory as well, the important difference being that the directions of the arcs must be followed in each of these walks. In particular, when referring to digraphs, the terms **directed path**, **directed cycle**, and **directed circuit** are synonymous with the terms **path**, **cycle**, and **circuit**. More formally, for vertices $u$ and $v$ in a digraph $D$, a **directed $u$-$v$ walk** $W$ (or simply a $u$-$v$ walk) in $D$ is a finite sequence

$$W : u = u_0, u_1, u_2, \ldots, u_k = v$$

of vertices, beginning with $u$ and ending with $v$ such that $(u_i, u_{i+1})$ is an arc for $0 \leq i \leq k - 1$. The number $k$ of occurrences of arcs (including repetition) in the walk $W$ is its **length**. Digraphs in which every vertex has positive outdegree must contain cycles (see Exercise 2.10).

**Theorem 2.2.** *If $D$ is a digraph such that* od $v \geq k \geq 1$ *for every vertex $v$ of $D$, then $D$ contains a cycle of length at least $k + 1$.*

 **2.2**  # Strongly connected digraphs

A digraph $D$ is **connected** (or **weakly connected**) if the underlying graph of $D$ is connected. A digraph $D$ is called **strongly connected** (or sometimes just **strong**) if for every pair $u$, $v$ of vertices, $D$ contains both a $u$-$v$ path and a $v$-$u$ path. While all digraphs of Figure 2.6 are connected, only $D_1$ is strongly connected.

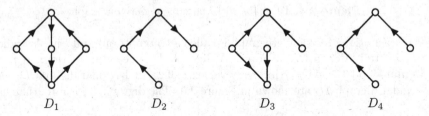

$D_1$                $D_2$                $D_3$                $D_4$

**Figure 2.6.** Connectedness properties of digraphs

Distance can be defined in digraphs as well. For vertices $u$ and $v$ in a digraph $D$ containing a $u$-$v$ path, the **directed distance** $\vec{d}(u, v)$ from $u$ to $v$ is the length of a shortest $u$-$v$ path in $D$. The distances $\vec{d}(u, v)$ and $\vec{d}(v, u)$ are defined for all pairs $u, v$ of vertices in a digraph $D$ if and only if $D$ is strongly connected. This distance is not a metric, in general. Although directed distance satisfies the triangle inequality, it is not symmetric unless $D$ is symmetric, in which case $D$ can be considered a graph.

The following theorem is the digraph analogue of Theorem 1.6, and its proof is analogous as well (see Exercise 2.14).

**Theorem 2.3.** *Let $u$ and $v$ be two vertices in a digraph $D$. For every $u$-$v$ walk $W$ in $D$, there exists a $u$-$v$ path $P$ such that every arc of $P$ belongs to $W$.*

Strongly connected digraphs are characterized in the following theorem.

**Theorem 2.4.** *A digraph $D$ is strongly connected if and only if $D$ contains a closed spanning walk.*

*Proof.* Assume that $W: u_1, u_2, \ldots, u_k, u_1$ is a closed spanning walk in $D$. Let $u, v \in V(D)$. Then $u = u_i$ and $v = u_j$ for some $i, j$ with $1 \le i, j \le k$ and $i \ne j$. Without loss of generality, assume that $i < j$. Then $W_1: u_i, u_{i+1}, \ldots, u_j$ is a $u_i$-$u_j$ walk in $D$ and $W_2: u_j, u_{j+1}, \ldots, u_k, u_1, \ldots, u_i$ is a $u_j$-$u_i$ walk in $D$. By Theorem 2.3, $D$ contains both a $u_i$-$u_j$ path and a $u_j$-$u_i$ path in $D$ and so $D$ is strongly connected.

Conversely, assume that $D$ is a nontrivial strongly connected digraph with $V(D) = \{v_1, v_2, \ldots, v_n\}$. Since $D$ is strongly connected, $D$ contains a $v_i$-$v_{i+1}$ path $P_i$ for $i = 1, 2, \ldots, n-1$ as well as a $v_n$-$v_1$ path $P_n$. Then the sequence $P_1, P_2, \ldots, P_n$ of paths produces a closed spanning walk in $D$. ∎

The **converse** $\vec{D}$ of a digraph $D$ is obtained from $D$ by reversing the direction of every arc of $D$. Thus, $D$ is strongly connected if and only if its converse $\vec{D}$ is strongly connected (see Exercise 2.15).

Recall that an orientation of a graph $G$ is a digraph obtained by assigning a direction to each edge of $G$. Certainly, if $G$ has a strong orientation, then $G$ must be connected. Also, if $G$ has a bridge, then it is impossible to produce a strong orientation of $G$. Robbins [202] showed that this condition is also sufficient for $G$ to have a strong orientation.

**Theorem 2.5 (Robbins' theorem).** *A graph has a strong orientation if and only if it is connected and bridgeless.*

*Proof.* We have already observed that if a graph has a strong orientation, then it is connected and bridgeless. For the converse, suppose $G$ is a connected, bridgeless graph. We wish to orient the edges of $G$ to produce a strongly connected digraph $D$.

Let $H$ be a maximal subgraph of $G$ that has a strong orientation (it may be the case that $H$ is a single vertex). We seek to show that $H = G$. We may assume that $H$ is an induced subgraph of $G$, as adding any missing edges, with any orientation, does not destroy the fact that $H$ is strongly connected.

Suppose to the contrary that $H \ne G$. Since $G$ is connected, there exists an edge $uv \in E(G)$ such that $u \in V(H)$ and $v \in V(G) \setminus V(H)$. Since $uv$ is not a bridge, there is a path from $v$ to $u$ in $G - uv$. Let $P$ be a shortest path from $v$ to any vertex of $H$ not containing the edge $uv$. By our choice of $P$, none of its edges lie in $H$. By orienting the edge $uv$ and the edges of $P$ to produce a directed path, a larger subgraph with a strong orientation is produced, contradicting the maximality of $H$. Thus, it must be that $H = G$, completing the proof. ∎

## 2.3   Tournaments

There are sporting events involving teams (or individuals) that require every two teams to compete against each other exactly once. This is referred to as a round-robin tournament. Round-robin tournaments give rise quite naturally to the class of digraphs called tournaments. Recall that a **tournament** is an orientation of a complete graph. Therefore, a tournament can be defined as a digraph such that for every pair $u, v$ of distinct vertices, exactly one of $(u, v)$ and $(v, u)$ is an arc. A tournament $T$ then models a round-robin tournament in which no ties are permitted. The vertices of $T$ are the teams in the round-robin tournament and $(u, v)$ is an arc in $T$ if team $u$ defeats team $v$.

Figure 2.7 shows two tournaments of order 3. In fact, these are the only two tournaments of order 3. The number of nonisomorphic tournaments increases sharply with their orders. For example, there is only one tournament of order 1 and one of order 2. As we just observed, the tournaments $T_1$ and $T_2$ in Figure 2.7 are the only two tournaments of order 3. There are four tournaments of order 4, 12 of order 5, 56 of order 6, and over 154 billion of order 12. (This is sequence A000568 in *The On-Line Encyclopedia of Integer Sequences* [229].)

**Figure 2.7.** The tournaments of order 3

Since the size of a tournament of order $n$ is $\binom{n}{2}$, it follows from Theorem 2.1 that

$$\sum_{v \in V(T)} \operatorname{od} v = \sum_{v \in V(T)} \operatorname{id} v = \binom{n}{2}.$$

A vertex $u$ in a tournament $T$ is a **king** in $T$ if, for every vertex $w$ different from $u$, either $(u, w)$ is an arc of $T$ or there is a vertex $v$ such that $(u, v)$ and $(v, w)$ are arcs of $T$. Landau [156] proved that every tournament has a king.

**Theorem 2.6.** *Every tournament has a king.*

*Proof.* Let $T$ be a tournament and let $u$ be a vertex having maximum outdegree in $T$. We show that $u$ is a king. Suppose, to the contrary, $u$ is not a king. Then there is a vertex $w$ in $T$ such that $(w, u)$ is an arc of $T$ and for every arc $(u, v)$ of $T$, it must be the case that $(w, v)$ is an arc of $T$ as well. Thus, $\operatorname{od} w \geq \operatorname{od} u + 1 > \operatorname{od} u$, producing a contradiction. Hence, $u$ is a king. ∎

## Transitive tournaments

A tournament $T$ is **transitive** if whenever $(u, v)$ and $(v, w)$ are arcs of $T$, then $(u, w)$ is also an arc of $T$. The tournament $T_2$ of Figure 2.7 is transitive while $T_1$ is not. The following result gives an elementary property of transitive tournaments. An **acyclic digraph** is a digraph having no cycles.

**Theorem 2.7.** *A tournament is transitive if and only if it is acyclic.*

*Proof.* Let $T$ be an acyclic tournament and suppose that $(u, v)$ and $(v, w)$ are arcs of $T$. Since $T$ is acyclic, $(w, u) \notin E(T)$. Therefore, $(u, w) \in E(T)$ and $T$ is transitive.

Conversely, suppose that $T$ is a transitive tournament and suppose, to the contrary, that $T$ contains a cycle, say $C = (v_1, v_2, \ldots, v_k, v_1)$, where $k \geq 3$. Since $(v_1, v_2)$ and $(v_2, v_3)$ are arcs of the transitive tournament $T$, it follows that $(v_1, v_3)$ is also an arc of $T$. If $k \geq 4$, then since $(v_1, v_3)$ and $(v_3, v_4)$ are arcs of $T$, it must be that $(v_1, v_4)$ is an arc of $T$. Continuing in this manner, we have that $(v_1, v_5)$, $(v_1, v_6)$, $\ldots$, $(v_1, v_k)$ are arcs of $T$. However, this contradicts the fact that $(v_k, v_1)$ is an arc of $T$. Thus, $T$ is acyclic. ∎

Although the number of nonisomorphic tournaments increases sharply as the number of vertices increases, transitive tournaments are unique.

**Theorem 2.8.** *For every positive integer $n$, there is exactly one transitive tournament of order $n$, up to isomorphism.*

*Proof.* The proof proceeds by induction on the order $n$ of the tournament. For $n = 1$, there is only one tournament of order $n$, and this tournament is also transitive. Thus, assume that $n \geq 2$ and that there is exactly one tournament of order $n - 1$. Let $T$ be a tournament of order $n$. By Theorem 2.6, $T$ has a king, say vertex $u$ is a king. Thus for every vertex $w$ different from $u$, either $(u, w)$ is an arc of $T$ or there exists a vertex $v$ such that $(u, v)$ and $(v, w)$ are arcs of $T$. However, if $(u, v)$ and $(v, w)$ are arcs of $T$, then, since $T$ is transitive, it must be the case that $(u, w)$ is an arc of $T$. Therefore, $u$ is adjacent to every vertex of $T$ and hence od $u = n - 1$. Consider the tournament $T - u$ of order $n - 1$ and note that, since od $u = n - 1$, $T - u$ is also transitive. Hence, by the induction hypothesis, $T - u$ is the unique tournament of order $n - 1$. Since $T$ is constructed from $T - u$ by joining $u$ to every vertex of $T - u$, it follows that $T$ is also unique. ∎

In light of Theorems 2.7 and 2.8, it follows that for every positive integer $n$, there is exactly one acyclic tournament of order $n$.

Although there is only one transitive tournament of each order $n$, in a certain sense, which we now describe, every tournament has the structure of a transitive tournament. Let $T$ be a tournament. We define a relation on $V(T)$ by saying that $u$ is related to $v$ if there is both a $u$-$v$ path and a $v$-$u$ path in $T$. This relation is an equivalence relation and, as such, this relation partitions $V(T)$ into equivalence classes $V_1, V_2, \ldots, V_k$ ($k \geq 1$). Let $S_i = T[V_i]$ for $i = 1, 2, \ldots, k$. Then each subdigraph $S_i$ is a strongly connected tournament and, indeed, is maximal with respect to the property of being strongly connected. The subdigraphs $S_1, S_2, \ldots, S_k$ are called the **strong components** of $T$. So the vertex sets of the strong components of $T$ produce a partition of $V(T)$.

Let $T$ be a tournament with strong components $S_1, S_2, \ldots, S_k$, and let $\widetilde{T}$ denote that digraph whose vertices $u_1, u_2, \ldots, u_k$ are in one-to-one correspondence with these strong components (where $u_i$ corresponds to $S_i$, $i = 1, 2, \ldots, k$) such that $(u_i, u_j)$ is an arc of $\widetilde{T}$, $i \neq j$, if and only if some vertex of $S_i$ is adjacent to some vertex of $S_j$. If $(u_i, u_j)$ is an arc of $\widetilde{T}$, then because $S_i$ and $S_j$ are distinct strong components of $T$, it follows that every vertex of $S_i$ is adjacent to *every* vertex of $S_j$. Hence, $\widetilde{T}$ is obtained by identifying the vertices of $S_i$ for $i = 1, 2, \ldots, k$. A tournament $T$ and its associated digraph $\widetilde{T}$ are shown in Figure 2.8.

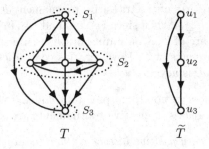

**Figure 2.8.** A tournament $T$ and its associated transitive tournament $\widetilde{T}$

Observe that for the tournament $T$ of Figure 2.8, $\widetilde{T}$ is itself a tournament, indeed a transitive tournament. That this always occurs follows from Theorem 2.9 (see Exercise 2.25).

**Theorem 2.9.** *If $T$ is a tournament with exactly $k$ strong components, then $\widetilde{T}$ is the transitive tournament of order $k$.*

Since for every tournament $T$, the tournament $\widetilde{T}$ is transitive, it follows that if $T$ is a tournament that is not strongly connected, then $V(T)$ can be partitioned as $\{V_1, V_2, \ldots, V_k\}$ ($k \geq 2$) such that $T[V_i]$ is a strongly connected tournament for each $i$, and if $v_i \in V_i$ and $v_j \in V_j$, where $i < j$, then $(v_i, v_j) \in E(T)$. This decomposition is often useful when studying the properties of tournaments that are not strongly connected.

We already noted that there are four tournaments of order 4. Of course, one of these is transitive, which consists of four trivial strong components $S_1, S_2, S_3, S_4$, where the vertex of $S_i$ is adjacent to the vertex of $S_j$ if and only if $i < j$. There are two tournaments of order 4 containing two strong components $S_1$ and $S_2$, depending on whether $S_1$ or $S_2$ is the strong component of order 3. (No strong component has order 2.) Since there are four tournaments of order 4, there is exactly one strongly connected tournament of order 4. These tournaments are depicted in Figure 2.9 where all arcs not depicted are directed downward.

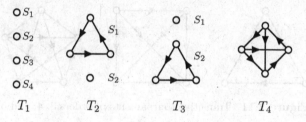

**Figure 2.9.** The four tournaments of order 4

We also stated that there are 12 tournaments of order 5. There are six tournaments of order 5 that are not strongly connected, as shown in Figure 2.10. Again all arcs not depicted are directed downward. Thus there are six strongly connected tournaments of order 5.

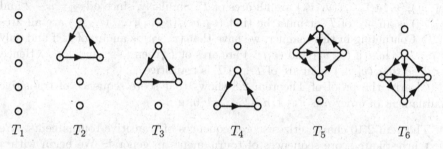

**Figure 2.10.** The six tournaments of order 5 that are not strongly connected

# 2.4 Score sequences

Suppose a tournament $T$ of order $n$ with vertex set $V(T) = \{v_1, v_2, \ldots, v_n\}$ represents a round-robin tournament between teams $v_1, v_2, \ldots, v_n$. If team $v_i$ defeats team $v_j$, then $(v_i, v_j)$ is an arc of $T$. The number of victories by team $v_i$ is the outdegree of $v_i$. For this reason, the outdegree of the vertex $v_i$ in a tournament is also referred to as the **score** of $v_i$. A sequence $s_1, s_2, \ldots, s_n$ of nonnegative integers is called a **score sequence of a tournament** if there exists a tournament $T$ of order $n$ whose vertices can be labeled $v_1, v_2, \ldots, v_n$ such that od $v_i = s_i$ for $i = 1, 2, \ldots, n$.

Figure 2.11 shows transitive tournaments of order $n$ for $n = 3, 4, 5$. Notice that the score sequence of each tournament of order $n$ in Figure 2.11 is $n-1, n-2, \ldots, 1, 0$. That this is always the case is proved next.

**Figure 2.11.** Transitive tournaments of orders 3, 4, and 5

**Theorem 2.10.** *Let $T$ be a tournament of order $n$. Then the score sequence of $T$ is $n-1, n-2, \ldots, 1, 0$ if and only if $T$ is transitive.*

*Proof.* Let $T$ be a tournament of order $n$. Suppose first that the score sequence of $T$ is $n-1, n-2, \ldots, 1, 0$. Without loss of generality, let $V(T) = \{v_1, v_2, \ldots, v_n\}$ with $\text{od}\, v_i = n - i$ for $i = 1, 2, \ldots, n$. Since $\text{od}\, v_1 = n - 1$, it follows that $(v_1, v_2), (v_1, v_3), \ldots, (v_1, v_n)$ are all arcs of $T$. Similarly, since $\text{od}\, v_2 = n - 2$ and $(v_1, v_2)$ is an arc of $T$, it must be that $(v_2, v_3), (v_2, v_4), \ldots, (v_2, v_n)$ are all arcs of $T$. Continuing in this manner, we have that $(v_j, v_k)$ is an arc of $T$ if and only if $j < k$. Thus, if $(v_i, v_j)$ and $(v_j, v_k)$ are arcs of $T$, then $i < j$ and $j < k$. Hence, $i < k$ so that $(v_i, v_k)$ is an arc of $T$ and $T$ is transitive.

Clearly, the proof of Theorem 2.8 shows that score sequence of transitive tournament of order $n$ is $n-1, n-2, \ldots, 1, 0$. ∎

Theorem 2.10 characterizes score sequences of transitive tournaments. We next investigate score sequences of tournaments in general. We begin with a theorem similar to Theorem 1.4.

**Theorem 2.11.** *A nondecreasing sequence $\pi : s_1, s_2, \ldots, s_n$ $(n \geq 2)$ of nonnegative integers is a score sequence of a tournament if and only if the sequence $\pi_1 : s_1, s_2, \ldots, s_{s_n}, s_{s_n+1} - 1, \ldots, s_{n-1} - 1$ is a score sequence of a tournament.*

*Proof.* Assume that $\pi_1$ is a score sequence of a tournament. Then there exists a tournament $T_1$ of order $n - 1$ having $\pi_1$ as a score sequence. Hence the vertices of $T_1$ can be labeled as $v_1, v_2, \ldots, v_{n-1}$ such that

$$\text{od}\, v_i = \begin{cases} s_i & \text{for } 1 \leq i \leq s_n \\ s_i - 1 & \text{for } i > s_n. \end{cases}$$

We construct a tournament $T$ by adding a vertex $v_n$ to $T_1$ where $v_n$ is adjacent to $v_i$ if $1 \leq i \leq s_n$ and $v_n$ is adjacent from $v_i$ otherwise. The tournament $T$ then has $\pi$ as a score sequence.

For the converse, we assume that $\pi$ is a score sequence. Hence there exist tournaments of order $n$ whose score sequence is $\pi$. Among all such tournaments, let $T$ be one such that $V(T) = \{v_1, v_2, \ldots, v_n\}$, $\text{od}\, v_i = s_i$ for $i = 1, 2, \ldots, n$ and the sum of the scores of the vertices adjacent from $v_n$ is minimum. We claim that $v_n$ is adjacent to vertices having scores $s_1, s_2, \ldots, s_{s_n}$. Assume, to the contrary, that $v_n$ is not adjacent to vertices having scores $s_1, s_2, \ldots, s_{s_n}$. Necessarily, then, there exist vertices $v_j$ and $v_k$ with $j < k$ and $s_j < s_k$ such that $v_n$ is adjacent to $v_k$ and $v_n$ is adjacent from $v_j$. Since the score of $v_k$ exceeds the score of $v_j$, there exists a vertex $v_t$ such that $v_k$ is adjacent to $v_t$, and $v_t$ is adjacent to $v_j$ as shown below.

Thus, a 4-cycle $C: v_n, v_k, v_t, v_j, v_n$ is produced. If we reverse the directions of the arcs of $C$, a tournament $T'$ is obtained also having $\pi$ as a score sequence as shown below.

However, in $T'$, the vertex $v_n$ is adjacent to $v_j$ rather than $v_k$. Hence the sum of the scores of the vertices adjacent from $v_n$ is smaller in $T'$ than in $T$, which is impossible. Thus, as claimed, $v_n$ is adjacent to vertices having scores $s_1, s_2, \ldots, s_{s_n}$. Then $T - v_n$ is a tournament having score sequence $\pi_1$. ∎

As an illustration of Theorem 2.11, we consider the sequence

$$\pi : 1, 2, 2, 3, 3, 4.$$

In this case, $s_n$ (actually $s_6$) has the value 4; thus, we delete the last term, repeat the first $s_n = 4$ terms, and subtract 1 from the remaining terms, obtaining

$$\pi_1' : 1, 2, 2, 3, 2.$$

Rearranging, we have

$$\pi_1 : 1, 2, 2, 2, 3.$$

Repeating this process twice more, we have

$$\pi_2' \;:\; 1, 2, 2, 1$$
$$\pi_2 \;:\; 1, 1, 2, 2$$
$$\pi_3 \;:\; 1, 1, 1.$$

The sequence $\pi_3$ is clearly a score sequence of a tournament. By Theorem 2.11, $\pi_2$ is as well, as are $\pi_1$ and $\pi$. We can use this information to construct a tournament with score sequence $\pi$. The sequence $\pi_3$ is the score sequence of the tournament $T_3$ of Figure 2.12. Proceeding from $\pi_3$ to $\pi_2$, we add a new vertex to $T_3$ and join it *to* two vertices of $T_3$ and *from* the other, producing a tournament $T_2$ with score sequence $\pi_2$. To proceed from $\pi_2$ to $\pi_1$, we add a new vertex to $T_2$ and join it *to* vertices having scores 1, 2, and 2 and *from* the remaining vertex of $T_2$, producing a tournament $T_1$ with score sequence $\pi_1$. Continuing in the same fashion, we finally produce a desired tournament $T$ with score sequence $\pi$ by adding a new vertex to $T_1$ and joining it *to* vertices having scores 1, 2, 2, and 3, and joining it *from* the other vertex.

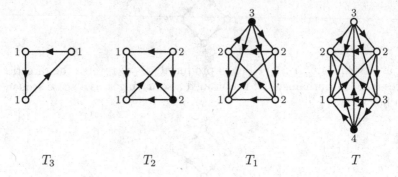

$$T_3 \qquad\qquad T_2 \qquad\qquad T_1 \qquad\qquad T$$

**Figure 2.12.** Construction of a tournament with a given score sequence

The following theorem by Landau [156] gives a nonconstructive criterion for a sequence of nonnegative integers to be the score sequence of a tournament, and the proof given here is due to Thomassen [231].

**Theorem 2.12.** *A nondecreasing sequence* $\pi : s_1, s_2, \ldots, s_n$ *of nonnegative integers is a score sequence of a tournament if and only if for each integer $k$ with* $1 \leq k \leq n$,

$$\sum_{i=1}^{k} s_i \geq \binom{k}{2}, \qquad (2.1)$$

*with equality when $k = n$.*

*Proof.* Suppose first that $\pi : s_1, s_2, \ldots, s_n$ is a score sequence of a tournament of order $n$. Then there exists a tournament $T$ with $V(T) = \{v_1, v_2, \ldots, v_n\}$ such that $\mathrm{od}\, v_i = s_i$ for $i = 1, 2, \ldots, n$. For an integer $k$ with $1 \leq k \leq n$ and $S = \{v_1, v_2, \ldots, v_k\}$, the subdigraph $T_1 = T[S]$ induced by $S$ is a tournament of order $k$ and size $\binom{k}{2}$. Since $\mathrm{od}_T v_i \geq \mathrm{od}_{T_1} v_i$ for $1 \leq i \leq k$, it follows that

$$\sum_{i=1}^{k} s_i = \sum_{i=1}^{k} \mathrm{od}_T v_i \geq \sum_{i=1}^{k} \mathrm{od}_{T_1} v_i = \binom{k}{2}.$$

We now verify the converse. Suppose that the converse is false. Then among all counterexamples for which $n$ is minimum, let $\pi : s_1, s_2, \ldots, s_n$ be one for which $s_1$ is minimum. Suppose first that there exists an integer $k$ with $1 \leq k \leq n-1$ such that

$$\sum_{i=1}^{k} s_i = \binom{k}{2}. \qquad (2.2)$$

Since $k < n$, it follows that $\pi_1 : s_1, s_2, \ldots, s_k$ is a score sequence of a tournament $T_1$ of order $k$.

Let $\tau : t_1, t_2, \ldots, t_{n-k}$ be the sequence, where $t_i = s_{k+i} - k$ for $i = 1, 2, \ldots, n - k$. Since

$$\sum_{i=1}^{k+1} s_i \geq \binom{k+1}{2},$$

it follows from (2.2) that

$$s_{k+1} = \sum_{i=1}^{k+1} s_i - \sum_{i=1}^{k} s_i \geq \binom{k+1}{2} - \binom{k}{2} = k.$$

Since $\pi$ is a nondecreasing sequence,

$$t_i = s_{k+i} - k \geq s_{k+1} - k \geq 0$$

for $i = 1, 2, \ldots, n - k$ and so $\tau$ is a nondecreasing sequence of nonnegative integers. We now show that $\tau$ satisfies (2.1).

For each integer $r$ with $1 \leq r \leq n - k$, we have

$$\sum_{i=1}^{r} t_i = \sum_{i=1}^{r} (s_{k+i} - k) = \sum_{i=1}^{r} s_{k+i} - rk = \sum_{i=1}^{r+k} s_i - \sum_{i=1}^{k} s_i - rk.$$

Since

$$\sum_{i=1}^{r+k} s_i \geq \binom{r+k}{2}$$

and

$$\sum_{i=1}^{k} s_i = \binom{k}{2},$$

it follows that

$$\sum_{i=1}^{r} t_i \geq \binom{r+k}{2} - \binom{k}{2} - rk = \binom{r}{2},$$

with equality for $r = n - k$. Thus, $\tau$ satisfies (2.1). Since $n - k < n$, there is a tournament $T_2$ of order $n - k$ having score sequence $\tau$.

Let $T$ be the tournament with $V(T) = V(T_1) \cup V(T_2)$ and

$$E(T) = E(T_1) \cup E(T_2) \cup \{(u, v) : u \in V(T_2), v \in V(T_1)\}.$$

Then $\pi$ is a score sequence for $T$, contrary to our assumption. Consequently,

$$\sum_{i=1}^{k} s_i > \binom{k}{2}$$

for $k = 1, 2, \ldots, n - 1$. In particular, $s_1 > 0$.

We now consider the sequence $\pi' : s_1 - 1, s_2, s_3, \ldots, s_{n-1}, s_n + 1$. Then $\pi'$ is a nondecreasing sequence of nonnegative integers satisfying (2.1). By the minimality of $s_1$, there is a tournament $T'$ of order $n$ having score sequence $\pi'$. Let $x$ and $y$ be vertices of $T'$ such that $\text{od}_{T'}\, x = s_n + 1$ and $\text{od}_{T'}\, y = s_1 - 1$. Since $\text{od}_{T'}\, x \geq \text{od}_{T'}\, y + 2$, there is a vertex $w \neq x, y$ such that $(x, w) \in E(T')$ and $(w, y) \in E(T')$. Thus, $P : x, w, y$ is a path in $T'$.

Let $T$ be a tournament obtained from $T'$ by reversing the directions of the arcs in $P$. Then $\pi$ is a score sequence for $T$, producing a contradiction. ∎

Harary and Moser [121] obtained a related characterization of sequences of nonnegative integers that are score sequences of strongly connected tournaments (see Exercise 2.45).

**Theorem 2.13.** *A nondecreasing sequence $\pi : s_1, s_2, \ldots, s_n$ of nonnegative integers is a score sequence of a strongly connected tournament if and only if*

$$\sum_{i=1}^{k} s_i > \binom{k}{2}$$

*for $1 \leq k \leq n - 1$ and*

$$\sum_{i=1}^{n} s_i = \binom{n}{2}.$$

*Furthermore, if $\pi$ is a score sequence of a strongly connected tournament, then every tournament with score sequence $\pi$ is strongly connected.*

# Exercises for Chapter 2

### Section 2.1.  Fundamentals

**2.1.** We have seen, in Exercise 1.44, that these does not exist a graph whose vertices have distinct degrees. Show that there exists a digraph of order 5 whose vertices have distinct outdegrees and distinct indegrees.

**2.2.** Does there exist a digraph of order 5 whose vertices have distinct outdegrees but the same indegree?

**2.3.** Determine all digraphs of order 4 and size 4.

**2.4.** Show that for every positive integer $k$, there exists a digraph of even order, half of whose vertices have outdegree $a$ and half have outdegree $b$ and $a - b = k$.

**2.5.** If all vertices of a digraph $D$ of order 5 have distinct outdegrees except for two vertices that have the same outdegree $a$, then what are the possible values of $a$?

**2.6.** Prove or disprove: No digraph contains an odd number of vertices of odd outdegree or an odd number of vertices of odd indegree.

**2.7.** Prove or disprove: If $D_1$ and $D_2$ are digraphs with $V(D_1) = \{u_1, u_2, \ldots, u_n\}$ and $V(D_2) = \{v_1, v_2, \ldots, v_n\}$ such that $\text{id}_{D_1} u_i = \text{id}_{D_2} v_i$ and $\text{od}_{D_1} u_i = \text{od}_{D_2} v_i$ for $i = 1, 2, \ldots, n$, then $D_1 \cong D_2$.

**2.8.** Prove that there exist regular tournaments of every odd order, but there are no regular tournaments of even order.

**2.9.** Let $T$ be a tournament with $V(T) = \{v_1, v_2, \ldots, v_n\}$. We know that

$$\sum_{i=1}^{n} \text{od}\, v_i = \sum_{i=1}^{n} \text{id}\, v_i = \binom{n}{2}.$$

   (a) Prove that $\sum_{i=1}^{n}(\text{od}\, v_i)^2 = \sum_{i=1}^{n}(\text{id}\, v_i)^2$.
   (b) Prove or disprove: $\sum_{i=1}^{n}(\text{od}\, v_i)^3 = \sum_{i=1}^{n}(\text{id}\, v_i)^3$.

**2.10.** Prove Theorem 2.2: If $D$ is a digraph such that $\text{od}\, v \geq k \geq 1$ for every vertex $v$ of $D$, then $D$ contains a cycle of length at least $k + 1$.

**2.11.** Prove that if $D$ is a digraph such that $\text{id}\, v \geq k \geq 1$ for every vertex $v$ of $D$, then $D$ contains a cycle of length at least $k + 1$.

**2.12.** Prove that a graph has an orientation with no directed paths of length 2 if and only if the graph is bipartite.

**2.13.** Prove that every digraph $D$ contains a set $S$ of vertices with the properties (1) no two vertices in $S$ are adjacent in $D$ and (2) for every vertex $v$ of $D$ not in $S$, there exists a vertex $u$ in $S$ such that $\vec{d}(u, v) \leq 2$.

## Section 2.2. Strongly connected digraphs

**2.14.** Prove Theorem 2.3: Let $u$ and $v$ be two vertices in a digraph $D$. For every $u$-$v$ walk $W$ in $D$, there exists a $u$-$v$ path $P$ such that every arc of $P$ belongs to $W$.

**2.15.** Show that a digraph $D$ is strongly connected if and only if its converse $\bar{D}$ is strongly connected.

**2.16.** Let $G$ be a nontrivial connected graph without bridges.
   (a) Show that for every edge $e$ of $G$ and for every orientation of $e$, there exists an orientation of the remaining edges of $G$ such that the resulting digraph is strongly connected.
   (b) Show that (a) need not be true if we begin with an orientation of two edges of $G$.

**2.17.** According to Robbins' theorem (Theorem 2.5), a nontrivial graph $G$ has a strong orientation if and only if $G$ is connected and contains no bridges.
  (a) Prove that if $G$ is a nontrivial connected graph with at most two bridges, then there exists an orientation $D$ of $G$ having the property that if $u$ and $v$ are any two vertices of $D$, there is either a $u$-$v$ path or a $v$-$u$ path.
  (b) Show that the statement (a) is false if $G$ contains three bridges.

**2.18.** Let $D$ be a digraph of order $n \geq 2$. Prove that if $\operatorname{od} v \geq (n-1)/2$ and $\operatorname{id} v \geq (n-1)/2$ for every vertex $v$ of $D$, then $D$ is strongly connected.

## Section 2.3.  Tournaments

**2.19.** Give an example of two nonisomorphic strongly connected tournaments of order 5.

**2.20.** How many tournaments of order 7 are there that are not strongly connected?

**2.21.** Determine those positive integers $n$ for which there exist regular tournaments of order $n$.

**2.22.** Give an example of two nonisomorphic regular tournaments of the same order.

**2.23.** Prove that a tournament $T$ is transitive if and only if $T$ does not contain a cycle of length 3.

**2.24.** Prove that a tournament $T$ of order $n$ is transitive if and only if the vertices of $T$ can be ordered as $v_1, v_2, \ldots, v_n$ so that $v_i$ is adjacent to $v_j$ if and only if $i < j$.

**2.25.** Prove Theorem 2.9: If $T$ is a tournament with exactly $k$ strong components, then $\widetilde{T}$ is the transitive tournament of order $k$.

**2.26.** Let $T$ be a tournament.
  (a) Show that if two vertices $u$ and $v$ have the same score in $T$, then $u$ and $v$ belong to the same strong component of $T$.
  (b) Prove that if $T$ is regular, then $T$ is strongly connected.

**2.27.** We have seen that there is exactly one transitive tournament of each order. A tournament of order $n \geq 3$ is defined to be **circular** if whenever $(u, v)$ and $(v, w)$ are arcs of $T$, then $(w, u)$ is an arc of $T$.
  (a) How many circular tournaments of order 3 are there?
  (b) Show that in a tournament of order 3 or more, every vertex, with at most two exceptions, has positive outdegree and positive indegree.
  (c) How many circular tournaments of order 4 or more are there?

**2.28.** For each positive integer $k$, there exist round-robin tournaments containing $2k$ teams with no ties permitted in which $k$ of these teams win $r$ games and the remaining $k$ of these teams win $s$ games for some $r$ and $s$ with $r \neq s$. What is the minimum value of $s$ for which this is possible?

**2.29.** Prove that if $u$ and $v$ are vertices of a tournament such that $\vec{d}(u,v) = k$, then $\operatorname{id} u \geq k - 1$.

**2.30.** For a tournament $T$ of order $n$, let

$$\Delta = \max\{\operatorname{od} v : v \in V(T)\} \text{ and } \delta = \min\{\operatorname{od} v : v \in V(T)\}.$$

Prove that if $\Delta - \delta < \frac{n}{2}$, then $T$ is strongly connected.

**2.31.** Let $(u,v)$ be an arc of a tournament $T$. Show that if $\operatorname{od} v > \operatorname{od} u$, then $(u,v)$ lies on a triangle of $T$.

**2.32.** Show that a tournament can contain three vertices of outdegree 1 but can never contain four vertices of outdegree 1.

**2.33.** Let $u$ and $v$ be two vertices in a tournament $T$. Prove that if $\vec{d}(u,v) = k \geq 2$, then $T$ contains a cycle of length $\ell$ for each integer $\ell$ with $3 \leq \ell \leq k + 1$.

**2.34.** Let $u$ and $v$ be two vertices in a tournament $T$. Prove that if $u$ and $v$ do not lie on a common cycle, then $\operatorname{od} u \neq \operatorname{od} v$.

**2.35.** Let $T$ be a tournament with the property that every vertex of $T$ belongs to a directed 3-cycle. Let $u$ and $v$ be distinct vertices of $T$. Prove that if $|\operatorname{od} u - \operatorname{od} v| \leq 1$, then $T$ contains both a directed $u$-$v$ path and a directed $v$-$u$ path.

**2.36.** Show that every vertex in a nontrivial regular tournament is a king.

**2.37.** A tournament $T$ of order $n$ can only be regular if $n$ is odd and so $\operatorname{od} v = (n-1)/2$ for every vertex $v$ of $T$. By Exercise 2.36, every vertex of $T$ is a king. Prove or disprove: There exists an even integer $n \geq 6$ such that for every tournament $T$ of order $n$ for which $\operatorname{od} v \geq (n-2)/2$ for each $v \in V(T)$, every vertex of $T$ is a king.

**2.38.** Show that there exists a tournament of order 4 having exactly three kings.

**2.39.** Show that there exists a tournament of order 5 having exactly four kings.

**2.40.** Show that there is an infinite class of tournaments in which every vertex except one is a king.

**2.41.** A vertex $z$ in a nontrivial tournament is called a **serf** if for every vertex $x$ distinct from $z$, either $x$ is adjacent to $z$ or $x$ is adjacent to a vertex that is adjacent to $z$. Prove that every nontrivial tournament has at least one serf.

## Section 2.4. Score sequences

**2.42.** Which of the following sequences are score sequences of tournaments? For each sequence that is a score sequence, construct a tournament having the given sequence as a score sequence.
  (a) $0, 1, 1, 4, 4$
  (b) $1, 1, 1, 4, 4, 4$
  (c) $1, 3, 3, 3, 3, 3, 5$
  (d) $2, 3, 3, 4, 4, 4, 4, 5$.

**2.43.** Show that if $\pi : s_1, s_2, \ldots, s_n$ is a score sequence of a tournament, then $\pi_1 : n - 1 - s_1, n - 1 - s_2, \ldots, n - 1 - s_n$ is a score sequence of a tournament.

**2.44.** What tournament $T$ of order $n$ has a score sequence $s_1, s_2, \ldots, s_n$ such that equality holds in (2.1) for every integer $k$ with $1 \le k \le n$?

**2.45.** Prove Theorem 2.13: a nondecreasing sequence $\pi : s_1, s_2, \ldots, s_n$ of nonnegative integers is a score sequence of a strongly connected tournament if and only if

$$\sum_{i=1}^{k} s_i > \binom{k}{2}$$

for $1 \le k \le n - 1$ and

$$\sum_{i=1}^{n} s_i = \binom{n}{2}.$$

Furthermore, if $\pi$ is a score sequence of a strongly connected tournament, then every tournament with score sequence $\pi$ is strongly connected.

**2.46.** Let $T$ be a tournament of order $n \ge 10$. Suppose that $T$ contains two vertices $u$ and $v$ such that when the arc joining $u$ and $v$ is removed, the resulting digraph $D$ contains neither a $u$-$v$ path nor a $v$-$u$ path. Show that $\mathrm{od}_D u = \mathrm{od}_D v$.

# Traversability 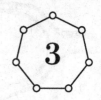 3

We begin this chapter by studying circuits that contain every edge (or arc) of a graph (or digraph). Such circuits, called eulerian circuits, are named after the famed mathematician Leonhard Euler (1707–1783). We also consider cycles that contain every vertex of a graph or digraph, called hamiltonian cycles and named after Sir William Rowan Hamilton (1805–1865).

##  3.1  Eulerian graphs and digraphs

In the 18th century, Königsberg (now known as Kaliningrad, a Russian exclave) was divided by the River Pregel and connected by seven bridges, which connected two islands in the river to each other and to the opposite banks as shown in Figure 3.1. It was a popular pastime among the residents of Königsberg to try to find a route that crossed each bridge exactly once. Euler proved that it was impossible to find such a route, a fact that many of the people of Königsberg had already suspected.

We denote the four regions of land by $A$, $B$, $C$, and $D$ as in Figure 3.1, and as Euler himself did. Euler observed that if it were possible to find a route that crossed each bridge exactly once, it could be represented by a sequence of letters, where each letter represents a land region and the letter immediately following it represents the land region reached after crossing a bridge. Since there are seven bridges that must be crossed precisely once, the sequence must have exactly eight letters. Euler also noticed that the letters $A$ and $B$, and the letters $A$ and $C$, must appear as consecutive terms in the sequence twice each. In addition, since five bridges lead to the island represented by $A$, the sequence must either start or end with $A$, and $A$ must appear a total of three times: once at the beginning or end, and twice to indicate entering and leaving the region. Similarly, each of the letters $B$, $C$, and $D$ must appear twice in the sequence. However, this means that the sequence would need nine letters, which is a contradiction. Therefore, it is impossible to find a route through Königsberg that crosses each bridge exactly once.

The Königsberg bridge problem has graphical overtones in many ways; indeed, Euler's representation of a route through Königsberg as a sequence of letters essentially encodes a walk in a graph. If the land regions of Königsberg are

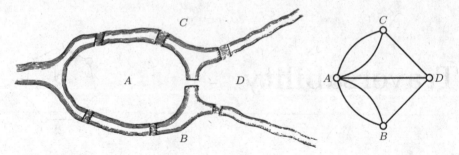

**Figure 3.1.** The bridges of Königsberg as Euler presented them (left) and the multigraph of Königsberg (right)

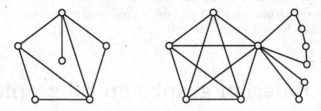

**Figure 3.2.** A graph with an eulerian trail but no eulerian circuit (left) and an eulerian graph (right)

represented by vertices and the bridges are represented by edges, then we obtain the multigraph shown on the right of Figure 3.1. The Königsberg bridge problem is then equivalent to the problem of determining whether this multigraph has a trail containing all of its edges. This problem suggests the following concepts.

An **eulerian trail** in a graph or multigraph $G$ is an *open* trail that contains every edge of $G$. Recall that trails are not allowed to repeat edges, so eulerian trails must actually contain every edge precisely once. An **eulerian circuit** in $G$ is a *closed* trail that contains every edge of $G$ (again, necessarily precisely once). A graph or multigraph that has an eulerian circuit is called **eulerian**. Note that since we allow the trivial circuit which has no edges, the graph $K_1$ is eulerian. Nontrivial examples are shown in Figure 3.2.

Simple characterizations of both eulerian (multi)graphs and (multi)graphs with eulerian trails exist. In fact, Euler [86] knew both characterizations in 1736, although complete proofs of these results were not given until the work of Hierholzer [127] in 1873.

**Theorem 3.1.** *A connected multigraph is eulerian if and only if every vertex has even degree.*

*Proof.* One direction is clear. If a vertex appears $k$ times in an eulerian circuit, then it must have degree $2k$. Thus every vertex of an eulerian multigraph must have even degree.

For the other direction, we proceed by induction on the number of edges of $G$ that every connected multigraph with all even degrees has an eulerian circuit. For the base case, if $G$ has 0 edges and is connected, then $G$ must be $K_1$, which is eulerian. Now suppose that $G$ is a connected multigraph with at least one edge and all degrees even, and that the result holds for all multigraphs with fewer edges.

Consider a trail $T: v_0, v_1, \ldots, v_k$ of maximal length in $G$. Suppose first that $v_k \neq v_0$. Then, since $v_k$ has even degree and every time $v_k$ is encountered on $T$ prior to the last requires two edges incident to $v_k$, there is an edge incident to $v_k$ that is not on $T$ and hence $T$ can be extended, a contradiction. Therefore $T$ must be closed, and it must be nontrivial because $G$ has at least one edge. If $T$ contains all of the edges of $G$, then we are done, so we may assume that the multigraph $G' = G - E(T)$ has at least one edge.

Since $T$ is closed, every vertex of $G$ lies in an even number of edges of $T$, so $G'$ also has all degrees even. Since $T$ is nontrivial, $G'$ has fewer edges than $G$. Since $G$ is connected and $T$ does not contain all of its edges, $G'$ must contain an edge $e$ incident with a vertex of $T$. By the induction hypothesis, we may conclude that the connected component of $G'$ that contains $e$ contains a nontrivial eulerian circuit $T'$, which must itself contain $e$. Thus by following $T$ until we reach a vertex incident with $e$, then following $T'$, and then finishing $T$, we obtain a longer trail in $G$ than $T$. As this would contradict our choice of $T$, it must be the case that $T$ is an eulerian circuit, as desired. ∎

Theorem 3.1 allows us to see easily that the multigraph of Königsberg shown on the right of Figure 3.1 is not eulerian; in fact, every vertex of this multigraph has odd degree. This theorem also leads quickly to a characterization of the connected multigraphs containing an eulerian trail.

**Theorem 3.2.** *A connected multigraph contains an eulerian trail if and only if precisely two of its vertices have odd degree. Furthermore, in this case all of its eulerian trails begin at one of these vertices and end at the other.*

*Proof.* Let $G$ be a connected multigraph with distinct vertex $u$ and $v$. Then $G$ contains an eulerian $u$-$v$ trail if and only if the multigraph obtained from $G$ by adding a new edge between $u$ and $v$ contains an eulerian circuit. The result now follows from Theorem 3.1. ∎

Using this result, we see that the multigraph of Königsberg from Figure 3.1 also does not have an eulerian trail.

##  Eulerian digraphs

Eulerian circuits and trails in graphs have natural analogues for digraphs. An **eulerian circuit** in a connected digraph $D$ is a circuit that contains every arc of $D$ (necessarily precisely once), while an **eulerian trail** in $D$ is an open trail that contains every arc of $D$. A connected digraph that contains an eulerian circuit is an **eulerian digraph**. The characterizations of eulerian digraphs and digraphs with eulerian trails are similar to the characterizations provided in the

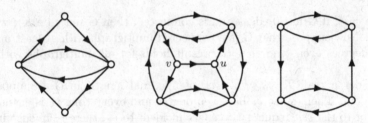

**Figure 3.3.** Eulerian circuits and trails in digraphs

undirected case by Theorems 3.1 and 3.2. As the proofs are also similar, we leave them for Exercises 3.14 and 3.16.

**Theorem 3.3.** *A weakly connected digraph is eulerian if and only if* $\operatorname{od} v = \operatorname{id} v$ *for every vertex* $v$.

**Theorem 3.4.** *A weakly connected digraph contains an eulerian trail if and only if it contains two vertices* $u$ *and* $v$ *such that*

$$\operatorname{od} u = \operatorname{id} u + 1 \quad and \quad \operatorname{id} v = \operatorname{od} v + 1,$$

*while* $\operatorname{od} w = \operatorname{id} w$ *for all other vertices* $w$. *Moreover, in this case all of its eulerian trails begin at* $u$ *and end at* $v$.

With the aid of Theorems 3.3 and 3.4, we see that the digraph on the left of Figure 3.3 contains an eulerian circuit, the digraph in the middle of this figure contains an eulerian $u$-$v$ trail, and the digraph on the right of this figure contains neither an eulerian circuit nor an eulerian trail.

## Veblen's theorem and cycle double covers

There are several other characterizations of eulerian graphs and digraphs. One of the oldest, due to Veblen [243] in 1912, characterizes these graphs in terms of edge-disjoint cycles (that is, cycles that do not share any edges). The analogous result for digraphs is also true (Exercise 3.17).

**Theorem 3.5 (Veblen's theorem).** *A connected graph is eulerian if and only if it contains a collection of pairwise edge-disjoint cycles that together contain all of the edges of the graph.*

*Proof.* One direction is trivial. If a graph contains such a collection of cycles, then every vertex has even degree and the graph is eulerian by Theorem 3.1.

We prove the other direction by induction on the number of edges of a connected eulerian graph $G$. If $G$ has no edges, then $G$ must be $K_1$, which does contain such a collection of cycles (the empty collection). Now suppose that $G$ has at least one edge, and that all eulerian graphs with fewer edges than $G$ have such a collection of cycles. Since $G$ is eulerian and has at least one edge, it is not

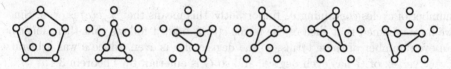

**Figure 3.4.** A cycle double cover of the Petersen graph

a tree. Let $C$ be any cycle of $G$ and define $H = G - E(C)$. Every component of $H$ has all degrees even. Therefore, by the induction hypothesis, each of these components contains a collection of pairwise edge-disjoint cycles such that every edge of the component lies on one of the cycles of the collection. Taking all of these collections of cycles together with $C$ gives the desired collection of cycles for $G$. ∎

A **cycle double cover** of a graph is a collection of (not necessarily distinct) cycles of the graph that together contain each of its edges precisely twice. Veblen's theorem (Theorem 3.5) implies that eulerian graphs always possess a cycle double cover, because we can take the double cover to consist of two copies of every cycle given by Veblen's theorem. Many graphs that are not eulerian also have cycle double covers, for example, the Petersen graph has a cycle double cover, consisting of six 5-cycles, as shown in Figure 3.4. On the other hand, it is easy to see that graphs with bridges cannot have cycle double covers. Szekeres [225] in 1973 and Seymour [218] in 1979 independently conjectured that this is the only obstruction to graphs having cycle double covers.

**Conjecture 3.6** (Cycle double cover conjecture). *Every bridgeless graph has a cycle double cover.*

Bondy [30] later proposed the stronger **small cycle double cover conjecture**, which states that we can always find a cycle double cover with fewer cycles than the graph has vertices.

## ⊶ The Toida–McKee theorem

A more recent characterization of eulerian graphs involving cycles is given next; the necessity is due to Toida [233] in 1973, while McKee [171] established the sufficiency in 1984. The digraph analogue of this result is false (Exercise 3.18).

**Theorem 3.7** (Toida–McKee theorem). *A connected graph is eulerian if and only if each of its edges lies on an odd number of cycles.*

*Proof.* We begin with the easier direction. Let $G$ be a connected graph in which every edge lies on an odd number of cycles, and take $v$ to be an arbitrary vertex of $G$. For each edge $e$ incident with $v$, denote by $c(e)$ the number of cycles containing $e$. Since every cycle containing $v$ must contain two edges incident to it, the sum $\sum c(e)$, taken over all edges $e$ incident with $v$, is equal to twice the number of cycles containing $v$. Importantly, this means that $\sum c(e)$ is even. Since we have assumed that each of the terms $c(e)$ is odd, it follows that there must be an even number of these terms, so the degree of $v$ is even. Since $v$ was arbitrary, every vertex of $G$ has even degree, and so $G$ is eulerian by Theorem 3.1.

The converse is more difficult. Suppose that $G$ is a connected graph with all degrees even, let $uv$ be an arbitrary edge of $G$, and consider the set of all $u$-$v$ trails in $G - uv$ in which $v$ appears precisely once (necessarily at the end). This is a finite set because trails are not allowed to repeat edges. Moreover, since $G$ is eulerian, $G - uv$ is connected, so there is at least one such trail.

There are an odd number of edges possible for the initial edge of such a trail because the degree of $u$ in $G - uv$ is odd. Suppose we choose $vw$ as the initial edge. There are then an odd number of choices for next edge, because $w$ is incident to an odd number of edges that are different from $uw$. We continue this process until we arrive at $v$. At each vertex other than $v$ in such a trail, there are an odd number of edges available along which to continue the trail, and thus we can never fail to reach $v$ eventually.

This shows that there are an odd number of $u$-$v$ trails in $G - uv$ in which $v$ appears precisely once. We now argue that there are an odd number of $u$-$v$ *paths* in $G - uv$ (in these paths, $v$ necessarily appears precisely once). Suppose that $T$ is such a $u$-$v$ trail that is not a path. This implies that $T$ contains some vertex at least twice. Let $x$ denote the first vertex encountered on this trail that is later re-encountered. Thus $T$ contains an $x$-$x$ circuit $C \colon x = v_1, v_2, \ldots, v_k, v_1 = x$, and so there is another $u$-$v$ trail $T'$ that is identical to $T$ except that it traverses the circuit $C$ in the reverse order $x = v_1, v_k, \ldots, v_2, v_1 = x$. This proves that the $u$-$v$ trails of $G - uv$ in which $v$ appears precisely once that are not $u$-$v$ paths occur in pairs. Therefore, there are an odd number of $u$-$v$ paths in $G - uv$. Since these paths are in bijection with cycles in $G$ that use the edge $uv$, the proof of this direction of the theorem is complete. $\blacksquare$

 ## 3.2  Hamiltonian graphs

A **hamiltonian path** is a path containing all of the vertices of a graph, while a **hamiltonian cycle** is a cycle containing all of the vertices. A graph itself is called **hamiltonian** if it has a hamiltonian cycle. The single-vertex graph $K_1$ is considered to be hamiltonian (with the trivial, one-vertex cycle as its hamiltonian cycle). Because of the similarity of their definitions, one might expect hamiltonian graphs to have a simple characterization similar to that of eulerian graphs. As we will see, this is not the case.

This property is named for the mathematician Sir William Rowan Hamilton. In 1857, before the property had ever been formally studied, Hamilton introduced a game in which players attempted to find a hamiltonian cycle in the graph shown in Figure 3.5, which is the graph of the dodecahedron. The Icosian Game, as

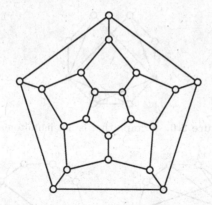

**Figure 3.5.** The graph of the dodecahedron

Hamilton called it, consisted of a wooden board with 20 pegs (the prefix icos- is Greek for 20), labeled by names of major cities at the time, and the player was to construct a round trip around the world in which each of the cities was visited precisely once. Since the graph of the dodecahedron is hamiltonian (as the reader is asked to verify in Exercise 3.25), Hamilton's Icosian Game can be won.

If $G$ is a hamiltonian graph, then $G$ is certainly connected and has no cut-vertices. This provides a necessary condition for a graph to be hamiltonian, but we can say more.

**Theorem 3.8.** *If $G$ is a hamiltonian graph, then*

$$k(G - S) \leq |S|$$

*for every nonempty subset $S \subseteq V(G)$, where $k(G - S)$ denotes the number of components of $G - S$.*

*Proof.* Suppose that $G$ is hamiltonian and that $S \subseteq V(G)$ is nonempty. Let $C: v_1, v_2, \ldots, v_n, v_1$ be a hamiltonian cycle of $G$. Because $S \neq \emptyset$, we may assume without loss of generality that $v_n \in S$. If $v_i$ is the last vertex of its component of $G - S$ that is encountered by $C$ (meaning that for all $i + 1 \leq j \leq n$, either $v_j$ lies in $S$ or in a different component of $G - S$ from $v_i$), then it must be the case that $v_{i+1} \in S$. This implies that $S$ must contain at least as many vertices as $G - S$ has components, as desired. ∎

Theorem 3.8 is most useful in its contrapositive formulation:

> *If there is a nonempty subset $S \subseteq V(G)$ such that the graph $G - S$ has more than $|S|$ components, then $G$ is not hamiltonian.*

For example, the graph shown in Figure 3.6 is not hamiltonian, because if we remove the two solid vertices, then we obtain a graph with $3 > 2$ components.

Chvátal [50] called a graph satisfying the necessary condition in Theorem 3.8 a **tough graph**. It can be shown, for example, that the Petersen graph is tough

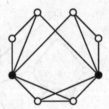

**Figure 3.6.** A graph that is not hamiltonian

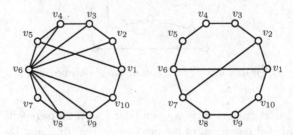

**Figure 3.7.** Proving that the Petersen graph is not hamiltonian

(see Exercise 3.30). On the other hand, we show below that the Petersen graph is not hamiltonian, so toughness is a necessary but not sufficient condition for a graph to be hamiltonian.

**Theorem 3.9.** *The Petersen graph is not hamiltonian.*

*Proof.* Suppose to the contrary that the Petersen graph did have a hamiltonian cycle, and label the vertices of the graph so that this cycle is

$$C: v_1, v_2, \ldots, v_{10}, v_1.$$

The Petersen graph is cubic, so every vertex must be incident to precisely one edge not on $C$.

Consider the vertex $v_1$. The Petersen graph does not contain 3- or 4-cycles, so $v_1$ cannot be adjacent to $v_3$, $v_4$, $v_8$, or $v_9$. Hence, $v_1$ must be adjacent to precisely one of $v_5$, $v_6$, or $v_7$. If $v_1$ was adjacent to $v_5$, as shown on the left of Figure 3.7, then the edge incident to $v_6$ that does not lie on $C$ must lie in a 3- or 4-cycle, a contradiction. Therefore, $v_1$ cannot be adjacent to $v_5$. By symmetry, $v_1$ also cannot be adjacent to $v_7$.

This leaves as the only possibility that $v_1$ is adjacent to $v_6$. Thus by a symmetrical argument, we must also have $v_2$ adjacent to $v_7$. However, in that case, $v_1, v_2, v_7, v_6, v_1$ would be a 4-cycle, as shown on the right of Figure 3.7, a contradiction. ■

## Sufficient conditions

Dirac [65] gave the first sufficient condition for a graph to be hamiltonian in 1952, and this initiated a long string of results.

**Theorem 3.10 (Dirac's theorem).** *Let $G$ be a graph with $n \geq 3$ vertices. If*

$$\deg v \geq n/2$$

*for every vertex $v$ of $G$, then $G$ is hamiltonian.*

*Proof.* Suppose that the statement is false, and choose an integer $n \geq 3$ such that there is a nonhamiltonian graph with $n$ vertices, all with degree at least $n/2$. Further let $G$ be such a graph with as many edges as possible. Because it is not hamiltonian, $G$ is certainly not a complete graph, so we can find two nonadjacent vertices and label them $v_1$ and $v_n$. By our choice of $G$, the graph $G + v_1 v_n$ has a hamiltonian cycle, and since $G$ is not hamiltonian, the hamiltonian cycle in $G + v_1 v_n$ must use the edge $v_1 v_n$. Label the rest of the vertices so that this hamiltonian cycle is $v_1, v_2, \ldots, v_n, v_1$.

The graph $G$ therefore contains the hamiltonian path, $v_1, v_2, \ldots, v_n$. Consider the $n - 1$ edges $v_i v_{i+1}$ of this path. For at least $n/2$ of these edges, we must have $v_1$ adjacent to $v_{i+1}$, and for at least $n/2$ of these edges, we must have $v_n$ adjacent to $v_i$. Therefore, there must be some edge $v_k v_{k+1}$ such that $v_1$ is adjacent to $v_{k+1}$ and $v_n$ is adjacent to $v_k$, as shown below.

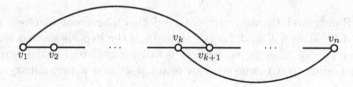

However, as this drawing shows, $G$ then contains the hamiltonian cycle

$$v_1, v_{k+1}, v_{k+2}, \ldots, v_n, v_k, v_{k-1}, \cdots, v_2, v_1,$$

and this contradiction completes the proof of the theorem. ∎

A careful inspection of our proof of Dirac's theorem shows that we only used the degree hypothesis to find the edge $v_k v_{k+1}$. In 1960, Ore [184] observed that this hypothesis can be weakened.

**Theorem 3.11 (Ore's theorem).** *Let $G$ be a graph with $n \geq 3$ vertices. If*

$$\deg u + \deg v \geq n$$

*for every pair of nonadjacent vertices $u$ and $v$, then $G$ is hamiltonian.*

*Proof.* The proof is identical to the proof of Dirac's theorem, with the exception of the argument for finding the edge $v_k v_{k+1}$. Consider again the $n-1$ edges $v_i v_{i+1}$ of the hamiltonian path $v_1, v_2, \ldots, v_n$. For at least $\deg v_1$ of these edges, we must have $v_1$ adjacent to $v_{i+1}$, and for at least $\deg v_n$ of these edges, we must have $v_n$ adjacent to $v_i$. Because our hypotheses imply that $\deg v_1 + \deg v_n \geq n$, it follows that there must be some edge $v_k v_{k+1}$ such that $v_1$ is adjacent to $v_{k+1}$ and $v_n$ is adjacent to $v_k$. This leads to the same contradiction as in the proof of Dirac's theorem, completing the proof. ∎

Ore's theorem implies a similar sufficient condition for the existence of hamiltonian paths.

**Corollary 3.12.** *Let $G$ be a graph with $n \geq 2$ vertices. If*

$$\deg u + \deg v \geq n - 1$$

*for every pair of nonadjacent vertices $u$ and $v$ of $G$, then $G$ contains a hamiltonian path.*

*Proof.* Suppose that the graph $G$ satisfies the hypotheses, and add a new vertex that is adjacent to every vertex of $G$ to obtain the graph $H = G \vee K_1$. Then $H$ satisfies the hypotheses of Ore's theorem, and so it is hamiltonian. By deleting the vertex that was added to form $H$, we obtain a hamiltonian path in $G$. ∎

 **The closure of a graph**

In 1976, Bondy and Chvátal [31] observed that the proof of Ore's theorem (Theorem 3.11) does not need the full strength of the hypothesis that the degree sum of each pair of nonadjacent vertices is at least the order of the graph. This observation leads to a more powerful result, but first a preliminary result is needed.

**Theorem 3.13.** *Let $u$ and $v$ be nonadjacent vertices in a graph $G$ of order $n$ satisfying*

$$\deg u + \deg v \geq n.$$

*Then $G + uv$ is hamiltonian if and only if $G$ is hamiltonian.*

*Proof.* Clearly $G + uv$ will be hamiltonian whenever $G$ is, so it suffices to prove the converse. Suppose that $G + uv$ is hamiltonian, and consider a particular hamiltonian cycle $C$ of $G + uv$. If $C$ does not use the edge $uv$, then $C$ is a hamiltonian cycle of $G$, and we are done. If $C$ does use the edge $uv$, then by removing this edge we obtain a hamiltonian $u$-$v$ path in $G$. We can now proceed as in the proof of Ore's theorem (Theorem 3.11) with $v_1 = u$ and $v_n = v$ to see that $G$ is hamiltonian, completing the proof. ∎

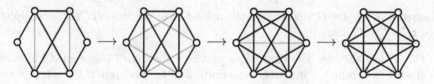

**Figure 3.8.** Constructing the closure of a graph of order 6; edges added in the next stage are indicated in gray

This result motivates our next definition. Let $G$ be a graph of order $n$. The **closure** $C(G)$ of $G$ is the graph obtained from $G$ by repeatedly joining pairs of nonadjacent vertices whose degree sum is at least $n$ until no such pair remains. An example of the transformation of a graph to its closure is shown in Figure 3.8. By Theorem 3.13, as we add these edges to the graph to form its closure, we neither create nor destroy any hamiltonian cycles.

The definition of the closure does not specify the order in which the edges should be added, which could lead to ambiguity and the possibility that a graph could have more than one closure. Before discussing closure further, we must prove that it is a **well-defined** operation, which means that the resulting graph does not depend on the order in which the edges are added. (If the operation was not well-defined, we would call it **ill-defined**.)

**Theorem 3.14.** *Closure is a well-defined operation on graphs.*

*Proof.* Let $G$ be a graph of order $n$, and suppose that both $H$ and $H'$ are obtained by repeatedly joining pairs of nonadjacent vertices whose degree sum is at least $n$, until no such pair remains (so, both $H$ and $H'$ are closures of $G$). We want to prove that $H = H'$. It suffices to prove that $H$ is a subgraph of $H'$, because then by repeating the argument with the roles of $H$ and $H'$ reversed, we will see that $H = H'$.

Suppose that $H$ is obtained from $G$ by adding the edges $e_1, e_2, \ldots, e_k$, in that order. We prove that if $e_1, e_2, \ldots, e_{i-1} \in E(H')$, then $e_i \in E(H')$. Suppose that $e_i = uv$. We know that in $G + e_1 + e_2 + \cdots + e_{i-1}$, the degree sum of $u$ and $v$ is at least $n$. Since we have assumed that $G + e_1 + e_2 + \cdots + e_{i-1}$ is a subgraph of $H'$, this means that the degree sum of $u$ and $v$ in $H'$ is also at least $n$. Therefore, the edge $e_i$ must have been added to $H'$ at some point, as desired. Since this is true for all $i$, it follows that every edge of $H$ is an edge of $H'$, completing the proof. ∎

Now that we know that closure is well-defined, we observe the following result by repeated application of Theorem 3.13.

**Theorem 3.15.** *A graph is hamiltonian if and only if its closure is hamiltonian.*

Since every complete graph with at least three vertices is hamiltonian, we immediately obtain the following.

**Theorem 3.16.** *Let $G$ be a graph with at least three vertices. If $C(G)$ is complete, then $G$ is hamiltonian.*

If a graph satisfies the conditions of Ore's theorem (Theorem 3.11), then its closure is complete, and so $G$ is hamiltonian by Theorem 3.16. Thus, Ore's theorem is an immediate corollary of Theorem 3.16. In fact, many sufficient conditions for a graph to be hamiltonian based on the degrees of the vertices of a graph can be deduced from Theorem 3.16, such as the following result of Chvátal [49].

**Theorem 3.17.** *Let $G$ be a graph of order $n \geq 3$. If the degree sequence $d_1 \leq d_2 \leq \cdots \leq d_n$ of $G$ satisfies $d_{n-i} \geq n - i$ whenever $d_i \leq i < n/2$, then $G$ is hamiltonian.*

*Proof.* Let $H = C(G)$. We will show that $H$ is complete which, by Theorem 3.16, will imply that $G$ is hamiltonian. Assume that $G$ satisfies the hypotheses of the theorem, but that $H$ is not complete. Let $u$ and $v$ be nonadjacent vertices of $H$ for which $\deg_H u + \deg_H v$ is as large as possible. Without loss of generality, we may assume that $k = \deg_H u \leq \deg_H v$, and since $u$ and $v$ are not adjacent in $H$, we have $\deg_H u + \deg_H v \leq n - 1$. Set $k = \deg_H u$, so we have $k \leq (n-1)/2 < n/2$ and $\deg_H v \leq n - k - 1$.

Because $\deg_H v \leq n - k - 1$, there are at least $k$ vertices that are not adjacent to $v$ in $H$. If $w$ is such a vertex, then by our choice of $u$ and $v$, we have

$$\deg_G w \leq \deg_H w \leq \deg_H u \leq k.$$

This shows that $G$ has at least $k$ vertices of degree at most $k$, so $d_k \leq k < n/2$.

Similarly, because $\deg_H u = k$, there are at least $n - k - 1$ vertices that are not adjacent to $u$ in $H$. If $w$ is one of these vertices, then by our choice of $u$ and $v$, we have

$$\deg_G w \leq \deg_H w \leq \deg_H v \leq n - k - 1.$$

This shows that $G$ has at least $n - k - 1$ vertices other than $u$ of degree at most $n - k - 1$. Since $u$ also has degree at most $n - k - 1$, we conclude that $d_{n-k} \leq n - k - 1$.

Therefore, we have shown that $d_k \leq k < n/2$ while $d_{n-k} < n-k$, contradicting our hypotheses. This contradiction proves that $H$ is complete, thereby completing the proof of the theorem. ∎

 **3.3   Hamiltonian digraphs**

A digraph is called **hamiltonian** if it contains a spanning cycle, which is itself called a **hamiltonian cycle**. A path that contains every vertex of a digraph is called a **hamiltonian path**.

Characterizing hamiltonian digraphs is at least as difficult as characterizing hamiltonian graphs, so there is no complete characterization in this case either. However, there are analogues of some of the sufficient conditions that we have seen in the undirected case. The proofs of these theorems are more difficult than their undirected counterparts and are therefore omitted.

One such result is due to Meyniel [176]. Readers interested in the proof of this result can refer to Bondy and Thomassen [32]. Two vertices $u, v$ of a digraph $D$ are considered **nonadjacent** if neither $(u, v)$ nor $(v, u)$ is an arc of $D$.

**Theorem 3.18 (Meyniel's theorem).** *Let $D$ be a strongly connected digraph with $n \geq 2$ vertices. If*

$$\deg u + \deg v \geq 2n - 1$$

*for every pair $u, v$ of nonadjacent vertices, then $D$ is hamiltonian.*

Meyniel's theorem on hamiltonian cycles in digraphs implies Ore's theorem on hamiltonian cycles in graphs in the following manner. Given a graph $G$ with the property that $\deg u + \deg v \geq n$ for every pair of nonadjacent vertices $u, v$, we can create a digraph $D$ on the same set of vertices by orienting each edge of $G$ in both directions. This digraph $D$ satisfies the conditions of Meyniel's theorem. Thus $D$ has a hamiltonian cycle, which corresponds to a hamiltonian cycle in $G$.

Meyniel's theorem also implies Camion's theorem (Theorem 3.23; see Exercise 3.39), and it generalizes the following two theorems on hamiltonian digraphs, by Woodall [258] and Ghouila-Houri [103] (see Exercises 3.41 and 3.42).

**Theorem 3.19 (Woodall's theorem).** *If $D$ is a digraph of order $n$ such that*

$$\operatorname{od} u + \operatorname{id} v \geq n$$

*for every pair $u, v$ of vertices with $(u, v) \notin E(D)$, then $D$ is hamiltonian.*

**Theorem 3.20 (Ghouila-Houri's theorem).** *If $D$ is a strongly connected digraph of order $n$ such that*

$$\deg v \geq n$$

*for each vertex $v$ of $D$, then $D$ is hamiltonian.*

## Tournaments

Tournaments cannot satisfy the hypotheses of Meyniel's, Woodall's, or Ghouila-Houri's theorems. However, in this well-studied case, much stronger results hold. The following fundamental result on the hamiltonicity of tournaments was first observed by Rédei [195] in 1934. This result is often considered to be the first theoretical result in the study of tournaments.

**Theorem 3.21.** *Every tournament contains a hamiltonian path.*

*Proof.* Consider a tournament $T$ of order $n$ and let $P\colon v_1, v_2, \ldots, v_k$ be a longest path in $T$. If $P$ is not a Hamiltonian path, then $k < n$ and there exists a vertex $v$ not on $P$. Since $P$ is a longest path, it follows that $(v, v_1), (v_k, v) \notin E(T)$, which implies that $(v_1, v), (v, v_k) \in E(T)$. This means that there exists a largest integer $i$ such that $(v_i, v) \in E(T)$. Thus, $(v, v_{i+1}) \in E(T)$ as shown in the diagram below.

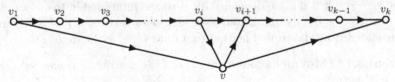

However, in this case the path

$$v_1, v_2, \ldots, v_i, v, v_{i+1}, \ldots, v_k$$

has a greater length than $P$. This contradicts our assumption that $P$ is a longest path in $T$, proving the theorem. ∎

For example, Figure 3.9 shows a tournament of order 5 consisting of three strong components $S_1$, $S_2$, and $S_3$, where $S_1$ and $S_3$ each consist of a single vertex and $S_2$ is a 3-cycle. This tournament has three hamiltonian paths, namely $P_1\colon u, v, w, x, y$, $P_2\colon u, w, x, v, y$, and $P_3\colon u, x, v, w, y$. This example in fact illustrates a result found independently by Rédei [195] and Szele [227] that strengthens Theorem 3.21: every tournament contains an odd number of hamiltonian paths.

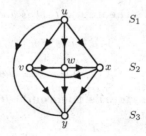

**Figure 3.9.** A tournament with three hamiltonian paths

It follows readily every transitive tournament contains precisely one hamiltonian path, but there are, not surprisingly, tournaments with many more hamiltonian paths. The next result we present, also due to Szele [227], establishes the existence of such tournaments.

**Theorem 3.22.** *For each integer $n \geq 2$, there exists a tournament of order $n$ with at least $n!/2^{n-1}$ hamiltonian paths.*

While every tournament has a hamiltonian path, not every tournament has a hamiltonian cycle. For example, the tournament shown in Figure 3.9 does not have a hamiltonian cycle, nor does any transitive tournament, because transitive tournaments are acyclic (Theorem 2.7). Recall that a digraph (or in particular, a tournament) is said to be strongly connected if there is a path between every pair of vertices in the digraph. If a tournament has a hamiltonian cycle, then it must be strongly connected, because one can use the arcs of such a cycle to reach any vertex from any other vertex. Camion [39] showed that the converse also holds, using a proof similar to that of Theorem 3.21.

**Theorem 3.23** (**Camion's theorem**). *A tournament is hamiltonian if and only if it is strongly connected.*

*Proof.* We have already observed that every hamiltonian tournament is strongly connected. The tournament with a single vertex is, trivially, both strongly connected and hamiltonian, while the tournament with two vertices is neither strongly connected nor hamiltonian. Thus, to establish the converse, let us assume that $T$ is a nontrivial strongly connected tournament. This means that $T$ contains cycles. Let $C$ be a cycle of maximum length in $T$. If $C$ contains all of the vertices of $T$, then $C$ is a hamiltonian cycle and we are done. Otherwise, let us suppose that $C$ is not hamiltonian and can be written as

$$C: v_1, v_2, \ldots, v_k, v_1$$

where $3 \leq k < n$. If $T$ contains a vertex $v$ that is adjacent to some vertex of $C$ and adjacent from some other vertex of $C$, then there must be a vertex $v_i$ of $C$ that is adjacent to $v$ such that $v_{i+1}$ is adjacent from $v$. In this case,

$$C': v_1, v_2, \ldots, v_i, v, v_{i+1}, \ldots, v_k, v_1$$

is a cycle whose length is greater than that of $C$, which leads to a contradiction. Therefore, every vertex of $T$ that is not on $C$ is either adjacent to every vertex of $C$ or adjacent from every vertex of $C$. Since $T$ is strongly connected, there must be vertices of each type.

Let $U$ be the set of all vertices of $T$ that are not on $C$ and such that each vertex of $U$ is adjacent to every vertex of $C$, and let $W$ be the set of those vertices of $T$ that are not on $C$ such that every vertex of $W$ is adjacent from each vertex of $C$ as shown in Figure 3.10. Then $U \neq \varnothing$ and $W \neq \varnothing$.

As $T$ is strongly connected, there exists a path from every vertex of $C$ to every vertex of $U$. Since no vertex of $C$ is adjacent to any vertex of $U$, there must be a vertex $w \in W$ that is adjacent to a vertex $u \in U$. However, this leads to a contradiction because then the cycle

$$C'': v_1, v_2, \ldots, v_k, w, u, v_1$$

has length greater than the length of $C$. ∎

It follows from Camion's theorem every vertex in a strongly connected tournament lies on a cycle (the hamiltonian cycle that the tournament must have). However, with a bit more work we can see that every vertex in a strongly connected tournament actually lies on a triangle as well.

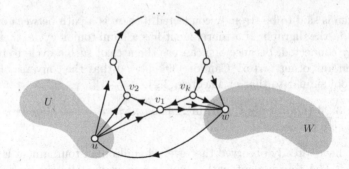

**Figure 3.10.** A situation that arises in the proofs of both Theorem 3.23 and Theorem 3.25.

**Corollary 3.24.** *Every vertex in a strongly connected tournament belongs to a triangle.*

*Proof.* Let $v$ be a vertex in a nontrivial strongly connected tournament $T$. By Theorem 3.23, $T$ is hamiltonian. Thus, $T$ contains a hamiltonian cycle $v = v_1, v_2, \ldots, v_n, v_1$. Since $v$ is adjacent to $v_2$ and adjacent from $v_n$, there exists a vertex $v_i$ with $2 \leq i < n$ such that $(v, v_i)$ and $(v_{i+1}, v)$ are arcs of $T$. Therefore, the triangle $v, v_i, v_{i+1}, v$ contains $v$. ∎

It is natural to wonder about longer cycles. It may be surprising to learn that if a tournament is hamiltonian, then it must possess significantly stronger properties. A digraph $D$ of order $n \geq 3$ is called **pancyclic** if it contains a cycle of every possible length, that is, $D$ contains a cycle of length $k$ for each $k = 3, 4, \ldots, n$. A digraph is called **vertex-pancyclic** if each vertex $v$ of $D$ lies on a cycle of every possible length. Harary and Moser [121] showed that every nontrivial strongly connected tournament is pancyclic, while Moon [178] went one step further by obtaining the following result.

**Theorem 3.25.** *Every strongly connected tournament is vertex-pancyclic.*

*Proof.* It suffices to consider the case where $T$ is a strongly connected tournament of order $n \geq 3$. Let $v_1$ be a vertex of $T$. We show that $v_1$ lies on an $k$-cycle for each $k = 3, 4, \ldots, n$. We proceed by induction on $k$. The base case $k = 3$ follows by Theorem 3.24. Now suppose that $v_1$ lies on an $k$-cycle $C: v_1, v_2, \ldots, v_k, v_1$ for $3 \leq k \leq n - 1$. We will show that $v_1$ lies on an $(k + 1)$-cycle.

First suppose that there is some vertex $v$ not on $C$ that is adjacent from at least one vertex of $C$ and adjacent to at least one vertex of $C$. Then, similarly to the proof of Theorem 3.21, for some index $i$, but $(v_i, v)$ and $(v, v_{i+1})$ are arcs of $T$ (where the indices are viewed modulo $k$). Thus, $v_1$ lies on the $(k + 1)$-cycle $v_1, v_2, \ldots, v_i, v, v_{i+1}, \ldots, v_k, v_1$ and we are done.

It remains to consider the case where there is no such vertex $v$. Then, as in the proof of Camion's theorem (Theorem 3.23), we let $U$ denote the set of vertices of $T$ that are not on $C$ and such that each vertex of $U$ is adjacent to every vertex of $C$, and we let $W$ be the set of vertices of $T$ that are not on $C$ and are adjacent from every vertex of $T$. Since $T$ is strongly connected, neither $U$ nor $W$ is empty, and there is a vertex $u \in U$ and a vertex $w \in W$ so that $(w, u) \in E(T)$, which is again the situation depicted in Figure 3.10. In this case we see that $v_1$ lies on the $(k+1)$-cycle $w, u, v_1, v_2, \ldots, v_{k-1}, w$, completing the proof. ∎

# 3.4   Highly hamiltonian graphs

A graph is **hamiltonian-connected** if for every pair $u$, $v$ of vertices, there is a hamiltonian $u$-$v$ path. Necessarily, every hamiltonian-connected graph of order 3 or more is hamiltonian, but the converse is not true. The cubic graph $C_3 \,\square\, K_2$ shown on the left of Figure 3.11 is hamiltonian-connected, while the 3-cube $C_4 \,\square\, K_2 = Q_3$ shown on the right of this figure is not hamiltonian-connected since there is no hamiltonian $u$-$v$ path (see Exercise 3.57).

The following theorem, which resembles Ore's theorem (Theorem 3.11) and is also due to Ore [186], provides a sufficient condition for a graph to be hamiltonian-connected.

**Theorem 3.26.** *Let $G$ be a graph of order $n$. If*

$$\deg u + \deg v \geq n + 1$$

*for every pair of nonadjacent vertices $u$ and $v$ of $G$, then $G$ is hamiltonian-connected.*

*Proof.* Consider two vertices $u$ and $v$ in graph $G$, and let $H$ be the graph obtained by adding a new vertex $z$ and connecting it to both $x$ and $y$. We now construct the closure $F = C(H)$. Since $\deg_H u + \deg_H v \geq n + 1$ for every two nonadjacent vertices $u, v \in V(G)$, we know that the induced subgraph $F[V(G)]$ is complete. Additionally, if $u$ is a vertex in $V(G)$ and $u \neq x, y$, then the sum of the degrees of $u$ and $z$ in $F$ is at least $n + 1$, so $uz \in E(F)$. This means that the entire graph $F$ is complete, so $H$ is hamiltonian by Theorem 3.16. As the degree of

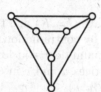 

**Figure 3.11.** Hamiltonian-connected and nonhamiltonian-connected graphs

vertex $z$ in $H$ is 2, any hamiltonian cycle $C$ in $H$ must contain the edges $xz$ and $yz$. Removing vertex $z$ and the edges incident to it from $C$ therefore gives us a hamiltonian path between $x$ and $y$ in $G$. ∎

A corollary similar to the sufficient condition for a graph to be hamiltonian given by Dirac's theorem (Theorem 3.10) now follows immediately.

**Corollary 3.27.** *If $G$ is a graph of order $n$ such that*

$$\deg v \geq (n+1)/2$$

*for every vertex $v$ of $G$, then $G$ is hamiltonian-connected.*

 **Panconnected and pancyclic graphs**

A graph $G$ of order $n$ is **panconnected** if it satisfies the following property: for every pair of distinct vertices $u$ and $v$ in $G$, there exists a path from $u$ to $v$ of length $\ell$ for every integer $\ell$ satisfying $d(u,v) \leq \ell \leq n-1$. Panconnected graphs are necessarily hamiltonian-connected, but the converse is not necessarily true, as shown in the following example.

For $k \geq 3$, let $G_k$ be the graph such that $V(G_k) = \{v_1, v_2, \ldots, v_{2k}\}$ and

$$
\begin{aligned}
E(G_k) \;=\; & \{v_i v_{i+1} : \; i = 1, 2, \ldots, 2k\} \\
& \bigcup \{v_i v_{i+3} : \; i = 2, 4, \ldots, 2k-4\} \\
& \bigcup \{v_1 v_3, v_{2k-2} v_{2k}\},
\end{aligned}
$$

where all subscripts are expressed modulo $2k$. Although for each pair $u, v$ of distinct vertices and for each integer $\ell$ satisfying $k \leq \ell \leq 2k-1$, the graph $G_k$ contains a $u$-$v$ path of length $\ell$, there is no $v_1$-$v_{2k}$ path of length $\ell$ if $1 < \ell < k$. Since $d(v_1, v_{2k}) = 1$, it follows that $G_k$ is not panconnected.

Williamson [256] obtained a sufficient condition for a graph to be panconnected in terms of its minimum degree, much like Dirac's theorem (Theorem 3.10).

**Theorem 3.28.** *If $G$ is a graph of order $n \geq 4$ such that*

$$\deg v \geq (n+2)/2$$

*for every vertex $v$ of $G$, then $G$ is panconnected.*

*Proof.* In the case where $n = 4$, we must have $G = K_4$ and the statement is true. We prove the statement for larger graphs by contradiction. Assume that there exists a graph $G$ of order $n \geq 5$ with minimum degree at least $(n+2)/2$ that is not panconnected; that is, there exist vertices $u$ and $v$ in $G$ such that there is no $u$-$v$ path of length $\ell$, where $\ell$ satisfies $d(u,v) \leq \ell \leq n-1$. Note that there is always a $u$-$v$ path of length $d(u,v)$, and that $G$ must be hamiltonian by Dirac's theorem (Theorem 3.10), so we may assume that $d(u,v) < \ell < n-1$.

Let $H = G - \{u, v\}$, which has order $n - 2 \geq 3$ and minimum degree at least $(n+2)/2 - 2 = (n-2)/2$. By Dirac's theorem (Theorem 3.10), the graph $H$ contains a hamiltonian cycle $C : v_1, v_2, \ldots, v_{n-2}, v_1$.

Consider a vertex $v_i$ in the cycle $C$, where $1 \leq i \leq n - 2$. If $uv_i \in E(G)$, then we must have $vv_{i+\ell-2} \notin E(G)$, where the subscripts are taken modulo $n - 2$. This is because otherwise, the path $u, v_i, v_{i+1}, \ldots, v_{i+\ell-2}, v$ would be a $u$-$v$ path of length $\ell$ in $G$. Therefore, for each vertex of $C$ that is adjacent to $u$ in $G$, there is a vertex of $C$ that is not adjacent to $v$ in $G$. Since the degree of $u$ in $G$ is at least $(n+2)/2$, we conclude that $u$ is adjacent to at least $n/2$ vertices of $C$. This means that the degree of $v$ in $G$ is at most $1 + (n - 2) - n/2 = n/2 - 1$, which is a contradiction. ∎

Having $\deg u + \deg v \geq n + 2$ for every pair of nonadjacent vertices $u$ and $v$ in a graph $G$ does not guarantee that the graph is panconnected. In fact, there does not exist a constant $c$ such that having $\deg u + \deg v \geq n + c$ for all pairs of nonadjacent vertices implies that the graph is panconnected.

To see this, consider a graph $G$ of order $n = 2c + 4$ for some $c \geq 2$, where $V(G) = \{w, z\} \cup W \cup Z$ and $|W| = |Z| = c + 1$ such that $G[W \cup Z] = K_{2c+2}$ and where $w$ is adjacent to every vertex of $W$, $z$ is adjacent to every vertex of $Z$, and $w$ and $z$ are adjacent to each other. For example, when $c = 2$, the graph $G$ is depicted below.

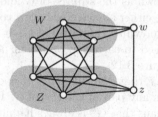

This graph $G$ satisfies the condition that $\deg u + \deg v \geq 3c + 4 = n + c$ for all pairs of nonadjacent vertices, but it is not panconnected, as $d(w, z) = 1$ but there is no $w$-$z$ path of length 2 in $G$.

According to Ore's theorem (Theorem 3.11), if a graph of order at least 3 satisfies $\deg u + \deg v \geq n$ for every pair of nonadjacent vertices $u$ and $v$, then it is hamiltonian. A result of Bondy [29] shows that under these hypotheses, with one exception, much more is true. A graph of order $n \geq 3$ is called **pancyclic** if it contains a cycle of every length between 3 and $n$.

**Theorem 3.29.** *Let $G$ be a graph of order $n \geq 3$. If*

$$\deg u + \deg v \geq n$$

*for every pair of nonadjacent vertices $u$ and $v$ of $G$, then either $G$ is pancyclic or $n$ is even and $G = K_{\frac{n}{2}, \frac{n}{2}}$.*

We close this section with a question posed by Nash-Williams [180] in 1971. He asked if Dirac's condition for hamiltonicity, that is, minimum degree at least $n/2$, guarantees at least $(n-2)/8$ edge-disjoint hamiltonian cycles in a graph $G$ of order $n$. Csaba, Kühn, Lo, Osthus, and Treglown [58] proved in 2016 that this is true for graphs that are large enough.

**Theorem 3.30.** *For all sufficiently large $n$, every graph $G$ of order $n$ and minimum degree at least $n/2$ contains at least $(n-2)/8$ edge-disjoint hamiltonian cycles.*

The number of edge-disjoint hamiltonian cycles guaranteed by Theorem 3.30 cannot, in general, be improved (see Exercise 3.58).

## 3.5    Graph powers

For any connected graph $G$ of order $n$, we can define a set of graphs based on the distances between vertices in $G$. The $k$th power of $G$, denoted as $G^k$, is a graph with the same set of vertices as $G$, but with an edge between two vertices $u$ and $v$ if and only if the distance between them in $G$ is between 1 and $k$. For example, the first power of $G$, $G^1$, is simply the original graph $G$. The second power of $G$, $G^2$, is also known as the **square** of $G$ and consists of edges between all pairs of vertices that are either adjacent or have a common neighbor in $G$. The third power of $G$, $G^3$, is known as the **cube** of $G$. It is important to note that graph powers are not related to cartesian products. A graph $G$ with its square and cube are shown in Figure 3.12.

Recall from Section 1.7 that the diameter of a connected graph $G$ is the maximum distance between any pair of vertices of $G$. If $k \geq \mathrm{diam}(G)$, then $G^k$ is a complete graph. This means that $G^k$ is always hamiltonian for sufficiently large values of $k$. It is interesting to consider the minimum value of $k$ such that $G^k$ is hamiltonian. Returning to the graphs in Figure 3.12, one can show that the square $G^2$ is not hamiltonian (Exercise 3.61), while the cube $G^3$ is hamiltonian. Thus, we need at least $k \geq 3$ to guarantee that $G^k$ is hamiltonian. It turns out that this is sufficient. In fact, a stronger result holds, which was discovered independently by Sekanina [217] and Karaganis [138].

**Theorem 3.31.** *If $G$ is a connected graph, then $G^3$ is hamiltonian-connected.*

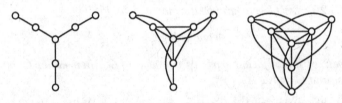

**Figure 3.12.** A graph $G$, its square $G^2$, and its cube $G^3$

*Proof.* It suffices to prove the theorem for trees, because if $T$ is a spanning tree of $G$ and $T^3$ is hamiltonian-connected, then $G^3$ is hamiltonian-connected. We proceed by induction on $n$, the order of the tree. The base case of $n = 1$ is clear, so assume that $T$ is a tree with $n \geq 2$ vertices and that for every tree with fewer vertices, the cube of the tree is hamiltonian-connected. Let $u$ and $v$ be arbitrary vertices of $T$. We need to show that $T^3$ contains a hamiltonian $u$-$v$ path.

Since $T$ is a tree, there is a unique $u$-$v$ path in $T$. Let $e = uw$ be the first edge of this path (thus we have $w = v$ if $u$ and $v$ are adjacent in $T$). The graph $T - e$ consists of two trees, one tree $T_u$ containing $u$ and the other tree $T_v$ containing $v$ and $w$. By the induction hypothesis, both $T_u^3$ and $T_v^3$ are hamiltonian-connected.

We now choose a vertex $u' \in V(T_u)$ in the following manner. If $T_u$ consists only of the vertex $u$, we set $u' = u$. Otherwise, we choose $u'$ to be any neighbor of $u$ in $T_u$. Thus $d_T(u', u) \leq 1$, and $T_u^3$ contains a hamiltonian $u$-$u'$ path, which we denote by $P_u$. Next we choose a vertex $w' \in V(T_v)$ in a similar but slightly different manner. If $T_v$ consists only of the vertex $v$, then we set $w' = v$. If $w \neq v$, then we set $w' = w$. Otherwise (if $w = v$, but $T_v$ contains more vertices than just $v$), we choose $w'$ to be any neighbor of $v$ in $T_v$. Therefore, no matter how $w'$ is chosen, we have $d_T(w, w') \leq 1$, and we know by our induction hypothesis that $T_v^3$ contains a hamiltonian $w'$-$v$ path, which we denote by $P_v$.

Now note that

$$d_T(u', w') \leq d_T(u', u) + d_T(u, w) + d_T(w, w') \leq 3.$$

Thus, $T^3$ contains an edge between $u'$ and $w'$. Therefore, the path formed by starting with $P_u$, then following the edge $u'w'$, and ending with $P_v$ is a hamiltonian $u$-$v$ path in $T^3$, proving the theorem. ∎

We have observed that not every connected graph with at least three vertices has a hamiltonian square (as shown in Figure 3.12). However, Nash-Williams and Beineke and Plummer separately conjectured that this is true for connected graphs with at least three vertices and without a cut-vertex. In 1974, Fleischner [93] proved this conjecture to be true. While Říha [247] and Georgakopoulos [102] have subsequently provided simplifications of Fleischner's original proof, all proofs of this result are quite lengthy and will not be presented here.

**Theorem 3.32 (Fleischner's theorem).** *If $G$ is a connected graph of order at least 3 without a cut-vertex, then $G^2$ is hamiltonian.*

A variety of results strengthening (but employing) Fleischner's theorem (Theorem 3.32) have been obtained. For example, Chartrand, Hobbs, Jung, Kapoor and Nash-Williams [44] strengthened the conclusion of Theorem 3.32 from hamiltonian to hamiltonian-connected.

**Theorem 3.33.** *If $G$ is a connected graph of order at least 3 without a cut-vertex, then $G^2$ is hamiltonian-connected.*

*Proof.* Suppose that $G$ is a connected graph of order at least 3 without a cut-vertex and let $u$ and $v$ be any two vertices of $G$. We need to show that there is a hamiltonian $u$-$v$ path.

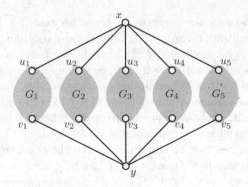

**Figure 3.13.** The graph constructed to prove Theorem 3.33

Let $G_1, G_2, \ldots, G_5$ be five distinct copies of $G$, and let $u_1, u_2, \ldots, u_5$ and $v_1, v_2,$ $\ldots, v_5$ be the vertices in these copies of $G$ that correspond to $u$ and $v$, respectively. Form a new graph $H$ by adding to the disjoint union $G_1 \cup G_2 \cup \cdots \cup G_5$ two new vertices $x$ and $y$ and ten new edges $xu_i$ and $yv_i$ for all $1 \leq i \leq 5$, as shown in Figure 3.13. Clearly $H$ is also connected, has order at least 3, and does not contain a cut-vertex. Therefore, $H^2$ is hamiltonian by Fleischner's theorem (Theorem 3.32).

Let $C$ be a hamiltonian cycle in $H^2$. In this cycle, the vertices $x$ and $y$ are each adjacent to vertices of at most two of the $G_i$. Therefore, there must be at least one $G_i$ that contains no vertex adjacent to either $x$ or $y$. Without loss of generality, suppose that this is $G_1$. Since $u_1$ and $v_1$ are the only vertices of $G_1$ adjacent to vertices outside $G_1$ in $H^2$, it follows that $C$ enters $G_1$ at either $u_1$ or $v_1$, traverses all of $G_1$, and then leaves $G_1$ at whichever of $u_1$ or $v_1$ it did not enter at. Therefore, $G_1^2$ has a hamiltonian $u_1$-$v_1$ path, which is equivalent to $G$ containing a hamiltonian $u$-$v$ path, as desired. ∎

# Exercises for Chapter 3

### Section 3.1. Eulerian graphs and digraphs

**3.1.** In present-day Königsberg (Kaliningrad), there are two additional bridges, one between regions B and C and one between regions B and D. Is it now possible to devise a route over all bridges of Königsberg without recrossing any of them?

**3.2.** Suppose that the Königsberg bridge problem had asked instead whether it was possible to take a route about Königsberg that crossed each bridge precisely twice. What would have been the answer in this case?

**3.3.** Let $F$ and $H$ be two disjoint connected noneulerian regular graphs and let $G = (F \cup H) \vee K_1$; that is, $G$ is obtained from $F$ and $H$ by adding a new vertex $v$ and joining $v$ to each vertex in $F$ and $H$. Prove that $G$ is eulerian.

**3.4.** Let $G$ be any connected non-eulerian graph. Let $H$ be the graph obtained from $G$ by adding a new vertex $v$ and joining it to every vertex that has odd degree in $G$. Prove that $H$ is eulerian.

**3.5.** Find a necessary and sufficient condition for the cartesian product $G \square H$ of two nontrivial connected graphs $G$ and $H$ to be eulerian.

**3.6.** Prove that if a graph of order $n \geq 6$ has an eulerian $u$-$v$ trail such that $\deg u - \deg v \geq n - 2$, then $n$ must be even.

**3.7.** Suppose that $G$ is an $r$-regular graph of order $n$ such that both $G$ and its complement $\overline{G}$ are connected. Is it possible that neither $G$ nor $\overline{G}$ is eulerian?

**3.8.** Show that if $T$ is a tree with at least one vertex of degree 2, then $\overline{T}$ is not eulerian.

**3.9.** Prove or disprove: Every eulerian bipartite graph has an even number of edges.

**3.10.** Let $G$ be a connected graph with precisely two vertices $u$ and $v$ of odd degree, where $\deg u \geq 3$ and $\deg v \geq 3$. Are there necessarily two edge-disjoint $u$-$v$ trails in $G$? Prove or disprove.

**3.11.** Let $G$ be a connected graph of order $n$ and size $m$ with $\delta(G) \geq 3$ such that every vertex of $G$ has odd degree.
   (a) Find a sharp upper bound (in terms of $m$ and $n$) for the size of an eulerian subgraph of $G$.
   (b) Give an example of such a graph $G$ where the maximum size of an eulerian subgraph of $G$ is less than the upper bound in (a).

**3.12.** Prove that every eulerian graph of odd order has three vertices of the same degree.

**3.13.** Prove that for each odd integer $n \geq 3$, there exists precisely one eulerian graph of order $n$ containing precisely three vertices of the same degree and at most two vertices of any other degree.

**3.14.** Prove Theorem 3.3: A weakly connected digraph is eulerian if and only if $\operatorname{od} v = \operatorname{id} v$ for every vertex $v$.

**3.15.** Prove that every weakly connected eulerian digraph is strongly connected.

**3.16.** Prove Theorem 3.4: A weakly connected digraph contains an eulerian trail if and only if it contains two vertices $u$ and $v$ such that

$$\operatorname{od} u = \operatorname{id} u + 1 \quad \text{and} \quad \operatorname{id} v = \operatorname{od} v + 1,$$

while $\operatorname{od} w = \operatorname{id} w$ for all other vertices $w$. Moreover, in this case all of its eulerian trails begin at $u$ and end at $v$.

**3.17.** Prove the digraph analogue of Veblen's theorem: a weakly connected digraph is eulerian if and only if it contains a collection of pairwise arc-disjoint directed cycles that together contain all of the arcs of the digraph.

**3.18.** Prove that the analogue of the Toida–McKee theorem does not hold for digraphs, by showing there is an eulerian orientation of $K_{4,2}$ in which one of its edges lies on an even number of directed cycles.

**3.19.** Prove that a graph has an eulerian orientation if and only if it is eulerian.

**3.20.** Suppose that $D$ is a connected digraph containing two vertices $u$ and $v$ such that $\operatorname{od} u = \operatorname{id} u + k$ and $\operatorname{id} v = \operatorname{od} v + k$ for some positive integer $k$, and $\operatorname{od} w = \operatorname{id} w$ for all other vertices $w$ of $D$. Prove that $D$ contains $k$ arc-disjoint $u$-$v$ paths.

**3.21.** Let $D$ be a digraph with an eulerian trail. Theorem 3.4 tells us that $D$ contains two vertices $u$ and $v$ such that $\operatorname{od} u = \operatorname{id} u + 1$ and $\operatorname{id} v = \operatorname{od} v + 1$, while $\operatorname{od} w = \operatorname{id} w$ for all other vertices $w$ of $D$.
  (a) Let $T'$ be a $u$-$x$ trail in $D$ that cannot be extended to a longer trail. Must $x = v$?
  (b) If $T$ is a $u$-$v$ trail in $D$, must $T$ be an eulerian trail?

**3.22.** Prove that a nontrivial connected digraph $D$ is eulerian if and only if $E(D)$ can be partitioned into subsets $E_i$ for $1 \le i \le k$, where the subdigraph $D[E_i]$ induced by the set $E_i$ is a cycle for each $i$.

**3.23.** Prove that if $D$ is a connected digraph such that

$$\sum_{v \in V(D)} |\operatorname{od} v - \operatorname{id} v| = 2t$$

for some $t \ge 1$, then $E(D)$ can be partitioned into subsets $E_i$, $1 \le i \le t$, so that the subgraph $G[E_i]$ induced by $E_i$ is an open trail for each $i$.

**3.24.** Let $D$ be a connected digraph of order $n$ with $V(D) = \{v_1, v_2, \ldots, v_n\}$. Prove that if $\operatorname{od} v_i \ge \operatorname{id} v_i$ for $1 \le i \le n$, then $D$ is eulerian.

### Section 3.2.  Hamiltonian graphs

**3.25.** Prove that the graph of the dodecahedron (shown in Figure 3.5) is hamiltonian by exhibiting a hamiltonian cycle.

**3.26.** Show that if $G$ is a graph containing a vertex that is adjacent to at least three vertices of degree 2, then $G$ is not hamiltonian.

**3.27.** Prove that bipartite graphs with an odd number $n \ge 3$ of vertices are not hamiltonian. Is the **Herschel graph** of Figure 3.14 hamiltonian?

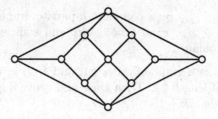

**Figure 3.14.** The Herschel graph

**3.28.** Prove that if $G$ and $H$ are hamiltonian graphs, then their cartesian product $G \square H$ is hamiltonian. Then apply this to show that the $n$-cube $Q_n$ is hamiltonian for all $n \geq 2$.

**3.29.** Show that the bound in Ore's theorem (Theorem 3.11) is sharp in the sense that for infinitely many integers $n$, there are nonhamiltonian graphs of order $n$ for which

$$\deg u + \deg v \geq n - 1$$

for every pair of nonadjacent vertices $u$ and $v$.

**3.30.** Prove that the Petersen graph is a tough graph.

**3.31.** A hamiltonian graph $G$ of order $n$ is **$k$-ordered hamiltonian** for an integer $k$ with $1 \leq k \leq n$ if for every ordered set $S = \{v_1, v_2, \ldots, v_k\}$ of $k$ vertices of $G$, there is a hamiltonian cycle of $G$ encountering these $k$ vertices of $S$ in the order listed.

  (a) Let $G$ be a graph of order $n \geq 3$ such that $\deg v \geq n/2$ for every vertex $v$ of $G$. Show that $G$ is 3-ordered hamiltonian.

  (b) Let $G$ be a graph of order $n \geq 4$ such that $\deg v \geq n/2$ for every vertex $v$ of $G$. Show that $G$ need not be 4-ordered hamiltonian.

  (c) Show that if $G$ is a 4-ordered hamiltonian graph, then $G$ is 3-connected.

**3.32.** Let $G$ be a graph of order $n \geq 2$, the degrees $d_i$ of whose vertices satisfy $d_1 \leq d_2 \leq \cdots \leq d_n$. Show that if there is no integer $k < (n+1)/2$ for which $d_k \leq k - 1$ and $d_{n+1-k} \leq n - k - 1$, then $G$ has a hamiltonian path. Then, apply this result to show that every self-complementary graph has a hamiltonian path.

**3.33.** Let $G$ be a bipartite graph with partite sets $U$ and $W$, and suppose that $|U| = |W| = k \geq 2$. Prove that if $\deg v > k/2$ for every vertex $v$ of $G$, then $G$ is hamiltonian.

**3.34.** Prove that if $G$ is a graph with $n \geq 3$ vertices and $m \geq \binom{n-1}{2} + 2$ edges, then $G$ is hamiltonian. What is the analogous result for hamiltonian paths?

**3.35.** Prove that if $T$ is a tree of order at least 4 that is not a star, then $\overline{T}$ contains a hamiltonian path.

**3.36.** Prove that the graph $K_{3r,2r,r}$ is hamiltonian for every positive integer $r$, but that there is no positive integer $r$ for which the graph $K_{3r+1,2r,r}$ is hamiltonian.

**3.37.** Let $G = K_{n_1, n_2, \ldots, n_k}$ be a complete $k$-partite graph of order at least 3, where $n_1 \geq n_2 \geq \cdots \geq n_k$. Find a necessary and sufficient condition for the graph $G$ to be hamiltonian.

**3.38.** Prove that if $G$ is a graph of order 101 and $\delta(G) = 51$, then every vertex of $G$ lies on a cycle of length 27. Then state and prove a generalization of this result to graphs of order $4k + 1$.

### Section 3.3.  Hamiltonian digraphs

**3.39.** Prove that Meyniel's theorem (Theorem 3.18) implies Camion's theorem (Theorem 3.23).

**3.40.** By Woodall's theorem (Theorem 3.19), if a digraph $D$ with $n$ vertices satisfies od $u +$ id $v \geq n$ whenever $(u, v) \notin E(D)$, then it is hamiltonian. Show that if a digraph $D$ with $n$ vertices satisfies od $u +$ id $v \geq n - 1$ whenever $(u, v) \notin E(D)$, then it must be strongly connected, but give an example to show that such a digraph may fail to be hamiltonian.

**3.41.** Prove that Meyniel's theorem (Theorem 3.18) implies Woodall's theorem (Theorem 3.19).

**3.42.** Prove that Meyniel's theorem (Theorem 3.18) implies Ghouila-Houri's theorem (Theorem 3.20).

**3.43.** Show that for infinitely many positive integers $n$, there exists a nonhamiltonian digraph $D$ of order $n$ such that od $v \geq (n-1)/2$ and id $v \geq (n-1)/2$ for every vertex $v$ of $D$.

**3.44.** Ghouila-Houri's theorem (Theorem 3.20) states that if $D$ is a strongly connected digraph of order $n$ and deg $v \geq n$ for every vertex $v$ of $D$, then $D$ is hamiltonian. Show that hypotheses that $D$ be strongly connected is necessary by exhibiting a digraph that satisfies the other hypothesis of the theorem but is not hamiltonian.

**3.45.** Prove that if $T$ is a tournament that is not transitive, then $T$ has at least three hamiltonian paths.

**3.46.** As mentioned after Theorem 3.21, every tournament has an odd number of hamiltonian paths.
  (a) If $T$ is a tournament of order 5 that is not strongly connected, then what is the maximum number of hamiltonian paths that $T$ can have?
  (b) If $T$ is a tournament of order 9 that has no strong components of order 5 or more, what the possible number of hamiltonian paths that $T$ can have?

**3.47.** A tournament $T$ of order 10 contains $k$ hamiltonian paths and consists of two strong components $S_1$ and $S_2$ of order 5. The strong component $S_1$ has $V(S_1) = \{v_1, v_2, \ldots, v_5\}$ and for $1 \leq i \leq 5$, $(v_i, v_j)$ is an arc of $S$ if $j = i + 1$ or $j = i + 2$ (addition modulo 5). Determine the number of hamiltonian paths in $S_2$ in terms of $k$.

**3.48.** Prove or disprove: If every vertex of a tournament $T$ belongs to a cycle, then $T$ is strongly connected.

**3.49.** Prove or disprove: Every arc of a strongly connected tournament $T$ lies on a hamiltonian cycle of $T$.

**3.50.** A digraph $D$ is **hamiltonian-connected** if for every pair $u, v$ of vertices of $D$, there exists a hamiltonian $u$-$v$ path. Prove or disprove: Every vertex-pancyclic tournament is hamiltonian-connected.

**3.51.** Prove that if a tournament $T$ contains a cycle of length $k$, then it also contains cycles of all lengths between 3 and $k$.

## Section 3.4. Highly hamiltonian graphs

**3.52.** Show that if $G$ is a graph of order $n \geq 4$ and size $m \geq \binom{n-1}{2} + 3$, then $G$ is hamiltonian-connected.

**3.53.** Give a proof by contradiction of Theorem 3.26 by first observing that $G - v$ is hamiltonian for every vertex $v$ of $G$.

**3.54.** Show that if $G$ is a graph of order $n \geq 3$ and $\delta(G) \geq n/2$, then every path of order at most $2\delta(G) - n + 1$ lies on a hamiltonian cycle.

**3.55.** Let $n$ and $k$ be positive integers such that $n \geq k + 2$. Prove that if $G$ is a graph of order $n$ such that $\deg u + \deg v \geq n + k - 1$ for every pair of nonadjacent vertices $u$ and $v$ of $G$, then every path of order at most $k$ lies on a hamiltonian cycle.

**3.56.** Give an example of a graph $G$ that is pancyclic but not panconnected.

**3.57.** Prove that no bipartite graph of order 3 or more is hamiltonian-connected, panconnected, or pancyclic.

**3.58.** Let $\{A, B\}$ be a partition of the vertex set of a graph $G$ of order $n = 8k + 2 \geq 10$ where $|A| = 4k$ and $|B| = 4k + 2$. The edge set of $G$ is defined by $G[A] = \overline{K}_{4k}$, $G[B] = (2k + 1)K_2$ and all possible edges joining a vertex of $A$ and a vertex of $B$.
  (a) Show that $G$ is hamiltonian.
  (b) Show that $G$ contains at most $\lfloor |B|/4 \rfloor = k = (n - 2)/8$ edge-disjoint hamiltonian cycles.

**3.59.** Determine the smallest integer $c$ such that if $G$ is a graph of order $n \geq 3$ and satisfies $\deg u + \deg v \geq n + c$ for every pair of nonadjacent vertices $u$ and $v$ of $G$, then $G$ is necessarily pancyclic.

**Section 3.5. Graph powers**

**3.60.** Show that if $G$ is a connected graph of diameter $\ell$ and $1 \le k \le \ell$, then $\mathrm{diam}(G^k) = \lceil \ell/k \rceil$.

**3.61.** Show that the graph $G^2$ of Figure 3.12 is not hamiltonian.

**3.62.** A graph $H$ is called a **square root** of a connected graph $G$ if $H^2 = G$.
  (a)  Give an example of a connected graph with two nonisomorphic square roots.
  (b)  Give an example of a connected graph with a unique square root.

**3.63.** Prove that every self-complementary graph has diameter 2 or 3. Use this to prove that if $G$ is a self-complementary graph of order at least 5, then $G^2$ is hamiltonian-connected.

**3.64.** Prove that if $v$ is any vertex of a connected graph $G$ of order at least 4, then $G^3 - v$ is hamiltonian.

**3.65.** Prove or disprove: If $G$ is hamiltonian, then $G^2$ is hamiltonian-connected.

# Connectivity

Connectivity is a fundamental concept in graph theory. Vertex connectivity quantifies how resilient a graph is to the removal of vertices, while edge-connectivity measures how resilient a graph is to the removal of edges. By a classic theorem of Menger, these measures of connectivity are both tied to the number of independent routes between pairs of vertices. Graphs that are connected but only barely are held together by their cut-vertices, and can be decomposed in a tree-like manner into blocks, which themselves are built from cycles by adding what are called ears. In this chapter, we will explore these and other aspects of graph connectivity.

##  4.1 Cut-vertices, bridges, and blocks

In Section 1.7, we defined a **cut-vertex** of a graph as a vertex whose removal increases the number of connected components. An edge whose removal has the same effect is called a **bridge**. Therefore, if $v$ is a cut-vertex of a graph $G$, then the graph $G - v$ obtained by removing $v$ from $G$ is disconnected. If $G$ is connected, then we can say more: a vertex $v$ is a cut-vertex of a connected graph $G$ if and only if $G - v$ is disconnected. We begin with a related equivalent condition. (The analogue of this result for graphs that are not necessarily connected is requested in Exercise 4.8.)

**Theorem 4.1.** *A vertex $v$ of a connected graph $G$ is a cut-vertex if and only if there are vertices $u$ and $w$ distinct from $v$ such that $v$ lies on every $u$-$w$ path.*

*Proof.* First suppose that $v$ is a cut-vertex of a connected graph $G$. As $G - v$ is disconnected, there are vertices $u$ and $w$ in $G - v$ that are not connected. It follows that $v$ must lie on every $u$-$w$ path of $G$. (In fact, this holds even if $G$ is not connected.)

For the converse, suppose that there are vertices are $u$ and $w$, distinct from $v$, so that $v$ lies on every $u$-$w$ path. Thus there are no $u$-$w$ paths in $G - v$, so $G - v$ is disconnected. Since $G$ is connected, this implies that $v$ is a cut-vertex. ∎

We have already given an alternative characterization of bridges with Theorem 1.8: an edge is a bridge if and only if it lies on no cycle. The edge analogue of Theorem 4.1 is also true, but for a fairly trivial reason (see Exercise 4.9).

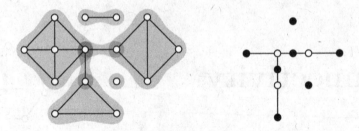

**Figure 4.1.** A graph with its blocks shaded, and the corresponding block-cutpoint graph, where blocks are indicated with solid circles and cut-vertices with hollow circles

## Blocks

A **block** of a graph $G$ is a maximal connected subgraph without cut-vertices. Note that blocks are defined to be subgraphs, and not sets of vertices, and also that blocks will necessarily be induced subgraphs. If $G$ itself has no cut-vertices, for example if $G$ is a cycle or a complete graph, then it is sometimes called a block itself (another term is **nonseparable**). For example, the graph on the left of Figure 4.1 is shown with its seven blocks shaded. Note that if $B$ is a block of the graph $G$, then $B$ has no cut-vertices itself, but it may contain vertices that are cut-vertices of $G$.

Every vertex of a graph lies in at least one block, but the blocks do not partition the vertices of a graph because they may intersect, albeit only barely.

**Theorem 4.2.** *Two distinct blocks of a graph may share at most one vertex.*

*Proof.* Suppose that $B_1$ and $B_2$ are distinct connected subgraphs of $G$ without cut-vertices, and that they share at least two vertices in common. We will show that $B_1 \cup B_2$ is also connected and also has no cut-vertices. This will prove the result because it implies that if $B_1$ and $B_2$ are distinct, then at least one of them is not maximal and thus not a block.

Because $B_1$ and $B_2$ share at least two vertices, $B_1 \cup B_2$ is connected. Consider an arbitrary vertex $v \in V(B_1 \cup B_2)$. We aim to show that $(B_1 \cup B_2) - v$ is connected. Because $B_1$ and $B_2$ share at least two vertices, there is some vertex $x \neq v$ with $x \in V(B_1) \cap V(B_2)$. Therefore, for every pair $u, w$ of vertices of $(B_1 \cup B_2) - v$, there are $u$-$x$ and $x$-$w$ walks in $(B_1 \cup B_2) - v$. By pasting these walks together, we see that there is a $u$-$w$ walk in $(B_1 \cup B_2) - v$, so $B_1 \cup B_2 - v$ is connected and thus $v$ is not a cut-vertex of $B_1 \cup B_2$. Since $v$ was arbitrary, this proves that $B_1 \cup B_2$ has no cut-vertices, as desired. ∎

In the opposite direction, every cut-vertex of $G$ lies in at least two blocks.

The blocks of a graph describe its rough structure, and we can visualize this structure with another graph. The **block-cutpoint graph** of a graph $G$ is the bipartite graph whose two partite sets are the cut-vertices and the blocks of $G$, respectively, and where the cut-vertex $v$ is adjacent to the block $B$ if and only if $v \in V(B)$. An example is shown on the right of Figure 4.1, where the blocks are indicated with solid circles and the cut-vertices with hollow circles.

**Theorem 4.3.** *The block-cutpoint graph of a graph $G$ is a forest. Moreover, if $G$ is connected, then its block-cutpoint graph is a tree.*

*Proof.* It suffices to prove that the block-cutpoint graph of a connected graph $G$ is a tree, as then the statement for general graphs follows. To this end, let $G$ be a connected graph and let $H$ be its block-cutpoint graph. We need to show both that $H$ is connected and that $H$ does not contain any cycles.

As every cut-vertex lies in at least one block in $G$, every vertex of $H$ that corresponds to a cut-vertex of $G$ is adjacent in $H$ to at least one block. Therefore, to establish connectivity, it suffices to show that given any two blocks $B_1$ and $B_2$ of $G$, we can find a path connecting the vertices corresponding to $B_1$ and $B_2$ in $H$. Let $v_1 \in V(B_1)$ and $v_2 \in V(B_2)$ be arbitrary. Because $G$ is connected, there is a $v_1$-$v_2$ path in $G$, and this corresponds to a path connecting the vertices corresponding to $B_1$ and $B_2$ in $H$.

To prove that the block-cutpoint graph $H$ is acyclic, assume for the sake of contradiction that $H$ contains a cycle. This cycle must contain at least two vertices that correspond to blocks of $G$; let $B_1$ and $B_2$ be two such blocks of $G$. Moreover, this cycle must also correspond to a cycle in $G$, which we denote by $C$. Since $C$ is a connected subgraph of $G$ without cut-vertices, there must be a block $B$ of $G$ with $V(C) \subseteq V(B)$. However, in this case $B$ must intersect both $B_1$ and $B_2$ in at least two vertices, and this is a contradiction to Theorem 4.2. ∎

An **end-block** of a graph $G$ is a block of $G$ that contains exactly one cut-vertex. For example, the graph shown in Figure 4.1 contains three end-blocks. If $G$ has at least one cut-vertex, then its block-cutpoint graph must contain at least two leaves. Each such leaf must correspond to a block of $G$ (it cannot correspond to a cut-vertex because every cut-vertex lies in at least two blocks), and the block it corresponds to must necessarily be an end-block of $G$. This proves the following.

**Theorem 4.4.** *Every graph with a cut-vertex has at least two end-blocks.*

The following related result can be proved by considering a longest path in the block-cutpoint graph (see Exercise 4.16).

**Theorem 4.5.** *If a connected graph $G$ has at least one cut-vertex, then it has a cut-vertex $v$ with the property that all but at most one of the blocks of $G$ containing $v$ are end-blocks.*

# 4.2   Vertex connectivity

Intuitively, a graph with a cut-vertex is not as connected as a graph without one. Vertex connectivity captures this difference by counting how many vertices must be removed to disconnect a graph. However, complete graphs cannot be disconnected no matter how many vertices we remove, so we need to introduce a convention to handle this case.

We say that a graph $G$ is **$k$-connected** if $|V(G)| \geq k + 1$ and $G - X$ is connected for every set $X \subseteq V(G)$ with $k - 1$ or fewer vertices. The (vertex) **connectivity** of a graph $G$, denoted by $\kappa(G)$, is the largest integer $k$ such that $G$ is $k$-connected.

For instance, the cycles $C_n$ for $n \geq 3$ have connectivity 2, while all trees with at least two vertices have connectivity 1, and the complete graph $K_n$ has connectivity $n - 1$. This leads to the odd fact that the single-vertex graph $K_1$ is connected but not 1-connected.

Therefore, a graph is 0-connected if and only if it is disconnected or $K_1$, and a graph is 1-connected if and only if it is connected and not $K_1$. If a graph $G$ is 2-connected, then it does not have a cut-vertex, so it is a block. However, the two concepts are not exactly equivalent, because $K_1$ and $K_2$ are blocks but not 2-connected.

A set $X \subseteq V(G)$ such that $G - X$ is disconnected is called a **vertex-cut** (another term is **separating set**). A vertex-cut $X \subseteq V(G)$ is said to be **minimum** if no vertex-cut is smaller. If $G$ is not a complete graph, then $\kappa(G)$ is equal to the cardinality of a minimum vertex-cut $X \subseteq V(G)$. We also consider a different notion: a vertex-cut $X \subseteq V(G)$ is **minimal** if there is no proper subset $Y \subsetneq X$ that is a vertex-cut. All minimum vertex-cuts are minimal, but not vice versa.

## Degree conditions

The minimum degree of a graph provides an upper bound on its connectivity.

**Theorem 4.6.** *For every graph $G$, we have $\kappa(G) \leq \delta(G)$.*

*Proof.* If $G$ is complete, then $\kappa(G) = n - 1 = \delta(G)$ by definition. If $G$ is not complete, then take a vertex $v \in V(G)$ with $\deg v = \delta(G)$ and note that $N(v)$ is a vertex-cut of $G$. Since $\kappa(G)$ is equal to the cardinality of a minimum vertex-cut of $G$, it is at most $\delta(G)$. ∎

Let $G$ be a graph with $n$ vertices. If $G$ is $k$-connected, then Theorem 4.6 shows that we must have $\deg v \geq k$ for every vertex $v$ of $G$, and this requires $G$ to have at least $\lceil nk/2 \rceil$ edges. Harary [120] gave a construction of $k$-connected graphs with precisely this many edges, for every value of $n$ and $k$ satisfying $2 \leq k \leq n - 1$. Examples of his construction are shown in Figure 4.2.

**Figure 4.2.** Harary graphs with seven vertices and connectivity $k = 2, 3, 4, 5, 6$, from left to right

Graphs with greater connectivity must have greater degrees, but the converse relationship is more complicated. There are several sufficient conditions of this type, the simplest being the following due to Chartrand and Harary [43].

**Theorem 4.7.** *Let $G$ be a graph with $n \geq k + 1$ vertices. If*

$$\deg v \geq \left\lceil \frac{n + k - 2}{2} \right\rceil$$

*for every vertex $v \in V(G)$, then $G$ is $k$-connected.*

*Proof.* We prove the contrapositive. Suppose that there exists a graph $G$, which has $n \geq k + 1$ vertices but is not $k$-connected. This implies that $G$ is not complete, so there is some vertex-cut $X \subseteq V(G)$ with $|X| = k - 1$. Since $G - X$ is a disconnected graph with $n - k + 1$ vertices, it has a component with at most $\lfloor (n - k + 1)/2 \rfloor$ vertices. Take $v$ to be any vertex in this component. Then $v$ is adjacent in $G$ only to other vertices of this component and to vertices of $X$. Thus

$$\deg v \leq \left( \left\lfloor \frac{n - k + 1}{2} \right\rfloor - 1 \right) + (k - 1) = \left\lfloor \frac{n + k - 3}{2} \right\rfloor,$$

proving the result. ∎

 **Paths and cycles in 2-connected graphs**

If $G$ is 2-connected, then it is not a tree, so it must contain a cycle. In fact, much more can be said.

**Theorem 4.8.** *A graph with at least three vertices is 2-connected if and only if every pair of vertices lie on a common cycle.*

*Proof.* First suppose that $G$ has at least three vertices and the property that every pair of vertices lie on a common cycle. Thus given any choice of distinct vertices $u, v, w \in V(G)$, there is a cycle containing $u$ and $w$, so $G - v$ has a $u$-$w$ path. Since $u$, $v$, and $w$ were arbitrary, $G$ cannot have a cut-vertex. This and the hypothesis that $G$ has at least three vertices imply that $G$ is 2-connected.

The converse is more significant. Let $G$ be a 2-connected graph. We prove, by induction on $k$, that every pair of vertices $u$ and $v$ in $G$ that are distance $k$ apart lie on a common cycle. The base case is $k = 1$, so $u$ and $v$ are adjacent.

We must have $\deg v \geq 2$ because $G$ is 2-connected, so $v$ has another neighbor $w$. Since $G - v$ is connected, it has a $u$-$w$ path $P$. Combining $P$ with the path $w, v, u$ gives us the desired cycle.

Now suppose that $k \geq 2$ and let $u = v_0, v_1, \ldots, v_k = v$ be a shortest $u$-$v$ path in $G$. Since $d(u, v_{k-1}) = k - 1$, there is a cycle $C$ containing $u$ and $v_{k-1}$ by induction. Because $G - v_{k-1}$ is connected, there is a path from $v$ to some vertex of $C$. Choose $P$ to be such a path of minimal possible length. Thus only the final vertex of $P$ lies on $C$. By combining $C$ and $P$, we obtain the desired cycle: starting at $u$, follow "one side" of $C$ to $v_{k-1}$, then follow the edge $v_{k-1}v_k$ to go to $v_k$, then follow $P$ to return to the "other side" of $C$, and finally follow $C$ back to $u$. ∎

This result has several consequences (see Exercises 4.21-4.24) that hint at the results of Section 4.4. For two distinct vertices $u$ and $v$ in a graph $G$, two $u$-$v$ paths are **internally disjoint** if they have only $u$ and $v$ in common.

**Corollary 4.9.** *A graph $G$ is 2-connected if and only if for every two distinct vertices $u$ and $v$ in $G$, there are two internally disjoint $u$-$v$ paths.*

 **Ear decompositions of 2-connected graphs**

The ear decomposition shows how every 2-connected graph can be built by starting with a cycle (the simplest type of 2-connected graph) and repeatedly adding paths. An **open ear** of a graph $G$ is a path $P$ in $G$ with distinct end-vertices and in which all non-end-vertices have degree 2 in $G$. Thus, every edge of $G$ is trivially an open ear. Given an open ear $P$ of a graph $G$ with end-vertices $u$ and $v$, we denote by $P^\circ$ the graph obtained by removing the end-vertices of $P$, so $P^\circ = P - u - v$. We then say that the graph $G$ is obtained by **adding an open ear** to the graph $G - P^\circ$.

An **open ear decomposition** of a graph $G$ is a sequence $G_0, G_1, \ldots, G_k$ of graphs where $G_0$ is a cycle and $G_i$ is obtained by adding an open ear to $G_{i-1}$ for all $1 \leq i \leq k$. An example is shown in Figure 4.3.

If $G$ can be obtained by adding an open ear to the graph $H$ and $H$ is 2-connected, then $G$ must be 2-connected as well. Since cycles are 2-connected,

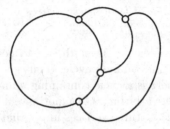

**Figure 4.3.** An example of an open ear decomposition

it follows that every graph that has an open ear decomposition is 2-connected. In 1932, Whitney [255] showed that the converse also holds, and thus every 2-connected graph has an open ear decomposition.

**Theorem 4.10 (Open ear decomposition).** *A graph $G$ is 2-connected if and only if it has an open ear decomposition. Moreover, if $G$ is 2-connected, then any cycle of $G$ can be taken as the initial cycle of such an open ear decomposition.*

*Proof.* We have already observed that every graph with an open ear decomposition is 2-connected, so now let $G$ be a 2-connected graph. This implies that $G$ is not a tree, so we can take $G_0$ to be some cycle of $G$. Now take $H$ to be a maximal subgraph of $G$ that has an open ear decomposition starting from $G_0$. Since $H$ contains $G_0$, it has at least three vertices, and since every edge of $G$ is an open ear, $H$ is an induced subgraph of $G$. We want to show that $H = G$.

Suppose, to the contrary, that $H \neq G$. Because $H$ is an induced subgraph of $G$, there must be some edge $uv \in E(G)$ with $u \in V(H)$ and $v \in V(G) \setminus V(H)$. Let $w$ be any vertex of $H$ other than $u$. Since $G$ is 2-connected, there must be a path from $v$ to $w$ in $G - u$. Starting at $u$ and then following this path until it returns to $H$ gives us an open ear that can be added to $H$. However, this contradicts the maximality of $H$, proving the theorem. ∎

# 4.3  Edge-connectivity

Thus far, we have measured the connectedness of a graph in terms of how many vertices must be removed to disconnect it, but we now consider removing edges. An **edge-cut** of a graph $G$ is a subset $F \subseteq E(G)$ such that $G - F$ is disconnected. As with vertex-cuts, a **minimum edge-cut** is one of the smallest possible size, while a **minimal edge-cut** is one for which no proper subset is an edge-cut. Again note that every minimum edge-cut is minimal, but not vice versa. The **edge-connectivity** of a graph $G$ of order at least 2, denoted by $\kappa'(G)$, is defined to be the cardinality of a minimum edge-cut of $G$. As with vertex-connectivity, we must make an exception and define the edge-connectivity of $K_1$ to be 0, but we do not need to make exceptions for larger complete graphs.

The set of edges incident with any vertex is an edge-cut, so it follows that

$$0 \leq \kappa'(G) \leq \delta(G) \leq n - 1$$

for every graph $G$ with $n$ vertices. A graph $G$ is **$k$-edge-connected** if $\kappa'(G) \geq k$.

In particular, a graph is 1-edge-connected if and only if it is not $K_1$ and is connected, while it is 2-edge-connected if and only if it is not $K_1$ and does not have a bridge. These observations allow us to see that vertex connectivity and edge-connectivity can differ for a graph. Indeed, every connected graph $G$ with a cut-vertex but no bridges satisfies $1 = \kappa(G) < 2 \leq \kappa'(G)$. The simplest such

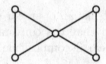

**Figure 4.4.** The bowtie graph

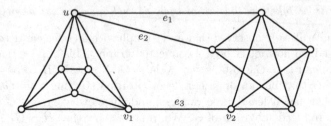

**Figure 4.5.** A graph with vertex connectivity 2 and edge-connectivity 3

graph is the **bowtie graph**, shown in Figure 4.4, which consists of two triangles joined at a common vertex.

A more complicated example is shown in Figure 4.5. Letting $G$ denote this graph, we see that $\kappa(G) = 2$ and $\kappa'(G) = 3$ because both $\{u, v_1\}$ and $\{u, v_2\}$ are minimum vertex-cuts, while $\{e_1, e_2, e_3\}$ is a minimum edge-cut.

It is frequently helpful to observe that minimal edge-cuts break a connected graph into precisely two components.

**Lemma 4.11.** *If $F$ is a minimal edge-cut of a connected graph $G$, then $G - F$ has precisely two components.*

*Proof.* Suppose that $F \subseteq E(G)$ is a minimal edge-cut of a connected graph $G$. Since $G$ is connected, $F$ must contain at least one edge, so we may choose an arbitrary edge $e \in F$. By the minimality of $F$, $F \setminus \{e\}$ is not an edge-cut, so $G - (F \setminus \{e\})$ is connected and $e$ is a bridge of this graph. It follows that $G - F$ has precisely two components, one for each vertex incident with $e$. ∎

Because minimum edge-cuts are also minimal edge-cuts, Lemma 4.11 allows us to conclude that if $F$ is a minimum edge-cut of a connected graph $G$, then $G - F$ has precisely two components.

We have $\kappa'(K_n) \leq n - 1$ for all complete graphs $K_n$, and it is not difficult to establish that equality holds.

**Theorem 4.12.** *For every positive integer $n$, we have $\kappa'(K_n) = n - 1$.*

*Proof.* Let $F$ be a minimum edge-cut of $K_n$. By Lemma 4.11, $K_n - F$ has precisely two components. Label these components $G_1$ and $G_2$, suppose that $G_1$ has $k$ vertices, and choose a vertex $u \in V(G_1)$ arbitrarily. Then, $F$ must contain all $n - k$ edges between $u$ and the vertices of $G_2$, and also at least one edge incident to each of the other $k - 1$ vertices of $G_1$, for a total of at least $n - 1$ edges. ∎

## ⋈ Whitney's inequalities

Whitney [255] was the first to observe that $\kappa'(G) \leq \delta(G)$ holds for every graph $G$. Additionally, Whitney proved the more profound inequality $\kappa(G) \leq \kappa'(G)$. The idea of the proof of this inequality is quite simple. Since the result is immediate if $G$ is complete (because $\kappa(K_n) = \kappa'(K_n) = n - 1$), we may assume that $G$ is not complete. Take a minimum edge-cut $F$ of $G$ and then choose, for each edge of the edge-cut, one of the vertices it joins to be part of a vertex-cut $X$. We must exercise some care, however, that we do not entirely remove one of the components of $G - F$, thereby accidentally making $G - X$ connected.

**Theorem 4.13.** *For every graph $G$,*

$$\kappa(G) \leq \kappa'(G) \leq \delta(G).$$

*Proof.* As observed above, it suffices to prove that $\kappa(G) \leq \kappa'(G)$, and it suffices to prove this in the case where $G$ is not complete. Let $F$ be a minimum edge-cut of $G$ (which exists because $G$ is not $K_1$), so $|F| = \kappa'(G)$. By Lemma 4.11, $G - F$ has precisely two components. Label these components $G_1$ and $G_2$.

First consider the case where every vertex in $G_1$ is adjacent to every vertex in $G_2$, and let $u$ be a vertex in $G_1$ and $v$ be a vertex in $G_2$. Since $F$ is an edge-cut, it must contain all edges of the form $xv$ for $x \in V(G_1)$ and all edges of the form $uy$ for $y \in V(G_2)$. The total number of these edges is $n - 1$, so $\kappa'(G) = n - 1 \geq \kappa(G)$ in this case.

Now suppose that we can find vertices $u \in V(G_1)$ and $v \in V(G_2)$ that are not adjacent. We can build a vertex-cut $X$ as follows. Take an edge $e \in F$. If $e$ is incident with $u$, meaning that $e = uy$ for some $y \in V(G_2)$, then we place $y$ in $X$. If $e$ is not incident with $u$, meaning that $e = xy$ for some $x \in V(G_1) \setminus u$ and $y \in V(G_2)$, then we place $x$ in $X$. This means that for every edge $e \in F$, one of its two ends belongs to $X$, so there are no edges between the remaining vertices of $G_1$ and $G_2$ in $G - X$. Furthermore, since $u, v \notin X$, $G_1 - X$ and $G_2 - X$ both contain vertices, and so $X$ is indeed a vertex-cut. Therefore, we have $\kappa(G) \leq |X| \leq |F| = \kappa'(G)$ in this case, which completes the proof of the theorem. ∎

We have already seen that we can have the strict inequality $\kappa(G) < \kappa'(G)$. To see that both inequalities of Theorem 4.13 can be strict simultaneously, we can appeal again to the graph of Figure 4.5 (which has minimum degree 4), or we can add a few vertices to the bowtie, as below.

In fact, given *any* integers $k$, $\ell$, and $d$ satisfying $1 \leq k \leq \ell \leq d$, there is a graph $G$ with $\kappa(G) = k$, $\kappa'(G) = \ell$, and $\delta(G) = d$ (see Exercise 4.45, which is originally due to Chartrand and Harary [43]).

There are several conditions that guarantee that one of the two inequalities in Whitney's inequalities are actually equalities. We present two.

**Theorem 4.14.** *For every cubic graph $G$, we have $\kappa(G) = \kappa'(G)$.*

*Proof.* Let $G$ be a cubic graph. By Whitney's inequalities, it suffices to prove that $\kappa'(G) \leq \kappa(G)$. If $G$ is $K_4$, then the result is clearly true, so we may assume that $G$ is not complete. Let $X$ be a minimum vertex-cut of $G$, and let $G_1$ and $G_2$ be two of the components of $G - X$.

Since $X$ is a minimum vertex-cut, every vertex $x \in X$ has at least one neighbor in $G_1$ and at least one neighbor in $G_2$. We build an edge-cut $F$ in the following manner. For a vertex $x \in X$, if there is only one edge joining $x$ to $G_1$, then we add that edge to $F$. Otherwise, there must be two edges joining $x$ to $G_1$ and one edge joining it to $G_2$, and we add the edge joining $x$ to $G_2$ to $F$.

To see that $F$ is an edge-cut, we need only note that in $G - F$, the vertices of $X$ that are adjacent to vertices of $G_1$ are only adjacent to vertices of $G_1$. Thus, there are no vertices of $X$ that can be used in a walk from $G_1$ to $G_2$, so $G - F$ is disconnected. This proves that $\kappa'(G) \leq |F| = |X| = \kappa(G)$. ∎

Recall from Section 1.7 that the diameter of a connected graph $G$ is the maximum distance between any pair of vertices of $G$. If $\text{diam}(G) = 1$, then $G$ is complete, and we have $\kappa'(G) = \delta(G)$. The following theorem of Plesník [192] extends this equality to graphs of diameter at most 2.

**Theorem 4.15.** *If $G$ is connected and has diameter 2, then $\kappa'(G) = \delta(G)$.*

*Proof.* Let $G$ be a graph of diameter 2 and take $F \subseteq E(G)$ to be a minimum edge-cut, so $\kappa'(G) = |F|$ because $G$ is not complete. Suppose that the components of $G - F$ are $G_1$ and $G_2$. It cannot be the case that both $G_1$ and $G_2$ have vertices $u$ and $v$, respectively, that are not incident with any edge of $F$, for then we would have $d(u, v) \geq 3$. Therefore, we may assume without loss of generality that every vertex of $G_1$ is incident with an edge of $F$. Let $u \in V(G_1)$ be arbitrary, and suppose that $u$ is incident with $k$ edges of $F$. Thus $u$ is adjacent to at least $\delta(G) - k$ other vertices of $G_1$. As every vertex of $G_1$ is incident with an edge of $F$, we see that $F$ has at least $\delta(G)$ edges: $k$ incident to $u$ and at least $\delta(G) - k$ incident to the vertices of $N(u) \cap G_1$. ∎

Exercise 4.46 asks the reader to show that Theorem 4.15 cannot be extended to graphs whose diameter exceeds 2. Together, Theorems 4.14 and 4.15 can be used to determine the connectivity of the Petersen graph (see Exercise 4.47).

# Ear decompositions of 2-edge-connected graphs

Whitney's inequalities tell us that every 2-connected graph is 2-edge-connected, but we have seen that not every 2-edge-connected graph is 2-connected (the simplest example being the bowtie). Thus, 2-edge-connected graphs do not

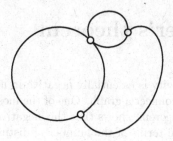

**Figure 4.6.** An example of a closed ear decomposition

necessarily have open ear decompositions. They do, however, possess a related but slightly weaker kind of decomposition, as was proved in 1939 by Robbins [202].

A **closed ear** of a graph $G$ is a cycle of $G$ in which every vertex except one has degree 2 in $G$. If $C$ is a closed ear of $G$ and $C'$ denotes the vertices of $C$ that have degree 2 in $G$, then we say that $G$ is obtained by **adding a closed ear** to the graph $G - C'$. A **closed ear decomposition** of a graph $G$ is a sequence $G_0, G_1, \ldots, G_k$ of graphs where $G_0$ is a cycle and $G_i$ is obtained by adding an open or a closed ear to $G_{i-1}$ for all $1 \leq i \leq k$. An example is shown in Figure 4.6.

The proof of the closed ear decomposition, below, is very similar to that of the open ear decomposition (Theorem 4.10), and also to our proof of Robbins' theorem (Theorem 2.5), see Exercise 4.49.

**Theorem 4.16** (Closed ear decomposition). *A graph $G$ is 2-edge-connected if and only if it has a closed ear decomposition. Moreover, if $G$ is 2-edge-connected, then any cycle of $G$ can be taken as the initial cycle of such a closed ear decomposition.*

*Proof.* We first note that every graph with a closed ear decomposition is 2-edge-connected. Suppose that $G$ is 2-edge-connected. As $G$ is not a tree, it has a cycle; let $G_0$ be any cycle of $G$. Let $H$ be a maximal subgraph of $G$ that has a closed ear decomposition starting from $G_0$. Since every edge of $G$ is an open ear, $H$ is an induced subgraph of $G$. If $H \neq G$, then there exists an edge $uv \in E(G)$ such that $u \in V(H)$ and $v \in V(G) \setminus V(H)$. Because $G$ is 2-edge-connected, there must be a path from $v$ to $u$ in $G - uv$. Starting at $u$ and following this path until it returns to $H$ either gives us an open ear (if it returns to a vertex of $H$ other than $u$) or a closed ear (if it returns to $u$). However, either of these cases would contradict the maximality of $H$, so we must have $H = G$, which proves the theorem. ∎

# 4.4   Menger's theorem

The definition of connectivity is essentially negative in nature, as it is a measure of how difficult it is to disconnect a graph. One of the most celebrated theorems of graph theory, Menger's theorem, shows that this negative definition is equivalent to a positive description in terms of the number of distinct paths connecting any pair of vertices. In this way, Menger's theorem provides an alternative, more intuitive way of understanding the connectivity of a graph.

Before stating and presenting a proof of Menger's theorem, some additional terminology is needed. We first introduce new terminology for paths. For two sets $A, B \subseteq V(G)$, an **$A$-$B$ path** is a path from a vertex in $A$ to a vertex in $B$ that passes through no other vertex of $A$ or $B$. In other words, the path $P : v_0, v_1, \ldots, v_k$ is an $A$-$B$ path if and only if $v_0 \in A$, $v_k \in B$, and $v_i \notin A \cup B$ for all $1 \leq i \leq k - 1$. Two paths are **disjoint** if they do not share any vertices, including end-vertices; this implies that they also do not share edges. A set of paths is **pairwise disjoint** if each pair of them is disjoint.

For two sets $A, B \subseteq V(G)$, an **$A$-$B$ separating set** is a set $X \subseteq V(G)$ such that there are no $A$-$B$ paths in $G - X$. Just as any walk between two vertices can be shortened to obtain a path between them (Theorem 1.6), any walk from $A$ to $B$ can be shortened to obtain an $A$-$B$ path. This justifies us calling these sets *separating*: if $X$ is an $A$-$B$ separating set, then there is no way to get from $A$ to $B$ in $G - X$.

Note that we do not forbid the separating set from containing vertices of $A$ or $B$; in particular, both $A$ and $B$ are $A$-$B$ separating sets. Also, it is essential for our inductive proof that we do not insist that $A$ and $B$ be disjoint, and every $A$-$B$ separating set $X$ must contain $A \cap B$, as otherwise $G - X$ would contain trivial single-vertex $A$-$B$ paths.

We may now state our first of several forms of Menger's theorem.

**Theorem 4.17 (Menger's theorem, set form).** *Let $G$ be a graph and let $A, B \subseteq V(G)$. The maximum cardinality of a collection of pairwise disjoint $A$-$B$ paths is equal to the minimum cardinality of an $A$-$B$ separating set.*

*Proof.* One direction of the theorem is trivial. If $G$ has a collection of $k$ pairwise disjoint $A$-$B$ paths, then every $A$-$B$ separating set must have cardinality at least $k$, because it must contain at least one vertex from each of these paths.

To prove the converse, let $G$ be a graph, let $A, B \subseteq V(G)$, and suppose that $k$ is the minimum cardinality of an $A$-$B$ separating set. We need to prove that $G$ has a collection of $k$ pairwise disjoint $A$-$B$ paths. We prove this by induction on the number of edges of $G$. If $G$ has no edges, then the smallest $A$-$B$ separating set is $A \cap B$, and $G$ also has $|A \cap B|$ trivial single-vertex $A$-$B$ paths (one for each vertex of $A \cap B$). This verifies the base case.

Now suppose that $G$ has at least one edge, and that the result holds for all graphs with fewer edges. Let $e = xy$ be an arbitrary edge of $G$. Take $X \subseteq V(G)$ to be an $A$-$B$ separating set in $G - e$ of minimum possible cardinality. If $|X| = k$, then our induction hypothesis implies that $G - e$ has a collection of $k$ pairwise disjoint $A$-$B$ paths, so $G$ does as well. On the other hand, we cannot have $|X| \leq k - 2$, because then $X \cup \{x\}$ would be an $A$-$B$ separating set of $G$ of cardinality less than $k$. We are therefore left with the case where $|X| = k - 1$.

Since $|X| < k$, there must be an $A$-$B$ path in $G - X$. However, there is no $A$-$B$ path in $G - X - e$, because $X$ is an $A$-$B$ separating set in $G - e$. Therefore, there must be an $A$-$B$ path in $G - X$ using the edge $e = xy$. Without loss of generality, we can assume that this path traverses the edge $e$ from $x$ to $y$. Let $X_A = X \cup \{x\}$ and $X_B = X \cup \{y\}$. In $G - e$, every $A$-$X_A$ separating set has cardinality at least $k$, as does every $X_B$-$B$ separating set. Thus by our induction hypothesis, $G - e$ contains a collection $P_A$ of $k$ pairwise disjoint $A$-$X_A$ paths and a collection $P_B$ of $k$ pairwise disjoint $X_B$-$B$ paths.

Since $P_A$ consists of $k$ disjoint paths ending at the $k$ vertices of $X_A$, one path of $P_A$ must end at each vertex of $X_A$. Similarly, one path of $P_B$ must begin at each vertex of $X_B$. Additionally, the paths of $P_A$ intersect the paths of $P_B$ only at $X$ because $G - X - e$ has no $A$-$B$ paths. To complete the proof, we can combine the paths of $P_A$ and $P_B$ in the following way. One path of $P_A$ ends at $x$, and this is combined with the path of $P_B$ that begins at $y$. Each of the other $k - 1$ paths of $P_A$ ends at a vertex of $X$ and is combined with the path of $P_B$ that starts at that vertex. This gives us a collection of $k$ disjoint $A$-$B$ paths in $G$, as desired. ∎

Menger published his result, which he called the $n$-arc theorem, as a lemma in a 1927 paper [173] on the topology of curves. In a paper published more than fifty years later [175], he discussed the history of the theorem, although he did not mention the hole in his original proof, which was first identified by Kőnig [150]. (A more accurate history of Menger's theorem was given by Schrijver in [213].) The first flawless proof seems to have been given by Menger's student Nöbeling and was published in Menger's 1932 book *Kurventheorie* [174]. Since then, dozens of proofs have been presented; the proof given above follows that of Göring [105].

## The vertex form of Menger's theorem

Theorem 4.17 is the original form of Menger's theorem, as he stated it. However, Menger's theorem has many variations, which are all named after Menger despite being proved by other people. We present the vertex form next, which was first observed in 1929 by Rutt [210].

The vertex form of Menger's theorem concerns a pair of vertices $u$ and $v$. We call $\{u\}$-$\{v\}$-paths simply $u$-$v$ **paths** and we call $\{u\}$-$\{v\}$ separating sets simply $u$-$v$ **separating sets**. The reader will note that the naive specialization of the set form of Menger's theorem to this context, by setting $A = \{u\}$ and $B = \{v\}$, is

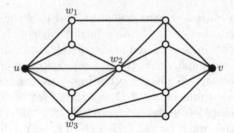

**Figure 4.7.** A graph illustrating the vertex form of Menger's theorem

not an impressive result, because both $\{u\}$ and $\{v\}$ are $u$-$v$ separating sets, and because all $u$-$v$ paths must intersect at both $u$ and $v$.

Thus for the vertex form of Menger's theorem, we change our requirements on both the separating sets and the paths. We require only that our paths be **internally disjoint**, meaning that they do not share any vertices other than their end-vertices. A set of paths is called **pairwise internally disjoint** if every pair of paths in the set is internally disjoint. We also require that our separating sets lie in $V(G) \setminus \{u, v\}$, to exclude these trivial cases. Note that if $u$ and $v$ are adjacent, then there cannot be any $u$-$v$ separating sets in $V(G) \setminus \{u, v\}$, thus the vertex form of Menger's theorem applies only when the two vertices are not adjacent.

**Theorem 4.18 (Menger's theorem, vertex form).** *Let $G$ be a graph and let $u, v \in V(G)$ be distinct, nonadjacent vertices. Then the maximum number of pairwise internally disjoint $u$-$v$ paths is equal to the minimum cardinality of a $u$-$v$ separating subset of $V(G) \setminus \{u, v\}$.*

As an example, consider the graph $G$ in Figure 4.7. There is a set of three vertices $X = \{w_1, w_2, w_3\}$ in $V(G) \setminus \{u, v\}$ that separates the vertices $u$ and $v$. It can be verified that no set with fewer than three vertices in $V(G) \setminus \{u, v\}$ separates $u$ and $v$, so by the vertex form of Menger's theorem, there is a set of three pairwise internally disjoint $u$-$v$ paths in $G$. We invite the reader to find three such paths. It follows from the vertex form of Menger's theorem that since there is a $u$-$v$ separating set in $V(G) \setminus \{u, v\}$ with three vertices, then there cannot be more than three pairwise internally disjoint $u$-$v$ paths. Conversely, if there are three pairwise internally disjoint $u$-$v$ paths, then there cannot be a $u$-$v$ separating set in $V(G) \setminus \{u, v\}$ with fewer than three vertices.

The set and vertex forms of Menger's theorem may look different, but they are actually equivalent. We prove the vertex form from the set form below, and Exercise 4.50 asks the reader to prove the set form from the vertex form.

*Proof of the vertex form of Menger's theorem from the set form.* Suppose that $u$ and $v$ are distinct, nonadjacent vertices of a graph $G$. As in the set form, there are two inequalities in Menger's theorem. The minimum cardinality of a $u$-$v$ separating subset of $V(G) \setminus \{u, v\}$ must be at least the maximum cardinality of a collection of pairwise internally disjoint $u$-$v$ paths, because any such separating set must contain at least one non-end-vertex from each of the paths in the collection.

To prove the more substantial inequality, let $A = N(u)$ and $B = N(v)$. Take $X \subseteq V(G) \setminus \{u, v\}$ to be a $u$-$v$ separating set of minimum possible cardinality $k$. Because $u$ and $v$ are not adjacent, $X$ is also a $N(u)$-$N(v)$ separating set, so the set form of Menger's theorem states that $G$ contains a collection of $k$ pairwise disjoint $N(u)$-$N(v)$ paths. By attaching the end-vertices $u$ and $v$ to these paths, we obtain a collection of $k$ pairwise internally disjoint $u$-$v$ paths. This establishes that the maximum number of pairwise internally disjoint $u$-$v$ paths is at least the minimum number of vertices in a $u$-$v$ separating set from $V(G) \setminus \{u, v\}$. ∎

## The global form of Menger's theorem

Both the set and vertex forms of Menger's theorem are local in the sense that they concern particular sets of vertices. As first observed by Whitney [255], the vertex form implies a global form, which can serve as an alternative and more positive definition of $k$-connectedness. Note that the $k = 2$ case of this result is equivalent to Corollary 4.9.

**Theorem 4.19 (Menger's theorem, global form).** *A graph with at least $k + 1$ vertices is $k$-connected if and only if for every pair of distinct vertices $u$ and $v$, there is a collection of $k$ pairwise internally disjoint $u$-$v$ paths.*

*Proof.* First, if a graph $G$ has at least $k + 1$ vertices and contains $k$ pairwise internally disjoint paths between every pair of distinct vertices, then the graph $G - X$ is connected for every set $X \subseteq V(G)$ with $|X| \le k - 1$, so $G$ is $k$-connected.

To prove the converse, suppose to the contrary that $G$ is $k$-connected, but that for some pair of distinct vertices $u, v \in V(G)$, we cannot find a collection of $k$ pairwise internally disjoint $u$-$v$ paths. If $u$ and $v$ are not adjacent, then the vertex form of Menger's theorem (Theorem 4.18) implies that there is a set $X \subseteq V(G) \setminus \{u, v\}$ with $|X| \le k - 1$ so that $u$ and $v$ are disconnected in $G - X$, but this is a contradiction to our hypothesis that $G$ is $k$-connected.

Thus, we may assume that $uv \in E(G)$. Consider the graph $G - uv$. Since we have assumed that we cannot find a collection of $k$ pairwise internally disjoint $u$-$v$ paths in $G$, we cannot find a collection of $k-1$ pairwise internally disjoint $u$-$v$ paths in $G - uv$. The vertex form of Menger's theorem (Theorem 4.18) then implies that there is a set $X \subseteq V(G) \setminus \{u, v\}$ with $|X| \le k - 2$ so that $u$ and $v$ are disconnected in $G - uv - X$. Because $G$ is $k$-connected, it has at least $k + 1$ vertices, so there is at least one vertex $w \in V(G - u - v - X)$. In $G - uv - X$, $w$ is disconnected from at least one of $u$ or $v$ (because there are no $u$-$v$ paths in this graph); without loss of generality, we may assume that $w$ is disconnected from $v$ in $G - uv - X$. However, this implies that $v$ is also disconnected from $w$ in $G - X - u$, and so $X \cup \{u\}$ is a vertex-cut of cardinality at most $k - 1$, contradicting our hypothesis that $G$ is $k$-connected. ∎

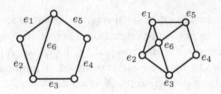

**Figure 4.8.** A graph (left) and its line graph (right)

 ## Edge forms of Menger's theorem

The versions of Menger's theorem we have seen so far consider separating sets of vertices, but they have analogues involving separating sets of edges. To derive these results from the vertex versions of Menger's theorem requires a construction. The **line graph** $L(G)$ of a graph $G$ is the graph whose vertices are the edges of $G$, and which has an edge $ef$ if and only if $e$ and $f$ are adjacent (that is, share a vertex) in $G$. Figure 4.8 shows an example.

For two distinct vertices $u$ and $v$ of a graph $G$, an edge-cut $F \subseteq E(G)$ is a **$u$-$v$ separating set** if $u$ and $v$ lie in different components of $G - F$. In our next form of Menger's theorem, note that we do not forbid the vertices $u$ and $v$ from being adjacent, and that our paths are **pairwise edge-disjoint**, meaning that no two of them may share an edge.

**Theorem 4.20 (Menger's theorem, edge form).** *Let $G$ be a graph and let $u, v \in V(G)$ be distinct vertices. Then the maximum number of pairwise edge-disjoint $u$-$v$ paths is equal to the minimum cardinality of a $u$-$v$ separating subset of $E(G)$.*

*Proof.* Let $A$ denote the set of edges incident to $u$ and let $B$ denote the set of edges incident to $v$. By the set form of Menger's theorem (Theorem 4.17) applied to the line graph $L(G)$, we see that the maximum number of pairwise disjoint $A$-$B$ paths in $L(G)$ is equal to the minimum cardinality of an $A$-$B$ separating subset of $V(L(G))$. These paths in $L(G)$ translate directly to pairwise edge-disjoint $u$-$v$ paths in $G$, and an $A$-$B$ separating subset of $V(L(G))$ is precisely a $u$-$v$ separating subset of $E(G)$. ■

There is also an edge analogue of the global form of Menger's theorem, which the reader is asked to prove in Exercise 4.51.

**Theorem 4.21 (Menger's theorem, global edge form).** *A graph with at least two vertices is $k$-edge-connected if and only if for every pair of distinct vertices $u$ and $v$, there are $k$ pairwise edge-disjoint $u$-$v$ paths.*

 ## The fan lemma and common cycles

Given a vertex $u$ and a set $B$ of vertices not containing $u$, a collection of $u$-$B$ paths is called a **$u$-$B$ fan** if each pair of these paths share only the vertex $u$ in common (so that the paths resemble a hand fan). For graphs with at least $k + 1$ vertices, the converse of the following result also holds (Exercise 4.56); both directions are originally due to Dirac [67].

**Theorem 4.22 (Fan lemma).** *If $G$ is a $k$-connected graph, then it contains a $u$-$B$ fan of cardinality at least $k$ for every vertex $u \in V(G)$ and every set $B \subseteq V(G) \setminus \{u\}$ with $|B| \geq k$.*

*Proof.* Suppose that $G$ is $k$-connected, that $u \in V(G)$, and that $B \subseteq V(G) \setminus \{u\}$ with $|B| \geq k$. Because $G$ is $k$-connected, any $N(u)$-$B$ separating set must have cardinality at least $k$. Thus, the set form of Menger's theorem (Theorem 4.17) shows that there are at least $k$ disjoint $N(u)$-$B$ paths. By prepending the vertex $u$ to each of these paths, we obtain a $u$-$B$ fan of cardinality at least $k$. ∎

By Theorem 4.8, every pair of vertices in a 2-connected graph lie on a common cycle of the graph. Using the fan lemma, Dirac [67] generalized this to $k$-connected graphs for all $k \geq 2$.

**Theorem 4.23.** *If $G$ is $k$-connected for some $k \geq 2$, then every $k$ vertices of $G$ lie on a common cycle.*

*Proof.* We prove the result by induction on $k$. The base case $k = 2$ is Theorem 4.8, but we can now derive it more simply from the global form of Menger's theorem (Theorem 4.19): for every pair of distinct vertices $u, v \in V(G)$, there are two pairwise internally disjoint $u$-$v$ paths, and these together form a cycle containing $u$ and $v$.

Now suppose that $k \geq 3$ and that the result holds for $k - 1$. Let $G$ be a $k$-connected graph and take $S \subseteq V(G)$ with cardinality $k$. Choose some vertex $u \in S$. By induction, since $G$ is also $(k - 1)$-connected, there is a cycle $C$ containing every vertex in $S \setminus \{u\}$. Set $B = V(C)$.

If $|B| = k - 1$, then the fan lemma (Theorem 4.22) guarantees the existence of a $u$-$B$ fan of cardinality at least $k - 1$. To obtain the desired cycle, we start at $u$, follow one path of this fan to $C$, traverse $C$, and then return to $u$ along another path of the fan.

We may therefore assume that $|B| \geq k$. Now the fan lemma (Theorem 4.22) guarantees the existence of a $u$-$B$ fan of cardinality at least $k$. The $k - 1$ vertices of $S \setminus \{u\}$ partition the cycle $C$ into $k - 1$ segments (each of which is a path). By the pigeonhole principle, at least two paths of our $u$-$B$ fan terminate in the same segment. Let $P$ and $Q$ be two such paths. To obtain the desired cycle, we start at $u$, follow $P$ to $C$, then traverse $C$ until we reach the end-vertex of $Q$, and then follow $Q$ (in reverse) to return to $u$. ∎

Note that the converse of Theorem 4.23 does not hold. For example, every three vertices of $C_4$ lie on a common cycle, but $C_4$ is not 3-connected.

# Exercises for Chapter 4

## Section 4.1. Cut-vertices, bridges, and blocks

**4.1.** Prove that vertices of degree 1 are never cut-vertices.

**4.2.** Show that if $G$ is a graph with $\delta(G) \geq 2$ containing a cut-vertex of degree 2, then $G$ has at least three cut-vertices.

**4.3.** Let $G$ be a graph with at least three vertices. Prove that if $e = uv$ is a bridge of $G$, then either $u$ or $v$ is a cut-vertex of $G$.

**4.4.** Prove that a graph in which all of the vertices have even degrees contains no bridges.

**4.5.** Prove that a 3-regular graph has a cut-vertex if and only if it has a bridge.

**4.6.** Prove that if $T$ is a tree of order at least 2, then $T$ contains a cut-vertex $v$ such that at most one vertex adjacent to $v$ is not a leaf.

**4.7.** Let $G$ be a connected graph, and let $e_1$ and $e_2$ be two edges of $G$. Prove that $G - e_1 - e_2$ has three components if and only if both $e_1$ and $e_2$ are bridges in $G$.

**4.8.** State and prove a version of Theorem 4.1 for graphs that are not necessarily connected.

**4.9.** Prove the edge analogue of Theorem 4.1: an edge $e$ of a connected graph $G$ is a bridge if and only if there are vertices $u$ and $w$ such that $e$ lies on every $u$-$w$ path.

**4.10.** If a connected graph has $k$ blocks and $\ell$ cut-vertices, what is the relationship between $k$ and $\ell$?

**4.11.** Prove that if $v$ is a cut-vertex of a graph $G$, then $\overline{G} - v$ is connected.

**4.12.** Prove or disprove: If $v$ is a cut-vertex of a connected graph $G$ and $H$ is a proper connected subgraph of order at least 3 containing $v$, then $v$ is a cut-vertex of $H$.

**4.13.** Prove or disprove: If $G$ is a connected graph and every proper connected induced subgraph of $G$ with at least three vertices contains a cut-vertex, then $G$ also contains a cut-vertex.

**4.14.** Prove or disprove: If $B$ is a block with three or more vertices in a connected graph $G$, then there is a cycle in $B$ that contains all the vertices of $B$.

**4.15.** Prove or disprove: If $G$ is a connected graph with cut-vertices, and $u$ and $v$ are antipodal vertices of $G$, then no block of $G$ contains both $u$ and $v$.

**4.16.** Prove Proposition 4.5: every connected graph $G$ with at least one cut-vertex contains a cut-vertex $v$ so that all but at most one of the blocks of $G$ containing $v$ are end-blocks.

**4.17.** Prove or disprove: If $B$ is a block of order 3 or more in a connected graph $G$, then there is a cycle in $B$ that contains all the vertices of $B$.

**4.18.** Let $G$ be a connected graph with cut-vertices. Prove that an orientation $D$ of $G$ is strongly connected if and only if the subdigraph of $D$ induced by the vertices of each block of $G$ is strongly connected.

## Section 4.2. Vertex connectivity

**4.19.** Let $G$ be a 2-connected graph. Prove that if $u, v \in V(G)$ such that $u$ is a cut-vertex of $G - v$, then $v$ is a cut-vertex of $G - u$.

**4.20.** Let $G$ be a nontrivial connected graph and let $u \in V(G)$. Prove that if $v$ is a vertex that is farthest from $u$ in $G$, then $v$ is not a cut-vertex of $G$.

**4.21.** Without appealing to Menger's theorem, prove Corollary 4.9: A graph $G$ is 2-connected if and only if for every two distinct vertices $u$ and $v$ in $G$, there are two internally disjoint $u$-$v$ paths.

**4.22.** Let $u$ and $v$ be distinct vertices of a 2-connected graph $G$. If $P$ is a $u$-$v$ path in $G$, does there necessarily exist another $u$-$v$ path $Q$ in $G$ that is internally disjoint from $P$?

**4.23.** Suppose that $G$ is a 2-connected graph with distinct vertices $u$ and $w$. Form a graph $H$ from $G$ by adding a new vertex $v$ and joining it to $u$ and $w$. Prove that $H$ is itself 2-connected.

**4.24.** Prove that if $U$ and $W$ are disjoint sets of vertices of a 2-connected graph $G$ of order at least 4 with $|U| = |W| = 2$, then $G$ contains two disjoint paths connecting the vertices of $U$ to the vertices of $W$.

**4.25.** Let $u$ and $v$ be distinct vertices of a 2-connected graph $G$. If $P$ is a $u$-$v$ path of $G$, does there necessarily exist a $u$-$v$ path $Q$ in $G$ such that $P$ and $Q$ are internally disjoint paths? Prove or give a counterexample.

**4.26.** Let $G$ and $H$ be graphs with $V(G) = \{v_1, v_2, \ldots, v_n\}$ and $V(H) = \{u_1, u_2, \ldots, u_n\}$, for some $n \geq 3$. Suppose further that the vertices $u_i$ and $u_j$ are adjacent in $H$ if and only if $v_i$ and $v_j$ belong to a common cycle in $G$. Characterize those graphs $G$ for which $H$ is complete.

**4.27.** Let us say that an **element** of a graph is a vertex or an edge. Prove that a graph $G$ with at least three vertices is 2-connected if and only if every pair of elements of $G$ lie on a common cycle of $G$.

**4.28.** Prove that if $G$ is a graph of order $n \geq 3$ with the property that $\deg u + \deg v \geq n$ for every pair of nonadjacent vertices $u$ and $v$ of $G$, then $G$ is 2-connected.

**4.29.** Prove that $\kappa(G \vee K_1) = \kappa(G) + 1$ for every graph $G$.

**4.30.** Prove the following, which is sometimes called the **expansion lemma**: If $G$ is formed from a $k$-connected graph $H$ by adding a vertex $u$ with at least $k$ neighbors in $H$, then $G$ is also $k$-connected.

**4.31.** Show for every $k$-connected graph $G$ and every tree $T$ of order $k + 1$ that there exists a subgraph of $G$ isomorphic to $T$.

**4.32.** Let $G$ be a noncomplete graph with $n$ vertices and connectivity $k$.
  (a) Prove that if $\deg v \geq (n + 2k - 2)/3$ for every vertex $v$ of $G$ and $X$ is a vertex-cut of minimum cardinality in $G$, then $G - X$ has exactly two components.
  (b) Prove that if $\deg v \geq (n + kt - t)/(t + 1)$ for some integer $t \geq 2$ and $X$ is a vertex-cut of minimum cardinality in $G$, then $G - X$ has at most $t$ components.

**4.33.** Verify that Theorem 4.7 is best possible by showing that for every positive integer $k$, there exists a graph $G$ of order $n \geq k + 1$ with $\delta(G) = \lceil (n + k - 3)/2 \rceil$ but $\kappa(G) < k$.

**4.34.** Prove that if a graph $G$ has $n$ vertices and $m$ edges, then $\kappa(G) \leq \lfloor 2m/n \rfloor$.

**4.35.** The **connection number** $\operatorname{con}(G)$ of a connected graph $G$ of order $n \geq 2$ is the smallest integer $k$ with $2 \leq k \leq n$ such that *every* induced subgraph of order $k$ in $G$ is connected. State and prove a theorem that gives a relationship between $\kappa(G)$ and $\operatorname{con}(G)$ for a graph $G$ of order $n$.

**4.36.** For an even integer $k \geq 2$, show that the minimum size of a $k$-connected graph of order $n$ is $kn/2$.

**4.37.** Prove or disprove: For any vertex $v$ of any graph $G$ of order $n \geq 2$, we have $\kappa(G - v) = \kappa(G)$ or $\kappa(G - v) = \kappa(G) - 1$.

**4.38.** Let $e$ be an edge of a $k$-connected graph $G$. Prove that $G - e$ is $(k - 1)$-connected.

**4.39.** Let $G_1$ and $G_2$ be $k$-connected graphs for some $k \geq 2$, and let $\mathcal{G}$ denote the set of all graphs obtained by adding $k$ edges between $G_1$ and $G_2$. Determine $\max\{\kappa(G) : G \in \mathcal{G}\}$.

## Section 4.3. Edge-connectivity

**4.40.** What are the possible cardinalities of minimal edge-cuts of $K_n$?

**4.41.** Determine the connectivity and edge-connectivity of each complete $k$-partite graph.

**4.42.** Prove that a graph $G$ with at least two vertices is $k$-edge-connected if and only if there exists no nonempty proper subset $X \subseteq V(G)$ such that the number of edges joining $X$ and $V(G) \setminus X$ is less than $k$.

**4.43.** Let $G$ be a graph. Prove that $\kappa'(G \vee K_1) \geq \kappa(G)' + 1$. Does equality always hold?

**4.44.** Suppose the graph $G$ has degree sequence $d_1 \geq d_2 \geq \cdots \geq d_n$. What is $\kappa'(G \vee K_1)$?

**4.45.** For every choice of positive integers $k$, $\ell$, and $d$ satisfying $1 \leq k \leq \ell \leq d$, construct a graph $G$ with $\kappa(G) = k$, $\kappa'(G) = \ell$, and $\delta(G) = d$.

**4.46.** Show that Theorem 4.15 cannot be extended to graphs whose diameter exceeds 2 by constructing an infinite sequence of graphs $G_k$ of diameter 3 with $\kappa'(G_k) < \delta(G_k)$.

**4.47.** Use Theorems 4.14 and 4.15 to determine the connectivity and edge-connectivity of the Petersen graph (depicted in Figure 1.7).

**4.48.** Show that if $G$ is a graph of order $n$ such that $\deg u + \deg v \geq n - 1$ for each pair $u$, $v$ of nonadjacent vertices, then $\kappa'(G) = \delta(G)$.

**4.49.** Use the closed ear decomposition (Theorem 4.16) to prove Robbins' theorem (Theorem 2.5): a graph has a strong orientation if and only if it is connected and bridgeless.

## Section 4.4. Menger's theorem

**4.50.** Prove that the vertex form of Menger's theorem (Theorem 4.18) implies the set form, Theorem 4.17.

**4.51.** Prove the global edge form of Menger's theorem (Theorem 4.21): A graph is $k$-edge-connected if and only if for every pair of distinct vertices $u$ and $v$, there are $k$ pairwise edge-disjoint $u$-$v$ paths.

**4.52.** Prove or disprove: If $G$ is a $k$-edge-connected graph and $v, v_1, v_2, \ldots, v_k$ are $k + 1$ vertices of $G$, then for $i = 1, 2, \ldots, k$, there exist $v$-$v_i$ paths $P_i$ such that each path $P_i$ contains exactly one vertex of $\{v_1, v_2, \ldots, v_k\}$, namely $v_i$, and for $i \neq j$, $P_i$ and $P_j$ are edge-disjoint.

**4.53.** Prove or disprove: If $G$ is a $k$-edge-connected graph with nonempty disjoint subsets $S_1$ and $S_2$ of $V(G)$, then there exist $k$ edge-disjoint paths $P_1, P_2, \ldots, P_k$ such that for each $i$, $P_i$ is a $u$-$v$ path for some $u \in S_1$ and some $v \in S_2$ for $i = 1, 2, \ldots, k$ and $|S_1 \cap V(P_i)| = |S_2 \cap V(P_i)| = 1$.

**4.54.** Show that $\kappa(Q_n) = \kappa'(Q_n) = n$ for all positive integers $n$.

**4.55.** Let $G$ be a graph of order $n$ with $\kappa(G) \geq 1$. Prove that

$$n \geq \kappa(G) \cdot (\operatorname{diam}(G) - 1) + 2.$$

**4.56.** Prove the following converse to the fan lemma (Theorem 4.22): If a graph $G$ with at least $k + 1$ vertices contains a $u$-$B$ fan of cardinality at least $k$ for every vertex $u \in V(G)$ and every set $B \subseteq V(G) \setminus \{u\}$ with $|B| \geq k$, then $G$ is $k$-connected.

**4.57.** A **chorded cycle** is a cycle $C$ (necessarily of length at least 4) together with an edge that joins two nonconsecutive vertices of $C$. Prove that every 3-connected graph contains a chorded cycle but that this need not be the case for a 2-connected graph.

**4.58.** Suppose that $G$ is a connected graph of order $n$ and that $k$ is an integer satisfying $1 \leq k \leq n - 3$. Suppose further that the maximum number of internally disjoint $u$-$v$ paths in $G$ is $k$ for every pair of nonadjacent vertices $u$ and $v$ of $G$. Prove that $G$ contains a vertex-cut with exactly $k + 1$ vertices.

**4.59.** Suppose that $G$ is a 3-connected graph. Prove that for every pair of vertices $u$ and $v$ of $G$, there exist two internally disjoint $u$-$v$ paths of different lengths in $G$. Then, show that this does not necessarily hold if $G$ is only 2-connected.

**4.60.** Suppose that $G$ is a $k$-connected graph of diameter $k$, where $k \geq 2$. Prove that $G$ contains $k + 1$ distinct vertices $v, v_1, v_2, \ldots, v_k$ and $k$ internally disjoint $v$-$v_i$ paths $P_1, P_2, \ldots, P_k$ such that each $P_i$ has length $i$.

**4.61.** Prove that if $G$ is a graph of order $n$ such that $\deg u + \deg v \geq n + k - 2$ for every pair of nonadjacent vertices $u$ and $v$ of $G$, then $G$ is $k$-connected. (The $k = 1$ case of this result is Exercise 1.59, and the $k = 2$ case is Exercise 4.28.)

# Planarity

**5**

In this chapter, we study planar graphs. These are graphs that can be drawn in the plane without any of their edges crossing. A formula developed by Euler plays a key role in our analysis of these graphs. We present two different characterizations of planar graphs, and discuss a necessary condition for a planar graph to be hamiltonian. Finally, we examine a parameter associated with nonplanar graphs.

## 5.1  Euler's formula

A **polyhedron** is a 3-dimensional object whose boundary consists of polygonal plane surfaces. These surfaces are typically called the **faces** of the polyhedron. The boundary of a face consists of the vertices and edges of the polygon, and the total number of faces in the polyhedron is commonly denoted by $F$, the total number of edges in the polyhedron by $e$ and the total number of vertices by $V$. The best known polyhedra are the so-called **platonic solids**: the **tetrahedron**, **cube (hexahedron)**, **octahedron**, **dodecahedron** and **icosahedron**. These are shown in the top row of Figure 5.1, and their values of $V$, $e$, and $F$ are shown below.

|  | $V$ | $e$ | $F$ |
|---|---|---|---|
| tetrahedron | 4 | 6 | 4 |
| cube | 8 | 12 | 6 |
| octahedron | 6 | 12 | 8 |
| dodecahedron | 20 | 30 | 12 |
| icosahedron | 12 | 30 | 20 |

Note that, for each of the platonic solids, the relationship $V - E + F = 2$ holds. Euler [87, 88] was evidently the first mathematician who observed that this relationship is true for any polyhedron, and hence it became known as:

**Euler's polyhedral formula.** *If a polyhedron has $V$ vertices, $e$ edges and $F$ faces, then*

$$V - E + F = 2.$$

Every polyhedron can be converted into a graph where the vertices and edges of the polyhedron are the vertices and edges of the graph. The graphs obtained

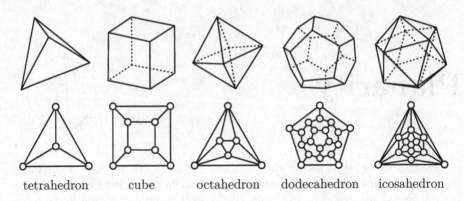

| tetrahedron | cube | octahedron | dodecahedron | icosahedron |

**Figure 5.1.** The five platonic solids

from the five platonic solids are shown in the second row of Figure 5.1. Note that
each graph depicted in Figure 5.1 has the property that no two edges cross, a
property in which we will be especially interested.

A graph $G$ is called a **planar graph** if $G$ can be drawn in the plane without
any two of its edges crossing, that is, edges intersect only at vertices. Such a
drawing is called an **embedding of $G$ in the plane**. In this case, the embedding
is a **planar embedding**. A graph $G$ that is already drawn in the plane in this
manner is a **plane graph**. Certainly then, every plane graph is planar and every
planar graph can be drawn as a plane graph. The graph $G_1 = K_{2,3}$ of Figure 5.2
is planar, although as drawn, it is not plane; however, $G_2 = K_{2,3}$ of Figure 5.2 is
both planar and plane. The graph $G_3 = K_{3,3}$ of Figure 5.2 is nonplanar as we
shall soon see.

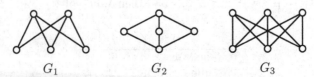

**Figure 5.2.** Planar, plane and nonplanar graphs

When the points in the plane that correspond to the vertices and edges of a
plane graph $G$ are removed from the plane, the resulting connected pieces of the
plane are the **regions** of $G$. One of the regions is unbounded and is called the
**exterior region** of $G$. The plane graph $H$ of Figure 5.3 has five regions, denoted
by $R_1$, $R_2$, $R_3$ and $R_4$, where $R_4$ is the exterior region.

For a region $R$ of a plane graph $G$, the vertices and edges incident with $R$
form a subgraph of $G$ called the **boundary** of $R$. Every edge of $G$ that lies on
a cycle belongs to the boundary of two regions of $G$, while every bridge of $G$
belongs to the boundary of a single region.

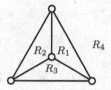

**Figure 5.3.** A plane graph $H$ and its regions

Observe that $n - m + r = 2$ for the graph $H$ of Figure 5.3 and the graph $G_2$ of Figure 5.2. In fact, this identity holds for every connected plane graph, and hence Euler's polyhedron formula is a special case of the next theorem, called Euler's formula.

**Theorem 5.1 (Euler's formula).** *For every connected plane graph with $n$ vertices, $m$ edges, and $r$ regions,*

$$n - m + r = 2.$$

*Proof.* We proceed by induction on the size $m$ of a connected plane graph. There is only one connected graph of size 0, namely $K_1$. In this case, $n = 1$, $m = 0$ and $r = 1$. Since $n - m + r = 2$, the base case of the induction holds.

Assume for a positive integer $m$ that if $H$ is a connected plane graph of order $n'$ and size $m'$, where $m' < m$ such that there are $r'$ regions, then $n' - m' + r' = 2$. Let $G$ be a connected plane graph of order $n$ and size $m$ with $r$ regions. We consider two cases.

*Case 1. Suppose $G$ is a tree.* In this case, $m = n - 1$ and $r = 1$. Thus $n - m + r = n - (n - 1) + 1 = 2$, producing the desired result.

*Case 2. Suppose $G$ is not a tree.* Since $G$ is connected and is not a tree, it follows by Theorem 1.8 that $G$ contains an edge $e$ that is not a bridge. Thus the edge $e$ is on the boundaries of two regions. So in $G - e$ these two regions merge into a single region. Since $G - e$ has order $n$, size $m - 1$ and $r - 1$ regions and $m - 1 < m$, it follows by the induction hypothesis that $n - (m - 1) + (r - 1) = 2$ and so $n - m + r = 2$. ∎

From Theorem 5.1, it follows that every two planar embeddings of a connected planar graph result in the same number of regions; thus one can speak of the number of regions of a connected planar graph. For planar graphs in general, we have the following result (see Exercise 5.2).

**Corollary 5.2.** *For every plane graph with $n$ vertices, $m$ edges, $r$ regions, and $k$ components,*

$$n - m + r = 1 + k.$$

If $G$ is a connected plane graph of order 4 or more, then the boundary of every region of $G$ must contain at least three edges. This observation is helpful in showing that, with respect to the order, its size cannot be too large.

**Theorem 5.3.** *If $G$ is a planar graph of order $n \geq 3$ and size $m$, then*

$$m \leq 3n - 6.$$

*Proof.* Since the size of every graph of order 3 cannot exceed 3, the inequality holds for $n = 3$. So we may assume that $n \geq 4$. Furthermore, we may assume that the planar graphs under consideration are connected, for otherwise edges can be added to produce a connected graph. Suppose that $G$ is a connected planar graph of order $n \geq 4$ and size $m$. Let $G$ be embedded in the plane with $r$ regions. By Euler's formula, $n - m + r = 2$. Let $R_1, R_2, \ldots, R_r$ be the regions of $G$ and let $m_i$ denote the number of edges on the boundary of $R_i$ $(1 \leq i \leq r)$. Then $m_i \geq 3$ for $1 \leq i \leq r$. Since each edge of $G$ is on the boundary of at most two regions of $G$, it follows that

$$3r \leq \sum_{i=1}^{r} m_i \leq 2m.$$

Hence,

$$6 = 3n - 3m + 3r \leq 3n - 3m + 2m = 3n - m$$

and thus $m \leq 3n - 6$. ∎

Thus, by Theorem 5.3, if $G$ is a graph of order $n \geq 5$ and size $m$ such that $m > 3n - 6$, then $G$ is nonplanar. Theorem 5.3 also has the following immediate consequence.

**Corollary 5.4.** *Every planar graph contains a vertex of degree 5 or less.*

*Proof.* The result is obvious for planar graphs of order 6 or less. Let $G$ be a graph of order $n$ and size $m$ all of whose vertices have degree 6 or more. Then $n \geq 7$ and

$$2m = \sum_{v \in V(G)} \deg v \geq 6n$$

and so $m \geq 3n$. By Theorem 5.3, $G$ is nonplanar. ∎

## The five regular polyhedra

We saw, by Euler's polyhedron formula, that if $V$, $e$ and $F$ are the number of vertices, edges and faces of a polyhedron, then

$$V - E + F = 2.$$

When dealing with a polyhedron $P$ (as well as the graph of the polyhedron $P$), it is customary to represent the number of vertices of degree $k$ by $V_k$ and number of faces bounded by a $k$-cycle ($k$-sided faces) by $F_k$. It follows then that

$$2E = \sum_{k \geq 3} kV_k = \sum_{k \geq 3} kF_k. \tag{5.1}$$

By Corollary 5.4, every polyhedron has at least one vertex of degree 3, 4 or 5. As an analogue to this result, we have the following.

**Theorem 5.5.** *At least one face of every polyhedron is bounded by a $k$-cycle for some $k$ where $k \in \{3, 4, 5\}$.*

*Proof.* Assume, to the contrary, that the statement is false. Then $F_3 = F_4 = F_5 = 0$. By Equation (5.1),

$$2E = \sum_{k \geq 6} k F_k \geq \sum_{k \geq 6} 6 F_k = 6 \sum_{k \geq 6} F_k = 6F.$$

Hence, $E \geq 3F$. Also,

$$2E = \sum_{k \geq 3} k V_k \geq \sum_{k \geq 3} 3 V_k = 3 \sum_{k \geq 3} V_k = 3V.$$

By Theorem 5.1, $V - E + F = 2$ and so $3V - 3E + 3F = 6$. Hence, $6 = 3V - 3E + 3F \leq 2E - 3E + E = 0$, which is a contradiction. ∎

A **regular polyhedron** is a polyhedron whose faces are bounded by congruent regular polygons and whose polyhedral angles are congruent. In particular, for a regular polyhedron, $F = F_s$ for some $s$ and $V = V_t$ for some $t$, where $s, t \in \{3, 4, 5\}$. For example, a cube is a regular polyhedron with $V = V_3$ and $F = F_4$. There are only four other regular polyhedra. These five regular polyhedra are the platonic solids we saw in Figure 5.1 and were known to be the only such polyhedra by the Greeks over two thousand years ago.

**Theorem 5.6.** *There are exactly five regular polyhedra.*

*Proof.* Let $P$ be a regular polyhedron and let $G$ be the associated plane graph. Then $V - E + F = 2$, where $V$, $e$ and $F$ denote the number of vertices, edges and faces of $P$ and the number of vertices, edges and regions of $G$, respectively. Therefore,

$$
\begin{aligned}
-8 &= 4E - 4V - 4F \\
&= 2E + 2E - 4V - 4F \\
&= \sum_{k \geq 3} k F_k + \sum_{k \geq 3} k V_k - 4 \sum_{k \geq 3} V_k - 4 \sum_{k \geq 3} F_k \\
&= \sum_{k \geq 3} (k - 4) F_k + \sum_{k \geq 3} (k - 4) V_k.
\end{aligned}
\tag{5.2}
$$

Since $G$ is regular, there exist integers $s$ and $t$ with $s, t \in \{3, 4, 5\}$ such that $F = F_s$ and $V = V_t$. Hence

$$-8 = (s - 4)F_s + (t - 4)V_t.$$

Moreover, $sF_s = 2E = tV_t$. If $s, t \geq 4$, then (5.2) yields $-8 = (s-4)F_s + (t-4)F_t \geq 0$, which is impossible. Hence either $s = 3$ or $t = 3$. This results in five possibilities for the pairs $s, t$.

*Case 1. Suppose $s = 3$ and $t = 3$.* Here we have

$$-8 = -F_3 - V_3 \quad \text{and} \quad 3F_3 = 3V_3;$$

so $F_3 = V_3 = 4$. Thus $P$ is the tetrahedron.

*Case 2.   Suppose $s = 3$ and $t = 4$.* Therefore,

$$-8 = -F_3 \quad \text{and} \quad 3F_3 = 4V_4.$$

Hence $F_3 = 8$ and $V_4 = 6$, implying that $P$ is the octahedron.

*Case 3. Suppose $s = 3$ and $t = 5$.* In this case,

$$-8 = -F_3 + V_5 \quad \text{and} \quad 3F_3 = 5V_5,$$

so $F_3 = 20$, $V_5 = 12$ and $P$ is the icosahedron.

*Case 4. Suppose $s = 4$ and $t = 3$.* We find here that

$$-8 = -V_3 \quad \text{and} \quad 4F_4 = 3V_3.$$

Thus $V_3 = 8$, $F_4 = 6$ and $P$ is the cube.

*Case 5. Suppose $s = 5$ and $t = 3$.* For these values,

$$-8 = F_5 - V_3 \quad \text{and} \quad 5F_5 = 3V_3.$$

Solving for $F_5$ and $V_3$, we find that $F_5 = 12$ and $V_3 = 20$, so $P$ is the dodecahedron.
∎

 **Maximal planar graphs**

A planar graph $G$ is **maximal planar** if the addition of any edge joining two nonadjacent vertices of $G$ results in a nonplanar graph. Necessarily then, if a maximal planar graph $G$ of order $n \geq 3$ and size $m$ is embedded in the plane resulting in $r$ regions, then the boundary of every region of $G$ is a triangle and so $3r = 2m$. It then follows by the proof of Theorem 5.3 that $m = 3n - 6$. All of the graphs shown in Figure 5.4 are maximal planar.

   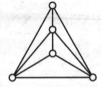 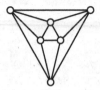

**Figure 5.4.** Maximal planar graphs

We now derive some results concerning the degrees of the vertices of a maximal planar graph.

**Theorem 5.7.** *If $G$ is a maximal planar graph of order 4 or more, then the degree of every vertex of $G$ is at least 3.*

*Proof.* Let $G$ be a maximal planar graph of order $n \geq 4$ and size $m$ and let $v$ be a vertex of $G$. Since $m = 3n - 6$, it follows that $G - v$ has order $n - 1$ and size $m - \deg v$. Since $G - v$ is planar and $n - 1 \geq 3$, it follows that

$$m - \deg v \leq 3(n - 1) - 6$$

and so $m - \deg v = 3n - 6 - \deg v \leq 3n - 9$. Thus, $\deg v \geq 3$. $\blacksquare$

Not only is the minimum degree of every maximal planar graph $G$ of order 4 or more at least 3, the graph $G$ is 3-connected (see Exercise 5.14). The next result gives a relationship among the degrees of the vertices in a maximal planar graph of order at least 4.

**Theorem 5.8.** *Let $G$ be a maximal planar graph of order $n \geq 4$ and size $m$ containing $n_i$ vertices of degree $i$ for $3 \leq i \leq \Delta = \Delta(G)$. Then*

$$3n_3 + 2n_4 + n_5 = 12 + n_7 + 2n_8 + \cdots + (\Delta - 6)n_\Delta.$$

*Proof.* Since $m = 3n - 6$, it follows that $2m = 6n - 12$. Therefore,

$$\sum_{i=3}^{\Delta} i n_i = \sum_{i=3}^{\Delta} 6n_i - 12$$

and so

$$\sum_{i=3}^{\Delta} (6 - i)n_i = 12. \tag{5.3}$$

Hence, $3n_3 + 2n_4 + n_5 = 12 + n_7 + 2n_8 + \cdots + (\Delta - 6)n_\Delta$. $\blacksquare$

## Discharging

We now describe a very important and useful technique, called **discharging**. The idea is to assign a "charge" to each vertex of a planar graph as well as **discharging rules** which indicate how charges can be redistributed among the vertices. For example, suppose $G$ is a maximal planar graph and that every vertex $v$ is assigned a **charge** of $6 - \deg v$. In particular, every vertex of degree 5 receives a charge of $+1$, every vertex of degree 6 receives a charge of 0, and every vertex of degree 7 or more receives a negative charge. By appropriately redistributing positive charges, some useful results can often be obtained. According to Equation (5.3) in the proof of Theorem 5.8, the sum of the charges of the vertices of a maximal planar graph of order 4 or more is 12. This is restated below.

**Theorem 5.9.** *If $G$ is a maximal planar graph of order $n \geq 4$, size $m$, and maximum degree $\Delta(G) = \Delta$ such that $G$ has $n_i$ vertices of degree $i$ for $3 \leq i \leq \Delta$, then*

$$\sum_{i=3}^{\Delta} (6 - i)n_i = 12.$$

According to Theorem 5.7, in a maximal planar graph of order 4 or more, the degree of every vertex is at least 3. Consequently, no vertex can contribute more than 3 to the sum in Theorem 5.9. Because this sum is 12, we have the following corollary, which is an extension of Corollary 5.4.

**Corollary 5.10.** *If $G$ is a maximal planar graph of order at least 4, then $G$ contains at least four vertices whose degrees are at most 5.*

Although the next theorem, due to Wernicke [253], can be proved using other methods, we will use the discharging method to illustrate this powerful technique.

**Theorem 5.11.** *If $G$ is a maximal planar graph of order 4 or more, then $G$ contains at least one of the following: (1) a vertex of degree 3, (2) a vertex of degree 4, (3) two adjacent vertices of degree 5, (4) two adjacent vertices, one of which has degree 5 and the other has degree 6.*

*Proof.* Assume, to the contrary, that there exists a maximal planar graph $G$ of order $n \geq 4$, where there are $n_i$ vertices of degree $i$ for $3 \leq i \leq \Delta = \Delta(G)$ such that $G$ contains none of (1)–(4). Thus, $\delta(G) = 5$. To each vertex $v$ of $G$ assign the charge $6 - \deg v$. Hence, each vertex of degree 5 receives a charge of $+1$, each vertex of degree 6 receives no charge, and each vertex of degree 7 or more receives a negative charge. By Theorem 5.9, the sum of the charges of the vertices of $G$ is

$$\sum_{i=3}^{\Delta}(6 - i)n_i = 12.$$

Let $G$ be embedded in the plane. For each vertex $v$ of degree 5 in $G$, redistribute its charge of $+1$ by moving a charge of $1/5$ to each of its five neighbors, resulting in $v$ now having a charge of 0. Hence, the sum of the charges of the vertices of $G$ remains 12. Since $G$ contains neither (3) nor (4), no vertex of degree 5 or 6 will have its charges increased. Consider a vertex $u$ with $\deg u = k \geq 7$. Thus $u$ received an initial charge of $6 - k$. Because no consecutive neighbors of $u$ in the embedding can have degree 5, the vertex $u$ can receive an added charge of $+1/5$ from at most $k/2$ of its neighbors. After the redistribution of charges, the new charge of $u$ is at most

$$6 - k + \frac{k}{2} \cdot \frac{1}{5} = 6 - \frac{9k}{10} < 0.$$

Hence, no vertex of $G$ now has a positive charge. This is impossible, however, since the sum of the charges of the vertices of $G$ is 12. ∎

Another result concerning maximal planar graphs that can be proved with the aid of the discharging method (see Exercise 5.18) is due to Franklin [97].

**Theorem 5.12.** *If $G$ is a maximal planar graph of order 4 or more, then $G$ contains at least one of the following: (1) a vertex of degree 3, (2) a vertex of degree 4, (3) a vertex of degree 5 that is adjacent to two vertices, each of which has degree 5 or 6.*

# 5.2 Characterizations of planarity

There are two graphs, namely $K_5$ and $K_{3,3}$, that play an important role in the study of planar graphs. We examine the planarity of $K_5$ and $K_{3,3}$ next.

**Theorem 5.13.** *The graph $K_5$ is nonplanar.*

*Proof.* The graph $K_5$ has order $n = 5$ and size $m = 10$. Since $m = 10 > 9 = 3n-6$, it follows by Theorem 5.3 that $K_5$ is nonplanar. ∎

Since any graph containing a nonplanar subgraph is itself nonplanar, it follows that once we know $K_5$ is nonplanar, we can conclude that $K_n$ is nonplanar for every integer $n \geq 5$. Of course, $K_n$ is planar for $1 \leq n \leq 4$.

Theorem 5.3 cannot be used to establish the nonplanarity of $K_{3,3}$ since $K_{3,3}$ has order $n = 6$ and size $m = 9$. On the other hand, we can use the fact that $K_{3,3}$ is bipartite to establish this property.

**Theorem 5.14.** *The graph $K_{3,3}$ is nonplanar.*

*Proof.* Suppose, to the contrary, that $K_{3,3}$ is planar. Then there is a plane embedding of $K_{3,3}$, say with $r$ regions. Thus, by Euler's formula, $n - m + r = 6 - 9 + r = 2$ and so $r = 5$. Let $R_1, R_2, \ldots, R_5$ be the five regions and let $m_i$ be the number of edges on the boundary of $R_i$ ($1 \leq i \leq 5$). Since $K_{3,3}$ is bipartite, $K_{3,3}$ contains no triangles and so $m_i \geq 4$ for $1 \leq i \leq 5$. Since every edge of $K_{3,3}$ lies on the boundary of a cycle, every edge of $K_{3,3}$ belongs to the boundary of two regions. Thus,

$$20 = 4r \leq \sum_{i=1}^{5} m_i = 2m = 18,$$

which is impossible. ∎

In the preceding section, we discussed several characteristics of planar graphs. However, a fundamental question remains: For a given graph $G$, how does one determine whether $G$ is planar or nonplanar? Of course, if $G$ can be drawn in the plane without any of its edges crossing, then $G$ is planar. On the other hand, if $G$ cannot be drawn in the plane without edges crossing, then $G$ is nonplanar. However, it may be very difficult to see how to draw a graph $G$ in the plane without edges crossing or to know that such a drawing is impossible. We saw from Theorem 5.3 that if $G$ has order $n \geq 3$ and size $m$ where $m > 3n - 6$, then $G$ is nonplanar. Also, as a consequence of Theorem 5.3, we saw in Corollary 5.4 that if $G$ contains no vertex of degree less than 6, then $G$ is nonplanar. Of course, in that case, $m \geq 3n$.

Any graph that is a subgraph of a planar graph must surely be planar. Equivalently, every graph containing a nonplanar subgraph must itself be nonplanar.

Thus, to show that a disconnected graph $G$ is planar, it suffices to show that each component of $G$ is planar. Hence, when considering planarity, we may restrict our attention to connected graphs. Since a connected graph $G$ is planar if and only if each block of $G$ is planar (see Exercise 5.1), it is sufficient to concentrate on 2-connected graphs only.

According to Theorems 5.13 and 5.14, the graphs $K_5$ and $K_{3,3}$ are nonplanar. Hence, if a graph $G$ should contain $K_5$ or $K_{3,3}$ as a subgraph, then $G$ is nonplanar. It is possible however for a graph $G$ to be nonplanar and yet contain no subgraph isomorphic to either $K_5$ or $K_{3,3}$. As we will see, such a graph must possess some subgraph that is "like" $K_5$ or $K_{3,3}$. For this, we will need the next definition.

A graph $H$ is a **subdivision** of a graph $G$ if either $H \cong G$ or $H$ can be obtained from $G$ by inserting vertices of degree 2 into some, all or none of the edges of $G$. Thus, for the graph $G$ of Figure 5.5, all of the graphs $H_1$, $H_2$ and $H_3$ are subdivisions of $G$. Indeed, $H_3$ is also a subdivision of $H_2$.

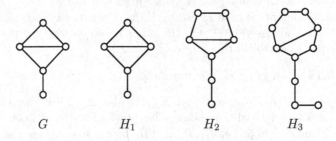

$\quad G \qquad\qquad H_1 \qquad\qquad H_2 \qquad\qquad H_3$

**Figure 5.5.** Subdivisions of a graph

Certainly, a subdivision $H$ of a graph $G$ is planar if and only if $G$ is planar. Therefore, $K_5$ and $K_{3,3}$ are nonplanar as is any subdivision of $K_5$ or $K_{3,3}$. These observations give the following result.

**Theorem 5.15.** *If $G$ contains a subgraph that is a subdivision of either $K_5$ or $K_{3,3}$, then $G$ is nonplanar.*

The remarkable feature about Theorem 5.15 is that its converse is also true. Hence, Theorem 5.15 and its converse provide a characterization of planar graphs, due to Kuratowski [155].

**Theorem 5.16 (Kuratowski's theorem).** *A graph $G$ is planar if and only if $G$ contains no subgraph that is a subdivision of $K_5$ or $K_{3,3}$.*

*Proof.* We have already noted the necessity of this condition for a graph to be planar. Hence, it remains to verify its sufficiency, namely that every graph containing no subgraph which is a subdivision of $K_5$ or $K_{3,3}$ is planar. Suppose that this statement is false. Then there is a nonplanar 2-connected graph $G$ of minimum size containing no subgraph that is a subdivision of $K_5$ or $K_{3,3}$. In fact, by Exercise 5.29, we may assume that such a graph is 3-connected.

Let $G$ be a 3-connected nonplanar graph of minimum size containing no subgraph that is a subdivision of $K_5$ or $K_{3,3}$. Let $e = uv$ be an edge of $G$. Then $H = G - e$ is planar. Since $G$ is 3-connected, $H$ is 2-connected. By Theorem 4.8, there is a cycle in $H$ containing both $u$ and $v$. Among all planar embeddings of $H$, choose one in which there is a cycle

$$C = (u = v_0, v_1, \ldots, v_\ell = v, \ldots, v_k = u)$$

containing $u$ and $v$ such that the number of regions interior to $C$ is maximized.

It is convenient to define two subgraphs of $H$. By the *exterior subgraph* of $H$ is meant the subgraph induced by those edges lying exterior to $C$ and the *interior subgraph* of $H$ is the subgraph induced by those edges lying interior to $C$. Both subgraphs exist, for otherwise the edge $e$ could be added either to the exterior or interior subgraph of $H$ so that the resulting graph (namely $G$) is planar.

No two distinct vertices of $\{v_0, v_1, \ldots, v_\ell\}$ or of $\{v_\ell, v_{\ell+1}, \ldots, v_k\}$ are connected by a path in the exterior subgraph of $H$, for otherwise there is a cycle in $H$ containing $u$ and $v$ and having more regions interior to it than $C$ has. Since $G$ is nonplanar, there must be a $v_s$-$v_t$ path $P$ in the exterior subgraph of $H$, where $0 < s < \ell < t < k$, such that $v_s$ and $v_t$ are the only vertices of $P$ that belong to $C$. This situation is pictured below.

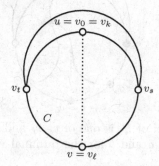

In fact, the path $P$ must be $(v_s, v_t)$; for otherwise, if there is an interior vertex $w$ on $P$, then, since $G$ is 3-connected, there are three internally disjoint paths from $w$ to $C$, creating a new cycle $C'$ containing $u$ and $v$ such that $C'$ has more regions interior to it than $C$ has, which is a contradiction.

Let $S$ be the set of vertices on $C$ different from $v_s$ and $v_t$, that is,

$$S = V(C) - \{v_s, v_t\},$$

and let $H_1$ be the component of $H - S$ that contains $P$. By the defining property of $C$, the subgraph $H_1$ cannot be moved to the interior of $C$ in a plane manner. This fact together with the fact that $G = H + e$ is nonplanar implies that the interior subgraph of $H$ must contain one of the following:

(1) A $v_a$-$v_b$ path with $0 < a < s$ and $\ell < b < t$ such that only $v_a$ and $v_b$ belong to $C$ as shown below.

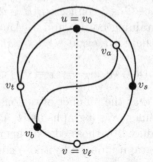

(2) A vertex $w$ not on $C$ that is connected to $C$ by three internally disjoint paths such that the terminal vertex of one such path $P'$ is one of $v_0, v_s, v_\ell$ and $v_t$. If, for example, the terminal vertex of $P'$ is $v_0$, then the terminal vertices of the other two paths are $v_a$ and $v_b$, where $s \leq a < \ell$ and $\ell < b \leq t$ where not both $a = s$ and $b = t$ occur as shown below.

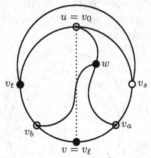

If the terminal vertex of $P'$ is one of $v_s, v_\ell$ and $v_t$, then there are corresponding bounds for $a$ and $b$ for the terminal vertices of the other two paths.

(3) A vertex $w$ not on $C$ that is connected to $C$ by three internally disjoint paths $P_1, P_2$ and $P_3$ such that the terminal vertices of these paths are three of the four vertices $v_0, v_s, v_\ell$ and $v_t$, say $v_0, v_\ell$ and $v_s$, respectively, together with a $v_c$-$v_t$ path $P_4$ ($v_c \neq v_0, v_\ell, w$), where $v_c$ is on $P_1$ or $P_2$ and $P_4$ is disjoint from $P_1, P_2$ and $C$ except for $v_c$ and $v_t$ as shown below.

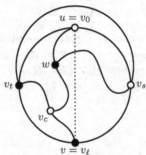

The remaining choices for $P_1, P_2$ and $P_3$ produce three analogous cases.

(4) A vertex $w$ not on $C$ that is connected to $v_0, v_s, v_\ell$ and $v_t$ by four internally disjoint paths as shown below.

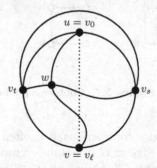

In the first three cases, there is a subgraph of $G$ that is a subdivision of $K_{3,3}$, while in the fourth case, there is a subgraph of $G$ that is a subdivision of $K_5$. This is a contradiction. ■

While it is rarely easy to use Kuratowski's theorem to test a graph for planarity, a number of efficient algorithms have been developed that determine whether a graph is planar, including linear-time algorithms, the first of which was obtained by Hopcroft and Tarjan [131].

As a consequence of Kuratowski's theorem, the 4-regular graph $G$ shown in Figure 5.6(a) is nonplanar since $G$ contains the subgraph $H$ in Figure 5.6(b) which is a subdivision of $K_{3,3}$.

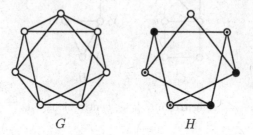

$G$                 $H$

**Figure 5.6.** A nonplanar graph

## Minors of graphs

There is another characterization of planar graphs closely related to that given by Kuratowski's theorem. Before presenting this theorem, it is useful to introduce some additional terminology. If two adjacent vertices $u$ and $v$ in a graph $G$ are identified, then we say that we have **contracted** the edge $uv$ (denoting the resulting vertex by $u$ or $v$). For the graph $G$ of Figure 5.7, the graph $G'$ is obtained by contracting the edge $uv$ in $G$ and $G''$ is obtained by contracting the edge $wy$ in $G'$.

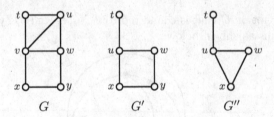

**Figure 5.7.** Contracting an edge

When dealing with edge contractions, it is often the case that we begin with a graph $G$, contract an edge in $G$ to obtain a graph $G'$, contract some edge in $G'$ to obtain another graph $G''$, and so on, until finally arriving at a graph $H$. Equivalently, $H$ can be obtained from $G$ by a succession of edge contractions if and only if the vertex set of $H$ is the set of elements in a partition $\{V_1, V_2, \ldots, V_k\}$ of $V(G)$ where each induced subgraph $G[V_i]$ is connected and $V_i$ is adjacent to $V_j$ ($i \neq j$) if some vertex in $V_i$ is adjacent to some vertex in $V_j$ in $G$. For example, in the graph $G$ of Figure 5.7, if we were to let

$$V_1 = \{t\}, \; V_2 = \{u, v\}, \; V_3 = \{x\} \text{ and } V_4 = \{w, y\},$$

then the resulting graph $H$ is shown in Figure 5.8.

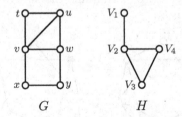

**Figure 5.8.** Edge contractions

A graph $H$ is called a **minor** of a graph $G$ if either $H \cong G$ or a graph isomorphic to $H$ can be obtained from $G$ by a succession of edge contractions, edge deletions and vertex deletions (in any order). Equivalently, $H$ is a minor of $G$ if $H \cong G$ or $H$ can be obtained from a subgraph of $G$ by a succession of edge contractions. Consequently, a graph $G$ is a minor of itself. The planar graph $H$ of Figure 5.8 is a minor of the planar graph $G$ of that figure. In fact, every minor of a planar graph is planar.

Consider next the graph $G$ of Figure 5.9, where $V_1 = \{t_1, t_2\}$, $V_2 = \{v\}$, $V_3 = \{u_1, u_2, u_3, u_4\}$, $V_4 = \{w_1, w_2, w_3\}$, $V_5 = \{s_1\}$, $V_6 = \{x_1, x_2\}$, $V_7 = \{s_2\}$ and $V_8 = \{z\}$. Then the graph $H$ of Figure 5.9 can be obtained from $G$ by successive edge contractions. Thus, $H$ is a minor of $G$. By deleting the edge $V_5 V_7$ and the vertices $V_7$ and $V_8$ from $H$ (or equivalently, deleting $V_7$ and $V_8$ from $H$), we see that $K_{3,3}$ is also a minor of $G$.

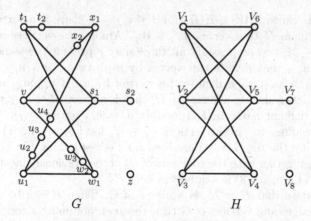

**Figure 5.9.** Minors of graphs

The example in Figure 5.9 serves to illustrate the following observation.

**Theorem 5.17.** *If a graph $G$ is a subdivision of a graph $H$, then $H$ is a minor of $G$.*

The converse of Theorem 5.17 is not true (see Exercise 5.30). The following is an immediate consequence of Theorem 5.16 and Theorem 5.17.

**Theorem 5.18.** *If $G$ is a nonplanar graph, then $K_5$ or $K_{3,3}$ is a minor of $G$.*

Our second characterization of planar graphs, due to Wagner [249], is given next and, perhaps unsurprisingly, involves minors of $K_5$ and $K_{3,3}$.

**Theorem 5.19 (Wagner's theorem).** *A graph $G$ is planar if and only if neither $K_5$ nor $K_{3,3}$ is a minor of $G$.*

In light of Theorem 5.17, to prove Theorem 5.19, we need only prove the following result.

**Theorem 5.20.** *Let $G$ be a graph.*
 (a) *If $G$ has $K_{3,3}$ as a minor, then $G$ contains a subdivision of $K_{3,3}$.*
 (b) *If $G$ has $K_5$ as a minor, then $G$ contains either a subdivision of $K_5$ or a subdivision of $K_{3,3}$.*

*Proof.* Suppose first that $H = K_{3,3}$ is a minor of $G$. The graph $H$ can be obtained by first deleting edges and vertices of $G$ (if necessary), obtaining a connected graph $G'$, and then by a succession of edge contractions in $G'$. We show, in this case, that $G'$ contains a subgraph that is a subdivision of $K_{3,3}$ and so $G'$, and $G$ as well, is nonplanar.

Denote the vertices of $H$ by $U_i$ and $W_i$ ($1 \leq i \leq 3$), where $\{U_1, U_2, U_3\}$ and $\{W_1, W_2, W_3\}$ are the partite sets of $H$. Since $H$ is obtained from $G'$ by a succession of edge contractions, the subgraphs

$$F_i = G'[U_i] \text{ and } H_i = G'[W_i] \text{ for } 1 \leq i \leq 3,$$

are connected. Since $U_iW_j \in E(H)$ for $1 \le i, j \le 3$, there is a vertex $u_{i,j} \in U_i$ that is adjacent in $H$ to a vertex $w_{i,j} \in W_j$. Among the vertices $u_{i,1}, u_{i,2}, u_{i,3}$ in $U_i$ ($1 \le i \le 3$), two or possibly all three may represent the same vertex. If $u_{i,1} = u_{i,2} = u_{i,3}$, then denote this vertex by $u_i$; if two of $u_{i,1}, u_{i,2}, u_{i,3}$ are the same, say $u_{i,1} = u_{i,2}$, then denote this vertex by $u_i$; if $u_{i,1}, u_{i,2}$ and $u_{i,3}$ are distinct, then let $u_i$ denote a vertex in $U_i$ that is connected to $u_{i,1}, u_{i,2}$ and $u_{i,3}$ by internally disjoint paths in $F_i$ (possibly $u_i = u_{i,j}$ for some $j$). We proceed in the same manner to obtain vertices $w_i \in W_i$ for $1 \le i \le 3$. The subgraph of $G$ induced by the previously described nine edges joining $U_1 \cup U_2 \cup U_3$ and $W_1 \cup W_2 \cup W_3$ together with the edge sets of all of the previously mentioned paths in $F_i$ and $H_j$ ($1 \le i, j \le 3$) is a subdivision of $K_{3,3}$.

Next, suppose that $H = K_5$ is a minor of $G$. Then $H$ can be obtained by first deleting edges and vertices of $G$ (if necessary), obtaining a connected graph $G'$, and then by a succession of edge contractions in $G'$. We show in this case that either $G'$ contains a subgraph that is a subdivision of $K_5$ or $G'$ contains a subgraph that is a subdivision of $K_{3,3}$.

We may denote the vertices of $H$ by $V_i$ ($1 \le i \le 5$), where $G_i = G'[V_i]$ is a connected subgraph of $G'$ and each subgraph $G_i$ contains a vertex that is adjacent to $G_j$ for each pair $i, j$ of distinct integers where $1 \le i, j \le 5$. For $1 \le i \le 5$, let $v_{i,j}$ be a vertex of $G_i$ that is adjacent to a vertex of $G_j$, where $1 \le j \le 5$ and $j \ne i$.

For a fixed integer $i$ with $1 \le i \le 5$, if the vertices $v_{i,j}$ ($i \ne j$) represent the same vertex, then denote this vertex by $v_i$. If three of the four vertices $v_{i,j}$ are the same, then we also denote this vertex by $v_i$. If two of the vertices $v_{i,j}$ are the same, the other two are distinct, and there exist internally disjoint paths from the coinciding vertices to the other two vertices, then we denote the two coinciding vertices by $v_i$. If the vertices $v_{i,j}$ are distinct and $G_i$ contains a vertex from which there are four internally disjoint paths (one of which may be trivial) to the vertices $v_{i,j}$, then denote this vertex by $v_i$. Hence, there are several instances in which we have defined a vertex $v_i$. Should $v_i$ be defined for all $i$ ($1 \le i \le 5$), then $G'$ (and therefore $G$ as well) contains a subgraph that is a subdivision of $K_5$.

We may assume then that for one or more integers $i$ ($1 \le i \le 5$), the vertex $v_i$ has not been defined. For each such $i$, there exist distinct vertices $u_i$ and $w_i$, each of which is connected to two of the vertices $v_{i,j}$ by internally disjoint (possibly trivial) paths, while $u_i$ and $w_i$ are connected by a path none of whose internal vertices are the vertices $v_{i,j}$ and where every two of the five paths have only $u_i$ or $w_i$ in common. If two of the vertices $v_{i,j}$ coincide, then we denote this vertex by $u_i$. If the remaining two vertices $v_{i,j}$ should also coincide, then we denote this vertex by $w_i$. We may assume that $i = 1$, that $u_1$ is connected to $v_{1,2}$ and $v_{1,3}$ and that $w_1$ is connected to $v_{1,4}$ and $v_{1,5}$, as described above. Denote the edge set of these paths by $E_1$.

We now consider $G_2$. If $v_{2,1} = v_{2,4} = v_{2,5}$, then let $w_2$ be this vertex and set $E_2 = \varnothing$; otherwise, there is a vertex $w_2$ of $G_2$ (which may coincide with $v_{2,1}, v_{2,4}$ or $v_{2,5}$) connected by internally disjoint (possibly trivial) paths to the distinct vertices in $\{v_{2,1}, v_{2,4}, v_{2,5}\}$. We then let $E_2$ denote the edge set of these paths. Similarly, the vertices $w_3$, $u_2$, and $u_3$ and the sets $E_3, E_4$, and $E_5$ are defined with the aid of the sets $\{v_{3,1}, v_{3,4}, v_{3,5}\}$, $\{v_{4,1}, v_{4,4}, v_{4,5}\}$ and $\{v_{5,1}, v_{5,2}, v_{5,3}\}$, respectively. The subgraph of $G'$ induced by the union of the sets $E_i$ and the edges $v_{i,j}v_{j,i}$ contains a subdivision of $K_{3,3}$ with partite sets $\{u_1, u_2, u_3\}$ and $\{w_1, w_2, w_3\}$. ∎

In the proof of Theorem 5.20, it was shown that if $K_5$ is a minor of a graph $G$, then $G$ contains a subdivision of $K_5$ or a subdivision of $K_{3,3}$. In other words, $G$ does not necessarily contain a subdivision of $K_5$. There is good reason for this, which is illustrated in the next example.

The Petersen graph $P$ has order $n = 10$ and size $m = 15$ and hence no conclusion can be drawn from Theorem 5.3 regarding the planarity or nonplanarity of $P$. Nevertheless, the Petersen graph is, in fact, nonplanar. Theorems 5.16 and 5.19 give two ways to establish this fact. Figures 5.10(a) and 5.10(b) show $P$ drawn in two ways. Since $P - x$ (shown in Figure 5.10(c)) is a subdivision of $K_{3,3}$, the Petersen graph is nonplanar. The partition $\{V_1, V_2, \ldots, V_5\}$ of $V(P)$ shown in Figure 5.10(d), where $V_i = \{u_i, v_i\}$, $1 \le i \le 5$, shows that $K_5$ in Figure 5.10(d) is a minor of $P$ and is therefore nonplanar. Since $P$ is a cubic graph, there is no subgraph of $P$ that is a subdivision of $K_5$, however.

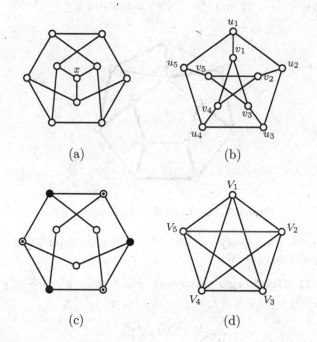

**Figure 5.10.** Showing that the Petersen graph is nonplanar

# 5.3    Hamiltonian planar graphs

In this section, we will prove a single result, namely a necessary condition for a planar graph to be hamiltonian.

Let $G$ be a hamiltonian planar graph of order $n$ with hamiltonian cycle $C$ and consider a planar embedding of $G$. An edge of $G$ not lying on $C$ is a **chord** of $G$. Every chord and every region of $G$ then lies interior to $C$ or exterior to $C$. For $i = 3, 4, \ldots, n$, let $r_i$ denote the number of regions interior to $C$ whose boundary contains exactly $i$ edges and let $r'_i$ denote the number of regions exterior to $C$ whose boundary contains exactly $i$ edges.

The plane graph $G$ of Figure 5.11 of order 12 is hamiltonian. The edges of the hamiltonian cycle $C = (v_1, v_2, \ldots, v_{12}, v_1)$ are drawn with bold lines. With respect to $C$, we have

$$r_3 = r'_3 = 1, \ r_4 = 3, \ r'_4 = 2, \ r_5 = r'_7 = 1,$$

while $r_i = 0$ for $6 \le i \le 12$ and $r'_i = 0$ for $i = 5, 6$ and $8 \le i \le 12$.

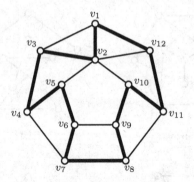

**Figure 5.11.** A hamiltonian planar graph

Next we have a necessary condition for a planar graph to be hamiltonian, discovered by Grinberg [108].

**Theorem 5.21 (Grinberg's theorem).** *For a plane graph $G$ of order $n$ with hamiltonian cycle $C$,*

$$\sum_{i=3}^{n}(i-2)(r_i - r'_i) = 0.$$

*Proof.* Suppose that $c$ chords of $G$ lie interior to $C$. Then $c+1$ regions of $G$ lie interior to $C$. Therefore,

$$\sum_{i=3}^{n} r_i = c+1 \text{ and so } c = \sum_{i=3}^{n} r_i - 1.$$

Let $N$ denote the result obtained by summing over all regions interior to $C$ the number of edges on the boundary of each such region. Then each edge on $C$ is counted once and each chord interior to $C$ is counted twice, that is,

$$N = \sum_{i=3}^{n} i r_i = n + 2c.$$

Therefore,

$$\sum_{i=3}^{n} i r_i = n + 2c = n + 2 \sum_{i=3}^{n} r_i - 2$$

and so

$$\sum_{i=3}^{n} (i - 2) r_i = n - 2.$$

Similarly,

$$\sum_{i=3}^{n} (i - 2) r_i' = n - 2.$$

Therefore, $\sum_{i=3}^{n} (i - 2)(r_i - r_i') = 0.$ ∎

The following observations are quite useful in applying Theorem 5.21. Let $G$ be a plane graph with hamiltonian cycle $C$. Furthermore, suppose that an edge $e$ of $G$ is on the boundary of two regions $R_1$ and $R_2$ of $G$. If $e$ is an edge of $C$, then one of $R_1$ and $R_2$ is in the interior of $C$ and the other is in the exterior of $C$. If, on the other hand, $e$ is not an edge of $C$, then $R_1$ and $R_2$ are either both in the interior of $C$ or both in the exterior of $C$.

In 1880, the British mathematician P. G. Tait conjectured that every 3-connected cubic planar graph is hamiltonian. This conjecture was disproved in 1946 by Tutte [236], who produced the graph $G$ in Figure 5.12 as a counterexample. In addition to disproving Tait's conjecture, Tutte [240] proved that every 4-connected planar graph is hamiltonian.

**Theorem 5.22.** *Every 4-connected planar graph is hamiltonian-connected.*

Since Theorem 5.21 gives a necessary condition for a planar graph to be hamiltonian, this theorem also provides a sufficient condition for a planar graph to be non-hamiltonian. We now see how Grinberg's theorem can be used to show that the plane graph of Figure 5.12 is not hamiltonian. This graph is called the **Tutte graph** and has a great deal of historical interest.

Assume, to the contrary, that the Tutte graph $G$ is hamiltonian. Then $G$ has a hamiltonian cycle $C$. Necessarily, $C$ must contain exactly two of the three edges $e$, $f_1$, and $f_2$. Suppose that $C$ contains $f_1$ and either $e$ or $f_2$. Similarly, $C$ must contain exactly two edges of the three edges $e'$, $f_2$ and $f_3$. Since we may assume that $C$ contains $f_2$, we may further assume that $e$ is not on $C$. Consequently, $R_1$ and $R_2$ lie interior to $C$.

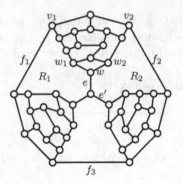

**Figure 5.12.** The Tutte graph

Let $G_1$ denote the component of $G-\{e, f_1, f_2\}$ containing $w$. Thus, $G_1$ contains a hamiltonian $v_1$-$v_2$ path $P'$. Therefore, $G_2 = G_1 + v_1v_2$ is hamiltonian and contains a hamiltonian cycle $C'$ consisting of $P'$ and $v_1v_2$. Applying Grinberg's theorem to $G_2$ with respect to $C'$, we obtain

$$1(r_3 - r_3') + 2(r_4 - r_4') + 3(r_5 - r_5') + 6(r_8 - r_8') = 0. \qquad (5.4)$$

Since $v_1v_2$ is on $C'$ and the exterior region of $G_2$ lies exterior to $C'$, it follows that

$$r_3 - r_3' = 1 - 0 = 1 \quad \text{and} \quad r_8 - r_8' = 0 - 1 = -1.$$

Therefore, from (5.4), we have

$$2(r_4 - r_4') + 3(r_5 - r_5') = 5.$$

Necessarily, both $ww_1$ and $ww_2$ are edges of $C'$ and so $r_4 \geq 1$, implying that either

$$r_4 - r_4' = 1 - 1 = 0 \quad \text{or} \quad r_4 - r_4' = 2 - 0 = 2.$$

If $r_4 - r_4' = 0$, then $3(r_5 - r_5') = 5$, which is impossible. On the other hand, if $r_4 - r_4' = 2$, then $3(r_5 - r_5') = 1$, which is also impossible. Hence, $G$ is not hamiltonian.

For many years, Tutte's graph was the only known example of a 3-connected cubic planar graph that was not hamiltonian. Much later, however, other such graphs were found; for example, Grinberg himself found another counterexample to Tait's conjecture (see Exercise 5.45).

 **5.4**  **The crossing number of a graph**

Nonplanar graphs cannot, of course, be embedded in the plane. Hence, whenever one attempts to draw a nonplanar graph in the plane, some of its edges must cross. This rather simple observation leads to the next concept.

The **crossing number** $\mathrm{cr}(G)$ of a graph $G$ is the minimum number of crossings (of its edges) among the drawings of $G$ in the plane. Before proceeding further, we comment on some assumptions we are making regarding all drawings under consideration. In particular, we assume that

- adjacent edges never cross
- two nonadjacent edges cross at most once
- no edge crosses itself
- no more than two edges cross at a point of the plane
- the (open) arc in the plane corresponding to an edge of the graph contains no vertex of the graph.

A few observations will prove useful. If $G$ is a subgraph of $H$, then $\mathrm{cr}(G) \leq \mathrm{cr}(H)$ while if $H$ is a subdivision of $G$, then $\mathrm{cr}(G) = \mathrm{cr}(H)$. A graph $G$ is planar if and only if $\mathrm{cr}(G) = 0$. In particular, if $G$ is a maximal planar graph of order $n \geq 3$ and size $m$, then $m = 3n - 6$ and $\mathrm{cr}(G) = 0 = m - 3n + 6$. If $m > 3n - 6$ and so $m - 3n + 6 > 0$, then $G$ is nonplanar and so $\mathrm{cr}(G) \geq 1$. In fact, the number $m - 3n + 6$ provides a lower bound for the crossing number of a graph of order $n \geq 3$ and size $m$.

**Theorem 5.23.** *If $G$ is a graph of order $n \geq 3$ and size $m$, then*

$$\mathrm{cr}(G) \geq m - 3n + 6.$$

*Proof.* Let $G$ be drawn in the plane with $\mathrm{cr}(G) = c$ crossings. At each crossing, a new vertex is introduced, producing a plane graph $H$ of order $n + c$ and size $m + 2c$. Since $H$ is planar, it follows by Theorem 5.3 that

$$m + 2c \leq 3(n + c) - 6.$$

Thus, $\mathrm{cr}(G) = c \geq m - 3n + 6.$ ∎

While the above lower bound for the crossing number of a graph can be useful in determining $\mathrm{cr}(G)$ for certain graphs $G$, this bound can differ significantly from $\mathrm{cr}(G)$. For example, for large integers $s$ and $t$, let $H = P_s \,\square\, P_t$. Then this planar graph $H$ can be embedded in the plane where one region has a $(2s + 2t - 4)$-cycle for its boundary and the boundary of all other regions are 4-cycles. Edges can be added to $H$ to produce a maximal planar graph $G$. By appropriately selecting nonadjacent vertices $x$ and $y$ of $G$, it follows that $\mathrm{cr}(G + xy) \geq 1$ by Theorem 5.23 but $G + xy$ can have a large crossing number.

##  Crossing numbers of complete graphs

One class of graphs whose crossing number has been a subject of study is complete graphs. By Theorem 5.23, it follows for $n \geq 3$ that

$$\mathrm{cr}(K_n) \geq \binom{n}{2} - 3n + 6 = \frac{(n-3)(n-4)}{2}. \tag{5.5}$$

Guy [115] discovered an even better lower bound for $\mathrm{cr}(K_n)$.

**Theorem 5.24.** *For $n \geq 5$, $\mathrm{cr}(K_n) \geq \frac{1}{5}\binom{n}{4}$.*

*Proof.* We proceed by induction on $n$. For $n = 5$, we have $\frac{1}{5}\binom{n}{4} = \frac{1}{5}\binom{5}{4} = 1$. Since $K_5$ is nonplanar, $\mathrm{cr}(K_5) \geq 1$. In fact, the drawing of $K_5$ in Figure 5.13 with one crossing shows that $\mathrm{cr}(K_5) = 1$.

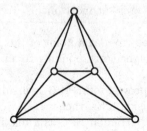

**Figure 5.13.** A drawing of $K_5$ with one crossing

Assume that $\mathrm{cr}(K_{n-1}) \geq \frac{1}{5}\binom{n-1}{4}$ for an integer $n \geq 6$. Let $K_n$ be drawn in the plane with $\mathrm{cr}(K_n)$ crossings. When a vertex of $K_n$ is deleted, a drawing of $K_{n-1}$ is obtained, where the number of crossings is at least $\mathrm{cr}(K_{n-1})$. Hence, the $n$ vertex-deleted subgraphs of $K_n$ produce $n$ drawings of $K_{n-1}$ having a total of at least $n \cdot \mathrm{cr}(K_{n-1})$ crossings.

A crossing in $K_n$ involves two nonadjacent edges, say $uv$ and $xy$. This crossing occurs in every vertex-deleted subgraph $K_{n-1}$ of $K_n$ except when $u, v, x$ or $y$ is deleted; that is, this crossing occurs in $n - 4$ subgraphs of $K_n$ isomorphic to $K_{n-1}$. Thus, the total number of crossings in these $n$ drawings of vertex-deleted subgraphs of $K_n$ is $(n - 4) \cdot \mathrm{cr}(K_n)$. Hence,

$$(n - 4) \cdot \mathrm{cr}(K_n) \geq n \cdot \mathrm{cr}(K_{n-1}) \geq \frac{n}{5}\binom{n-1}{4}.$$

Therefore,

$$\mathrm{cr}(K_n) \geq \frac{n}{5(n-4)}\binom{n-1}{4} = \frac{1}{5}\binom{n}{4},$$

as desired. ∎

The lower bounds for $\mathrm{cr}(K_n)$ given in (5.5) and in Theorem 5.24 are the same when $n = 5$ and $n = 6$, while

$$\frac{1}{5}\binom{n}{4} > \frac{(n-3)(n-4)}{2}$$

when $n \geq 7$. Thus the lower bound $\mathrm{cr}(K_n) \geq \frac{1}{5}\binom{n}{4}$ is an improvement over that given in (5.5). When $n = 6$, these bounds state that $\mathrm{cr}(K_6) \geq 3$. The drawing of $K_6$ with three crossings in Figure 5.14 shows that $\mathrm{cr}(K_6) = 3$.

It has been shown by Blažek and Koman [27] and Guy [114], among others, that for complete graphs,

$$\mathrm{cr}(K_n) \leq \frac{1}{4}\left\lfloor \frac{n}{2} \right\rfloor \left\lfloor \frac{n-1}{2} \right\rfloor \left\lfloor \frac{n-2}{2} \right\rfloor \left\lfloor \frac{n-3}{2} \right\rfloor. \tag{5.6}$$

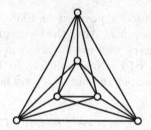

**Figure 5.14.** A drawing of $K_6$ with three crossings

Guy conjectured, in fact, that the upper bound in (5.6) is, in fact, the crossing number of $K_n$ for all positive integers $n$, and this is known to be true for some small values of $n$.

**Theorem 5.25.** *For* $1 \le n \le 12$,

$$\mathrm{cr}(K_n) = \frac{1}{4} \left\lfloor \frac{n}{2} \right\rfloor \left\lfloor \frac{n-1}{2} \right\rfloor \left\lfloor \frac{n-2}{2} \right\rfloor \left\lfloor \frac{n-3}{2} \right\rfloor.$$

Guy [115] established Theorem 5.25 for $1 \le n \le 10$, while Pan and Richter [187] verified Theorem 5.25 for $n = 11$ and $n = 12$. For $n > 12$, less is known. For example, McQuillan, Pan and Richter [172] showed that $\mathrm{cr}(K_{13}) \in \{219, 221, 223, 225\}$. Asymptotic evidence of the truth of Guy's conjecture was provided by Balogh, Lidický and Salazar [17] who showed that

$$\mathrm{cr}(K_n) > (0.985) \left( \frac{1}{4} \left\lfloor \frac{n}{2} \right\rfloor \left\lfloor \frac{n-1}{2} \right\rfloor \left\lfloor \frac{n-2}{2} \right\rfloor \left\lfloor \frac{n-3}{2} \right\rfloor \right)$$

for sufficiently large $n$.

## Crossing numbers of complete bipartite graphs

We now turn to the crossing number of complete bipartite graphs. The problem of determining $\mathrm{cr}(K_{s,t})$ has a memorable history. This problem is sometimes referred to as **Turán's brick-factory problem,** named for the Hungarian mathematician Paul Turán (1910–1976).

Born in Budapest, Hungary, Paul Turán displayed remarkable mathematical ability at a very early age. Turán was one of many young students in Budapest who studied graph theory under Dénes König. Turán met Paul Erdős in September 1930 and became and remained friends with him. Turán received a Ph.D. in 1935 from the University of Budapest, with a dissertation in number theory. By the end of 1935, Turán had seven papers in print. Despite his outstanding record at such an early age, Turán had great difficulty securing a faculty position because of his Jewish heritage. He could only make a living as a private mathematics tutor although he continued his research.

While Turán finally secured a position in 1938 (as a school teacher), his personal situation had grown worse. After the German invasion of Poland which began World War II, Hungary was not involved in the war at first but was nevertheless greatly influenced by Nazi policies. In 1940 Turán was sent to a labor camp. Indeed, Turán was in and out of several labor camps during the war. Turán himself [235] wrote:

> *We worked near Budapest, in a brick factory. There were some kilns where the bricks were made and some open storage yards where the bricks were stored. All the kilns were connected by rail with all the storage yards. The bricks were carried on small wheeled trucks to the storage yards. All we had to do was to put the bricks on the trucks at the kilns, push the trucks to the storage yards, and unload them there. We had a reasonable piece rate for the trucks, and the work itself was not difficult; the trouble was only at the crossings. The trucks generally jumped the rails there, and the bricks fell out of them; in short this caused a lot of trouble and loss of time which was precious to all of us. We were all sweating and cursing at such occasions, I too; but nolens volens the idea occurred to me that this loss of time could have been minimized if the number of crossings of the rails had been minimized. But what is the minimum number of crossings? I realized after several days that the actual situation could have been improved, but the exact solution of the general problem with s kilns and t storage yards seemed to be very difficult ... the problem occurred to me again ... at my first visit to Poland where I met Zarankiewicz. I mentioned to him my 'brick-factory'-problem ... and Zarankiewicz thought to have solved (it). But Ringel found a gap in his published proof, which nobody has been able to fill so far – in spite of much effort. This problem has also become a notoriously difficult unsolved problem ....*

Zarankiewicz [264] thought he had proved that

$$\mathrm{cr}(K_{s,t}) = \left\lfloor \frac{s}{2} \right\rfloor \left\lfloor \frac{s-1}{2} \right\rfloor \left\lfloor \frac{t}{2} \right\rfloor \left\lfloor \frac{t-1}{2} \right\rfloor \tag{5.7}$$

but, in actuality, he had only verified that the right-hand expression of (5.7) is an upper bound for $\mathrm{cr}(K_{s,t})$. As it turned out, both Kainen and Ringel found flaws in Zarankiewicz's argument. Hence, (5.7) remains only a conjecture, now called Zarankiewicz's conjecture. Due to the combined work of Kleitman [147] and Woodall [259], we do know that Zarankiewicz's conjecture is true for the following values of $s$ and $t$.

**Theorem 5.26.** *If $s$ and $t$ are positive integers with $s \leq t$ and either (1) $s \leq 6$ or (2) $s = 7$ and $t \leq 10$, then*

$$\mathrm{cr}(K_{s,t}) = \left\lfloor \frac{s}{2} \right\rfloor \left\lfloor \frac{s-1}{2} \right\rfloor \left\lfloor \frac{t}{2} \right\rfloor \left\lfloor \frac{t-1}{2} \right\rfloor .$$

It follows, therefore, from Theorem 5.26 that

$$\mathrm{cr}(K_{3,t}) = \left\lfloor \frac{t}{2} \right\rfloor \left\lfloor \frac{t-1}{2} \right\rfloor, \quad \mathrm{cr}(K_{4,t}) = 2 \left\lfloor \frac{t}{2} \right\rfloor \left\lfloor \frac{t-1}{2} \right\rfloor,$$

$$\mathrm{cr}(K_{5,t}) = 4 \left\lfloor \frac{t}{2} \right\rfloor \left\lfloor \frac{t-1}{2} \right\rfloor, \quad \text{and} \quad \mathrm{cr}(K_{6,t}) = 6 \left\lfloor \frac{t}{2} \right\rfloor \left\lfloor \frac{t-1}{2} \right\rfloor$$

for all $t$. For example, $\mathrm{cr}(K_{3,3}) = 1$, $\mathrm{cr}(K_{4,4}) = 4$, $\mathrm{cr}(K_{5,5}) = 16$, $\mathrm{cr}(K_{6,6}) = 36$ and $\mathrm{cr}(K_{7,7}) = 81$. A drawing of $K_{4,4}$ with four crossings is shown in Figure 5.15.

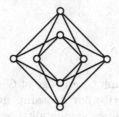

**Figure 5.15.** A drawing of $K_{4,4}$ with four crossings

Asymptotic evidence regarding Zarankiewicz's conjecture was given by de Klerk, Pasechnik and Schrijver [62] who showed that

$$\mathrm{cr}(K_{s,t}) > (0.8594) \left( \left\lfloor \frac{s}{2} \right\rfloor \left\lfloor \frac{s-1}{2} \right\rfloor \left\lfloor \frac{t}{2} \right\rfloor \left\lfloor \frac{t-1}{2} \right\rfloor \right)$$

for $s \geq 9$ and $t$ sufficiently large.

As would be expected, the situation regarding crossing numbers of complete $k$-partite graphs, $k \geq 3$, is even more complicated. For the most part, only bounds and highly specific results have been obtained in these cases. On the other hand, some of the proof techniques employed have been enlightening. As an example, the following result of White [254, p. 67] establishes the crossing number of $K_{2,2,3}$.

**Theorem 5.27.** *The crossing number of $K_{2,2,3}$ is 2.*

*Proof.* The graph $K_{2,2,3}$ has order 7 and size 16. Suppose that $\mathrm{cr}(K_{2,2,3}) = c$. Since $K_{3,3}$ is nonplanar and $K_{3,3}$ is a subgraph of $K_{2,2,3}$, it follows that $K_{2,2,3}$ is nonplanar and so $c \geq 1$. Let $K_{2,2,3}$ be drawn in the plane with $c$ crossings. At each crossing we introduce a new vertex, producing a connected plane graph $G$ of order $n = 7 + c$ and size $m = 16 + 2c$. By Theorem 5.3, $m \leq 3n - 6$.

Let $u_1 u_2$ and $v_1 v_2$ be two (nonadjacent) edges of $K_{2,2,3}$ that cross in the given drawing, giving rise to a new vertex. If $G$ is a maximal planar graph, then $C = (u_1, v_1, u_2, v_2, u_1)$ is a cycle of $G$, implying that the subgraph induced by $\{u_1, u_2, v_1, v_2\}$ in $K_{2,2,3}$ is $K_4$. However, $K_{2,2,3}$ contains no such subgraph; thus, $G$ is not a maximal planar graph and so $m < 3n - 6$. Therefore,

$$16 + 2c < 3(7 + c) - 6,$$

from which it follows that $c \geq 2$. The inequality $c \leq 2$ follows from the fact that there exists a drawing of $K_{2,2,3}$ with two crossings (see Figure 5.16). ∎

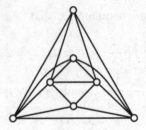

**Figure 5.16.** A drawing of $K_{2,2,3}$ with two crossings

 Fáry's theorem

In a planar embedding of a graph $G$, an edge of $G$ can be any curve, including a straight-line segment. The **rectilinear crossing number** $\overline{\mathrm{cr}}(G)$ of a graph $G$ is the minimum number of crossings among all those drawings of $G$ in the plane in which each edge is a straight-line segment. Since the crossing number $\mathrm{cr}(G)$ considers all drawings of $G$ in the plane (not just those for which edges are straight-line segments), we have the obvious inequality

$$\mathrm{cr}(G) \leq \overline{\mathrm{cr}}(G). \tag{5.8}$$

Clearly, $\overline{\mathrm{cr}}(G) \geq 0$ for every planar graph $G$. However, an interesting feature of planar graphs is that they can be embedded in the plane so that every edge is a straight-line segment. Such an embedding is referred to as a **straight-line embedding**. This result is known as Fáry's theorem but was proved independently not only by Fáry [91] but by Stein [221] and Wagner [248] as well.

**Theorem 5.28 (Fáry's theorem).** *If $G$ is a planar graph, then*

$$\overline{\mathrm{cr}}(G) = 0.$$

*Proof.* If the rectilinear crossing number of every maximal planar graph is 0, then the rectilinear crossing number of every planar graph is 0. Hence, it suffices to prove the theorem for maximal planar graphs. This result is obvious for $K_1$ and $K_2$.

We prove by induction on $n \geq 3$ that for every maximal plane graph $G$ of order $n$, the boundary of whose exterior region contains the vertices $u, v$ and $w$, there exists a straight-line embedding of $G$, each region of which has the same boundary as the given planar embedding of $G$, and whose exterior region has boundary vertices $u, v$ and $w$. The result is certainly true for $n = 3$ and $n = 4$. Assume that the statement is true for all maximal plane graphs of order $k$ for some integer $k \geq 4$. Let $G$ be a maximal plane graph of order $k+1$ whose exterior region has boundary vertices $u, v$ and $w$.

By Corollary 5.10, $G$ contains a vertex $x \notin \{u, v, w\}$ such that $\deg x = r$ and $3 \leq r \leq 5$. Let $N_G(x) = \{x_1, x_2, \ldots, x_r\}$. Remove $x$ from $G$ and let $R$ be the region of $G - x$ whose boundary vertices are $N_G(x)$. Add $r - 3$ edges to the region $R$ of $G - x$ so that a maximal plane graph $G'$ results. Since $G'$ is a maximal plane graph of order $k$, it follows by the induction hypothesis that there is a straight-line embedding of $G'$ resulting in a graph $G''$ each region of which has the same boundary and whose exterior region has boundary vertices $u, v$ and $w$.

Now remove the $r - 3$ edges that were added to $G - x$ to produce a straight-line embedding $G^*$ of $G - x$ such that the boundary of the region $R^*$ with boundary vertices $N_G(x)$ is an $r$-gon, where $3 \leq r \leq 5$. If the $r$-gon is convex, then the vertex $x$ can be added anywhere in $R^*$ and joined to the vertices of $N_G(x)$ by straight-line segments, producing a straight-line embedding of $G$.

Suppose that the $r$-gon $P$ is not convex and so $r = 4$ or $r = 5$. The nonconvex 4- and 5-gons are shown in Figure 5.17 (a) and Figure 5.18 (a). We then triangulate $P$ by adding $r - 3$ straight-line segments. If $r = 4$, then one straight-line segment is added as shown in Figure 5.17 (b). Place a new vertex $x$ on this segment and remove this segment. Straight-line segments can then be drawn from $x$ to each of the vertices $x_i$ $(1 \leq i \leq 4)$, as shown in Figure 5.17 (c), producing a straight-line embedding of $G$.

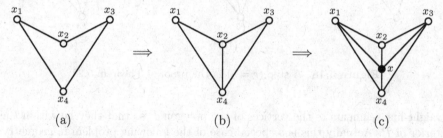

**Figure 5.17.** A step $(r = 4)$ in the proof of Theorem 5.28

If $r = 5$, then two straight-line segments can be added to triangulate $P$, producing three triangles as shown in Figure 5.18 (b) for each of the nonconvex 5-gons. One of the five vertices of $P$, say $x_1$, lies on each of these three triangles. One of these triangles, say $T$, has exactly one edge on the boundary of $P$. Since a triangle is convex, a straight-line segment can be drawn in $T$. Again place a new vertex $x$ on this segment and remove the segment and the two edges of $T$ not lying on the boundary of $P$. Straight-line segments can then be drawn from $x$ to each of $x_2, x_3, x_4$ and $x_5$, as shown in Figure 5.18 (c), producing a straight-line embedding of $G$. ∎

## The art gallery problem

By Fáry's theorem, there is a straight-line embedding of every planar graph. In the proof of this theorem, we used the fact that within the interior of each triangle, quadrilateral and pentagon, a point can be placed that can be joined by

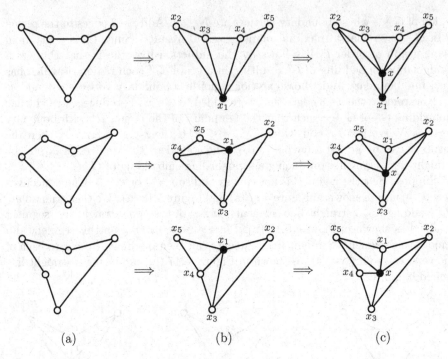

**Figure 5.18.** A step ($r = 5$) in the proof of Theorem 5.28

straight-line segments to the vertices of the polygon $P$ so that they lie within the interior of $P$. Actually this is a special case of the following problem in geometry:

**The art gallery problem.** *Suppose that a certain art gallery consists of a single large room with $n$ walls on which paintings are hung. What is the minimum number of security guards that must be hired and stationed in the gallery to guarantee that for every painting hung on a wall there is a guard who has straight line vision of the artwork?*

This problem was posed in 1973 by the geometer Victor Klee after a discussion with Vašek Chvátal. It was shown by Chvátal [51], with a simpler proof by Fisk [92], that no more than $\lfloor n/3 \rfloor$ guards are needed and that examples exist where $\lfloor n/3 \rfloor$ guards are required. In the case of an art gallery with three, four or five walls, only one guard need be hired, which is the fact used in the proof of Theorem 5.28.

It has been conjectured that $\mathrm{cr}(K_{s,t}) = \overline{\mathrm{cr}}(K_{s,t})$ for all positive integers $s$ and $t$. In the case of complete graphs, not only is $\overline{\mathrm{cr}}(K_n) = \mathrm{cr}(K_n)$ for $1 \leq n \leq 4$ (when $K_n$ is planar), it is also known that $\overline{\mathrm{cr}}(K_n) = \mathrm{cr}(K_n)$ for $5 \leq n \leq 7$ and $n = 9$. However, $\mathrm{cr}(K_8) = 18$ and $\overline{\mathrm{cr}}(K_8) = 19$ (see Guy [115]), so strict inequality in (5.8) is indeed a possibility. Furthermore, Brodsky, Durocher and Gethner [36] showed that $\overline{\mathrm{cr}}(K_{10}) = 62$, while $\mathrm{cr}(K_{10}) = 60$. Determining the rectilinear crossing number of $K_n$ has attracted a lot of interest over the last several decades and has some surprising connections to other branches of mathematics. Aichholzer maintains a website [2] where exact values of $\overline{cr}(K_n)$ for all $n$ with $3 \leq n \leq 27$ and $n = 30$ are given along with the best-known upper bounds in many other cases. In particular, Cetina, Hernández-Vélez, Leaños and Villalobos [42] showed that $\overline{cr}(K_{30}) = 9726$.

From an algorithmic point of view, Garey and Johnson [101] showed that determining whether $cr(G) \leq k$ for a given graph $G$ is NP-complete while Bienstock [25] showed that determining whether $\overline{cr}(G) \leq k$ is NP-hard. In computational complexity theory, the NP-hard (non-deterministic polynomial-time hard) problems are those problems that are informally "at least as hard as the hardest problems in NP". More precisely, a problem $H$ is **NP-hard** when every problem $L$ in NP can be reduced in polynomial time to $H$; that is, assuming a solution for $H$ takes 1 unit time, a solution for $H$ can be used to solve $L$ in polynomial time. As a consequence, finding a polynomial time algorithm to solve any NP-hard problem would give polynomial time algorithms for all the problems in NP. Note that a problem in NP that is also NP-hard is NP-complete.

# Exercises for Chapter 5

### Section 5.1. Euler's formula

**5.1.** Prove that a graph is planar if and only if each of its blocks is planar.

**5.2.** Prove Corollary 5.2: the crossing number of $K_{2,2,3}$ is 2.

**5.3.** Give an example of a graph $G$ of order 8 such that $G$ and $\overline{G}$ are planar.

**5.4.** Prove that if $G$ is a planar graph of order 11, then $\overline{G}$ is nonplanar.

**5.5.** Prove that the order of every 3-regular planar graph containing no triangle or 4-cycle is at least 20.

**5.6.** Show that the Petersen graph is nonplanar.

**5.7.** Show that if $G$ is a planar graph containing no vertex of degree less than 5, then $G$ contains at least 12 vertices of degree 5.

**5.8.** Give an example of a planar graph that contains no vertex of degree less than 5.

**5.9.** If the boundary of every interior region of a plane graph $G$ of order $n$ and size $m$ is a triangle and the boundary of the exterior region is a $k$-cycle ($k \geq 3$), express $m$ in terms of $n$ and $k$.

**5.10.** A cubic polyhedron $P$ has only 5-sided, 6-sided and 7-sided faces. Determine a formula for the number of 5-sided faces in $P$.

**5.11.** Give an example of two non-isomorphic maximal planar graphs of the same order.

**5.12.** Prove that there exists only one 4-regular maximal planar graph.

**5.13.** Prove that a planar graph of order $n \geq 3$ and size $m$ is maximal planar if and only if $m = 3n - 6$.

**5.14.** Prove that every maximal planar graph of order 4 or more is 3-connected.

**5.15.** In a planar embedding of a maximal planar graph $G$ of order 6, a vertex is placed in each interior region of $G$ and joined to the vertices on its boundary, producing a graph $H$. Prove or disprove: $H$ is hamiltonian.

**5.16.** Determine all maximal planar graphs $G$ of order 3 or more such that the number of regions in a planar embedding of $G$ equals its order.

**5.17.** A nontrivial tree $T$ of order $n$ has the property that $\overline{T}$ is a maximal planar graph.
  (a) What is $n$?
  (b) Give an example of a tree $T$ with this property.

**5.18.** Use the discharging method to prove Theorem 5.12: if $G$ is a maximal planar graph of order 4 or more, then $G$ contains at least one of the following: (1) a vertex of degree 3, (2) a vertex of degree 4, (3) a vertex of degree 5 that is adjacent to two vertices, each of which has degree 5 or 6.

**5.19.** If every vertex $v$ of a nontrivial tree $T$ is given a charge of $2 - \deg v$, then what is the sum of the charges of the vertices of $T$?

## Section 5.2.  Characterizations of planarity

**5.20.** Let $U$ be a subset of the vertex set of a graph $G$, and suppose that $H = G[U]$ is an induced subgraph of $G$ of order $k$. Let $H'$ be the graph obtained by inserting a vertex of degree 2 into every edge of $H$. Let $G'$ be the graph obtained by inserting a vertex of degree 2 into every edge of $G$.
  (a) Prove or disprove: If $G$ is planar, then $H'$ is planar.
  (b) What is $G'[U]$?

**5.21.** Let T be a tree of order at least 4, and let $e_1, e_2, e_3 \in E(\overline{T})$. Prove that $T + e_1 + e_2 + e_3$ is planar.

**5.22.** Let $T$ be a tree of order at least 5, and let $e_1, e_2, \cdots, e_5 \in E(\overline{T})$. Let $G = T + \{e_1, e_2, \cdots, e_5\}$. Prove that if $G$ does not contain a subdivision of $K_{3,3}$, then $G$ is planar.

**5.23.** Consider the graph $K_4 \,\square\, K_2$.
(a) Determine the order $n$, the size $m$ and the number $3n - 6$ for $K_4 \,\square\, K_2$.
(b) What does the information in (a) say about the planarity of $K_4 \,\square\, K_2$?
(c) Is $K_4 \,\square\, K_2$ planar or nonplanar?

**5.24.** Determine all graphs $G$ of order $n \geq 5$ and size $m = 3n - 5$ such that for each edge $e$ of $G$, the graph $G - e$ is planar.

**5.25.** Let $S_{a,b}$ denote the double star in which the degrees of the two vertices that are not end-vertices are $a$ and $b$. Determine all pairs $a, b$ of integers such that $\overline{S}_{a,b}$ is planar.

**5.26.** A nonplanar graph $G$ of order 7 has the property that $G - v$ is planar for every vertex $v$ of $G$.
(a) Show that $G$ does not contain $K_{3,3}$ as a subgraph.
(b) Give an example of a graph with this property.

**5.27.** Determine all integers $n \geq 3$ such that $\overline{C}_n$ is planar.

**5.28.** Determine all integers $n \geq 3$ such that $C_n^2$ is nonplanar.

**5.29.** Prove that if $G$ is a nonplanar graph of minimum size containing no subgraph that is a subdivision of $K_5$ or $K_{3,3}$, then $G$ is 3-connected.

**5.30.** Show that the converse of Theorem 5.17 is not, in general, true.

**5.31.** It has been observed that if a graph $H$ is a minor of a planar graph, then $H$ is planar. Prove or disprove: If a minor $H$ of a graph $G$ is planar, then $G$ is planar.

**5.32.** Let $G$ be the graph shown in Figure 5.19.
(a) Show that $G$ contains $K_{3,3}$ as a subgraph.
(b) Show that $G$ does not contain a subdivision of $K_5$ as a subgraph.
(c) Show that $K_5$ is a minor of $G$.

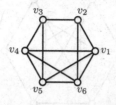

**Figure 5.19.** The graph $G$ in Exercise 5.32

**5.33.** Let $G$ be a 4-regular graph of order 10 and size $m$. What can be deduced about the planarity of $G$ by comparing the numbers $m$ and $3n - 6$?

**5.34.** Prove or disprove: There is a planar 2-connected 4-regular graph of order 10.

**5.35.** Show that the 2-connected 4-regular graph $H$ of order 10 shown in Figure 5.20 does not contain $K_5$ as a subgraph but does contain $K_5$ as a minor. What can you conclude from this?

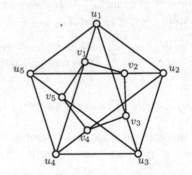

**Figure 5.20.** A 2-connected 4-regular graph of order 10

**5.36.** Prove that if $G$ is a graph of order $n \geq 5$ and size $m \geq 3n - 5$, then $G$ need not contain $K_5$ as a subgraph but must contain a subgraph with minimum degree 4.

**5.37.** What is the minimum possible order of a graph $G$ containing only vertices of degree 3 and degree 4 and an equal number of each such that $G$ contains a subdivision of $K_5$?

**5.38.** Consider the graph $H$ of Figure 5.21.
  (a) Does $H$ contain a subdivision of $K_5$ or a subdivision of $K_{3,3}$?
  (b) Does $H$ contain $K_5$ or $K_{3,3}$ as a minor?
  (c) Is $H$ planar or nonplanar?

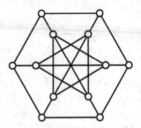

**Figure 5.21.** The graph $H$ in Exercise 5.38

**5.39.** A graph $G$ is **outerplanar** if there exists a planar embedding of $G$ so that every vertex of $G$ lies on the boundary of the exterior region. Prove the following:
  (a) A graph $G$ is outerplanar if and only if the join $G \vee K_1$ is planar.
  (b) A graph $G$ is outerplanar if and only if $G$ contains no subgraph that is a subdivision of $K_4$ or $K_{2,3}$.
  (c) The size of every outerplanar graph of order $n \geq 2$ is at most $2n - 3$.

**5.40.** Prove that every nontrivial outerplanar graph contains at least two vertices of degree 2 or less.

**5.41.** Determine all connected graphs $G$ of order $n \geq 4$ such that $G \vee K_1$ is outerplanar.

**5.42.** For a positive integer $k$, a graph $G$ of order $n > k$ and size $m$ is said to have property $\pi_k$ if (1) $m = kn - \binom{k+1}{2}$ and (2) for every induced subgraph $H$ of order $p$ and size $q$ in $G$, where $k \leq p < n$, it follows that $q \leq kp - \binom{k+1}{2}$.
  (a) Show that $\delta(G) \geq k$.
  (b) Show that $\omega(G) \leq k + 1$.
  (c) What familiar class of graphs has property $\pi_2$? Show that there is a graph having property $\pi_2$ that does not belong to this class.
  (d) What familiar class of graphs has property $\pi_3$? Show that there is a graph having property $\pi_3$ that does not belong to this class.

## Section 5.3. Hamiltonian planar graphs

**5.43.** Show, by applying Theorem 5.21, that $K_{2,3}$ is not hamiltonian.

**5.44.** Show, by applying Theorem 5.21, that each of the graphs in Figure 5.22 is not hamiltonian.

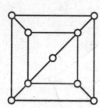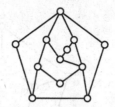

**Figure 5.22.** Graphs in Exercise 5.44

**5.45.** Show, by applying Theorem 5.21, that the **Grinberg graph** in Figure 5.23 is not hamiltonian.

**5.46.** Show, by applying Theorem 5.21, that the **Herschel graph** in Figure 5.24 is not hamiltonian.

**5.47.** Show, by applying Theorem 5.21, that no hamiltonian cycle in the graph of Figure 5.25 contains both the edges $e$ and $f$.

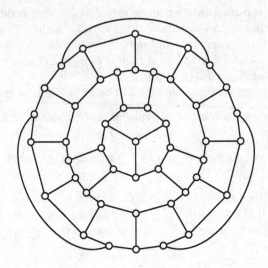

**Figure 5.23.** The Grinberg graph (see Exercise 5.45)

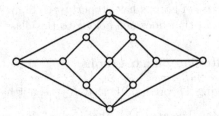

**Figure 5.24.** The Herschel graph (see Exercise 5.46)

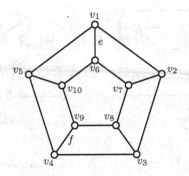

**Figure 5.25.** A hamiltonian planar graph in Exercise 5.47

## Section 5.4. The crossing number of a graph

**5.48.** Draw $K_7$ in the plane with nine crossings.

**5.49.** Determine $cr(K_{3,3})$ without using Theorem 5.26.

**5.50.** Show that $cr(K_{5,5}) \le 16$.

**5.51.** Determine $cr(K_{2,2,2})$.

**5.52.** Determine $cr(K_{1,2,3})$.

**5.53.** Show that $2 \le cr(C_3 \,\square\, C_3) \le 3$.

**5.54.** It is known that $cr(W_4 \,\square\, K_2) = 2$, where $W_4$ is the wheel $C_4 \vee K_1$ of order 5. Draw $W_4 \,\square\, K_2$ in the plane with two crossings.

**5.55.** Prove or disprove: If $G$ is a nonplanar graph containing an edge $e$ such that $G - e$ is planar, then $cr(G) = 1$.

**5.56.** Prove that $\overline{cr}(C_3 \,\square\, C_t) = t$ for $t \ge 3$.

**5.57.** Give an example of a straight-line embedding of a maximal planar graph of order exceeding 4 containing exactly four vertices of degree 5 or less.

**5.58.** Let $G$ be a connected planar graph. Prove or disprove: If $cr(G \,\square\, K_2) = 0$, then $G$ is outerplanar.

# Coloring

**6**

The four color problem, now known as the four color theorem, is one of the oldest and most well-known problems in graph theory. This problem sparked the study of graph coloring, which has been a major area of research in graph theory for nearly two centuries. In this chapter, we explore vertex and edge colorings of graphs, both planar and nonplanar, and also discuss snarks and perfect graphs.

## 6.1   Vertex coloring

A **coloring** of a graph $G$ is an assignment of colors—typically positive integers—to the vertices of $G$, one color to each vertex. A coloring is **proper** if every pair of adjacent vertices receive different colors. A proper coloring that uses $k$ or fewer colors is called a **$k$-coloring**, and a graph is called **$k$-colorable** if it has a $k$-coloring.

The minimum $k$ for which a graph $G$ is $k$-colorable is called the (vertex) **chromatic number** of $G$ and is denoted by $\chi(G)$. If $G$ is a graph with $\chi(G) = k$, then $G$ is said to be **$k$-chromatic**. Certainly, if $H$ is a subgraph of $G$, then $\chi(H) \leq \chi(G)$. Thus a subgraph of a $k$-chromatic graph is necessarily $k$-colorable, but need not be $k$-chromatic.

For a few families of graphs, the chromatic number is easy to determine. For example, $\chi(K_n) = n$ for every positive integer $n$. For cycles, the situation is only slightly more complicated: $\chi(C_{2k}) = 2$ while $\chi(C_{2k-1}) = 3$ for every integer $k \geq 2$ (Exercise 6.1).

The graph shown in Figure 6.1 is 3-colorable, as demonstrated by the proper 3-coloring shown in the figure. To show that this graph is 3-*chromatic*, we need to prove that it is not 2-colorable, which follows from the fact that it contains $C_5$ as a subgraph. This argument is a common one: to show that $\chi(G) = k$, we usually exhibit a proper $k$-coloring of $G$ and then argue that any proper coloring of $G$ must use at least $k$ colors.

Characterizations of $k$-chromatic graphs are simple to derive for $k = 1$ and $k = 2$. A graph is 1-chromatic if and only if it has no edges, and similarly, a graph is 2-chromatic if and only if it is bipartite. However, determining whether a given graph is $k$-colorable is NP-complete for all $k \geq 3$. Therefore, in general,

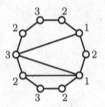

**Figure 6.1.** A 3-chromatic graph

we must be satisfied with finding bounds for the chromatic number rather than computing it exactly.

Two simple bounds on the chromatic number of a graph can be obtained from the sizes of certain special subgraphs. A **clique** is a complete subgraph in a graph, and the **clique number** of a graph $G$, denoted by $\omega(G)$, is defined as the order of the largest clique in $G$. This parameter provides a crude lower bound on the chromatic number of a graph: $\chi(G) \geq \omega(G)$ for all graphs $G$, since each vertex of a clique must be assigned a different color.

An **independent set** is a set of vertices of a graph such that no two of its members are adjacent. The maximum number of vertices that can be included in an independent set of a graph $G$ is called the (vertex) **independence number** and denoted by $\alpha(G)$. This parameter can also be used to bound the chromatic number. Given a proper coloring of a graph, the set of vertices assigned the same color is called a **color class**. Each color class must be an independent set of vertices. Therefore, the chromatic number of a graph can also be defined as the minimum number of independent sets into which its vertex set can be partitioned. Exercise 6.4 asks the reader to prove that

$$\frac{n}{\alpha(G)} \leq \chi(G) \leq n + 1 - \alpha(G)$$

for every graph $G$ with order $n$ and independence number $\alpha(G)$.

The maximum degree of a graph $G$, denoted by $\Delta(G)$, also features in a bound on the chromatic number of $G$ through the following result.

**Theorem 6.1.** *For every graph $G$,*

$$\chi(G) \leq 1 + \Delta(G).$$

*Proof.* To prove the result, we proceed by induction on the order of $G$. If the order of $G$ is 1, the result is trivial. Now, suppose that $G$ has order at least 2 and that the result holds for all graphs of lesser order. Let $\Delta = \Delta(G)$ and take $v \in V(G)$ to be arbitrary. By the induction hypothesis, the graph $G - v$ has a proper coloring with at most $1 + \Delta$ colors. Since $\deg v \leq \Delta$, there are at most $\Delta$ colors that $v$ cannot be given (those already assigned to its neighbors). Therefore, $G$ has a proper coloring with at most $1 + \Delta$ colors, completing the proof. ∎

Note that Theorem 6.1 is best possible for odd cycles and complete graphs.

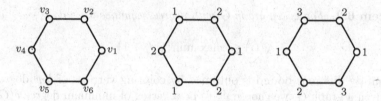

**Figure 6.2.** Greedy colorings of $C_6$

## Greedy coloring

While the inductive proof we have given for Theorem 6.1 is fine, there is a different perspective that often yields more insight. In a **greedy coloring**, we start by considering all of the vertices as uncolored. We then color the vertices one at a time in some predetermined order, assigning each vertex the smallest positive integer color that we have not already used to color one of its neighbors. (The colors are always positive integers in greedy colorings.)

For example, consider the graph $C_6$ with its vertices labeled as shown on the left of Figure 6.2. If we greedily color the vertices of this graph in the order $v_1$, $v_2$, $v_3$, $v_4$, $v_5$, $v_6$, then we obtain the proper 2-coloring shown in the center of Figure 6.2. Since we know that $\chi(C_6) = 2$, this is the best we could have done. By changing the vertex order, we can do worse, however. If we greedily color the vertices of this graph in the order $v_1$, $v_2$, $v_4$, $v_3$, $v_5$, $v_6$, then we obtain the proper 3-coloring shown on the right of Figure 6.2.

For every graph $G$, there is an ordering of its vertices for which the greedy coloring produces an optimal coloring (Exercise 6.6). On the other hand, with suboptimal vertex orders, greedy colorings can easily use too many colors (see Exercise 6.7). Although we do not study it further, we mention that the **Grundy number** of a graph is the maximum number of colors used to color it by a greedy coloring.

An important observation about greedy colorings is that the color assigned to a vertex is at most 1 more than the number of its neighbors that have already been colored. Thus when greedily coloring a graph $G$, we never need more than $1 + \Delta(G)$ colors, giving another proof of Theorem 6.1.

This idea can be refined a bit. Suppose we greedily color the vertices of a graph $G$ in the order $v_1, v_2, \ldots, v_n$. Letting $d_i = \deg v_i$, so $d_1, d_2, \ldots, d_n$ is the degree sequence of $G$, we see that when we come to color $v_i$, at most $\min\{i-1, d_i\}$ of its neighbors will already have been colored. Therefore, the maximum color that we could need to use to color $v_i$ is $\min\{i, d_i + 1\}$, and so the greedy coloring with this vertex order will use at most $\min\{i, d_i + 1\}$ colors, where the maximum is over all $1 \leq i \leq n$. This quantity is minimized when we color the vertices of high degree first, which is known as the **largest-first ordering**. We therefore obtain the following strengthening of Theorem 6.1 due to Welsh and Powell [252].

**Theorem 6.2.** *For every graph $G$ with degree sequence $d_1 \geq d_2 \geq \cdots \geq d_n$, we have*

$$\chi(G) \leq \max_{1 \leq i \leq n} \min\{i, d_i + 1\}.$$

A subtly different bound is obtained by coloring vertices of *low* degree *last*. Thus given a graph $G$, we choose $v_n$ to be a vertex of minimum degree $\delta(G)$. We then choose $v_{n-1}$ to be a vertex of minimum degree in $G - v_n$, and we continue in this manner. The resulting ordering is called the **smallest-last ordering**, and leads us to the following bound first observed (in a different manner) by Szekeres and Wilf [226].

**Theorem 6.3.** *For every graph $G$,*

$$\chi(G) \leq 1 + \max_{H \subseteq G} \delta(H),$$

*where the maximum is taken over all subgraphs $H$ of $G$.*

*Proof.* We prove the result by induction on the order of $G$. If the order of $G$ is 1, then the result is trivial. Now suppose that $G$ has order at least 2 and that the result holds for all graphs of lesser order. Set

$$d = \max_{H \subseteq G} \delta(H)$$

and choose $v_n \in V(G)$ with $\deg v_n = \delta(G) \leq d$. By induction, we have

$$\chi(G - v_n) \leq 1 + \max_{H \subseteq G - v_n} \delta(H) \leq 1 + d.$$

Choose a proper $(1 + d)$-coloring of $G - v_n$. Because $\deg v_n \leq d$, this coloring may be extended to a proper $(1 + d)$-coloring of $G$, proving the result. ∎

Greedy coloring with the smallest-last ordering is optimal on forests, as the bound in Theorem 6.3 shows, because every forest has a vertex of degree at most 1 (an isolated vertex or a leaf).

## Brooks' theorem

Theorem 6.1 states that the chromatic number of a graph $G$ is at most $1 + \Delta(G)$, and we know that this is best possible for odd cycles and complete graphs. However, for all other graphs, we can improve this bound by choosing a different vertex ordering. The key idea is that the greedy coloring assigns the color $1 + \Delta(G)$ to a vertex only if that vertex has $\Delta(G)$ neighbors *and* all of those neighbors have already been assigned different colors.

Suppose that we can order the vertices of $G$ as $v_1, v_2, \ldots, v_n$ so that for all $1 \leq i \leq n - 1$, the vertex $v_i$ has an uncolored neighbor (one in the set $\{v_{i+1}, \ldots, v_n\}$), and so that when it comes time to color $v_n$, at least two of its neighbors (all of which will have been colored by that point) share a color. Such a vertex ordering leads to a proof of the following 1941 theorem of Brooks [37]. The proof we present is due to Zając [262]; several other proofs are surveyed by Cranston and Rabern [57].

**Theorem 6.4 (Brooks' theorem).** *For every connected graph $G$ that is not an odd cycle or a complete graph,*

$$\chi(G) \leq \Delta(G).$$

*Proof.* The proof is by induction on the order of $G$, the base case where $G$ is a single vertex holding trivially. Now suppose that $G$ is a connected graph that is neither an odd cycle nor complete, and set $\Delta = \Delta(G)$. We begin by eliminating two cases.

Suppose that $G$ is not $\Delta$-regular, and let $v$ be a vertex with $\deg v < \Delta$. If $G - v$ is an odd cycle, then it follows that $\Delta = 3$ and $G$ is 3-colorable. If $G - v$ is a complete graph, then it must have precisely $\Delta$ vertices, and we have $\chi(G) = \Delta$ since $\deg v < \Delta$. As we are done in both of these cases, we may assume that $G - v$ is neither an odd cycle nor a complete graph, so it is $\Delta$-colorable by induction. Since $\deg v < \Delta$, we can extend any $\Delta$-coloring of $G - v$ to a $\Delta$-coloring of $G$. Therefore, we may assume that $G$ is $\Delta$-regular.

If $\Delta = 2$, then since $G$ is connected and $\Delta$-regular, it is a cycle. Since we have assumed that $G$ is not an odd cycle, it must be an even cycle and thus we have $\chi(G) = 2 = \Delta$, as desired. We may now assume that $\Delta \geq 3$.

Choose a vertex $v_2 \in V(G)$ arbitrarily. Because $G$ is $\Delta$-regular, $|N(v_2)| = \Delta$. If $G[N[v_2]]$ were complete, then it would have to be all of $G$ because $G$ is $\Delta$-regular and connected, and this would contradict our assumption that $G$ is not complete. Thus we can find two vertices $v_1, v_3 \in N(v_2)$ that are not adjacent. Extend the path $v_1, v_2, v_3$ to a maximal path

$$v_1, v_2, v_3, \ldots, v_r.$$

First suppose that $r = n$ (so, the path chosen above is hamiltonian). Because $\Delta \geq 3$, the vertex $v_2$ is adjacent to some vertex $v_k$ for $k \geq 4$. Greedily color the vertices of $G$ in the order

$$v_1, v_3, v_4, \ldots, v_{k-1}, v_n, v_{n-1}, \ldots, v_k, v_2.$$

Every vertex except $v_2$ has a neighbor that appears after it in this order, so none of them will require the color $1 + \Delta$. Moreover, since $v_1$ and $v_3$ are not adjacent, both will be given the color 1, and so when we come to $v_2$, it will not require the color $1 + \Delta$. Thus we have proved that in this case, with this vertex order, the greedy coloring uses at most $\Delta$ colors, as desired.

Next suppose that $r < n$, so not all of the vertices are on this maximal path. It must be the case that $N(v_r) \subseteq \{v_1, v_2, \ldots, v_{r-1}\}$, as otherwise we could extend the path. Choose $i$ to be the least index of a neighbor of $v_r$ in this set, so $v_i, v_{i+1}, \ldots, v_r, v_i$ is a cycle containing every vertex in $N(v_r)$. Label this cycle $C$.

Because $r < n$, $C$ is not all of $G$, and since $G$ is connected, $C$ must contain some vertex adjacent to $G - C$. Choose $k$ to be the greatest index of a vertex $v_k$ in $C$ that is adjacent to $G - C$. Note that $k \leq r - 1$, because $v_r$ is not adjacent to any vertex of $G - C$. Thus $v_{k+1}$ is defined and is not adjacent to $G - C$ by our choice of $k$.

By our induction hypothesis, there is a proper coloring of $G - C$ with the colors $\{1, 2, \ldots, \Delta\}$. To extend this coloring to all of $G$, we first give $v_{k+1}$ the same color as any neighbor of $v_k$, which we can do because $v_{k+1}$ is not adjacent to any vertex of $G - C$. We then greedily color the rest of $C$ in the order

$$v_{k+2}, v_{k+3}, \ldots, v_r, v_i, v_{i+1}, \ldots, v_k.$$

While greedily coloring in this order, every vertex before $v_k$ has a neighbor that occurs after it, so the color $1 + \Delta$ cannot be required until we come to color $v_k$. Then when we do come to color $v_k$, we do not need to use the color $1 + \Delta$ because $v_k$ will have at least two neighbors of the same color, as its neighbor $v_{k+1}$ shares a color with (at least) one of its neighbors in $G - C$. ∎

The bound given by Brooks' theorem is not particularly good for some graphs; for example, we have $\chi(K_{1,n}) = 2$, but $\Delta(K_{1,n}) = n$.

 **6.2   Edge coloring**

An **edge coloring** of a graph $G$ is an assignment of colors to the edges of $G$, one color to each edge. An edge coloring is **proper** if every pair of adjacent edges receive different colors. A proper edge coloring that uses $k$ or fewer colors is called a **$k$-edge-coloring**, and a graph is called **$k$-edge-colorable** if it has a $k$-edge-coloring.

As with vertex colorings, we are often interested in edge colorings that use as few colors as possible. The **edge chromatic number** (or **chromatic index**) $\chi'(G)$ of a graph $G$ is the minimum positive integer $k$ for which $G$ is $k$-edge colorable. In Figure 6.3, a 3-edge-coloring of a graph $G$ is given. Consequently, $\chi'(G) \leq 3$. On the other hand, since $G$ contains three mutually adjacent edges, at least three distinct colors are required in any edge coloring of $G$ and so $\chi'(G) \geq 3$. Therefore, $\chi'(G) = 3$.

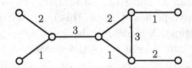

**Figure 6.3.** A 3-edge coloring of a graph

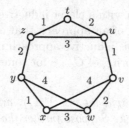

**Figure 6.4.** A graph with edge chromatic number 4

In a given edge coloring of a graph, a set consisting of all those edges assigned the same color is again referred to as a **color class**. Each color class, then, has the property that no two edges are adjacent; such a set of edges is, in general, called an **independent set** of edges. Thus the edge chromatic number of a graph can be defined as the minimum number of independent sets into which the edge set can be partitioned. The maximum number of edges in an independent set of edges of a graph $G$ is called the **edge independence number** of $G$ and is denoted by $\alpha'(G)$. It follows (see Exercise 6.11), that if $G$ is a graph of size $m$, then

$$\chi'(G) \geq m/\alpha'(G).$$

Since every edge coloring of a graph $G$ must assign distinct colors to adjacent edges, we must use at least $\deg v$ colors to color the edges incident to any vertex $v$. Therefore, we always have

$$\chi'(G) \geq \Delta(G).$$

In the graph $G$ of order $n = 7$ and size $m = 10$ shown in Figure 6.4, we have $\chi'(G) \geq \Delta(G) = 3$. In fact, we can see that $\alpha'(G) = 3$ (one always has $\alpha'(G) \leq n/2$, which in this case gives an upper bound of $7/2$, and an independent set of three edges is readily apparent). This gives the improved lower bound $\chi'(G) \geq m/\alpha'(G) = 10/3$. In the other direction, Figure 6.4 depicts a 4-edge-coloring, so $\chi'(G) \leq 4$. Since $\chi'(G)$ must be an integer, we conclude that $\chi'(G) = 4$.

## Vizing's theorem

As the graph in Figure 6.4 demonstrates, the obvious lower bound $\chi'(G) \geq \Delta(G)$ is not always best possible. However, a remarkable 1964 result of Vizing [244] (independently announced around the same time by Gupta [112]) shows that this lower bound is never far from the truth.

**Theorem 6.5 (Vizing's theorem).** *For every graph $G$,*

$$\chi'(G) \leq 1 + \Delta(G).$$

We prove Vizing's theorem with a clever inductive argument, which is adapted from Schrijver's version [214] of an approach that originated with Ehrenfeucht, Faber, and Kierstead [73]. This inductive approach relies on the following lemma to extend a $(\Delta + 1)$-edge-coloring of $G - v$ for some vertex $v$ to a $(\Delta + 1)$-edge-coloring of $G$.

**Lemma 6.6.** *Let $G$ be a graph, let $v \in V(G)$, and suppose that $k \geq 1$ is an integer for which $\chi'(G - v) \leq k$. Suppose further that $v$ and all its neighbors have degree at most $k$ and that $v$ has at most one neighbor of degree precisely $k$. Then $\chi'(G) \leq k$.*

Before proving the Lemma 6.6, we show how it implies Vizing's theorem.

*Proof of Vizing's theorem (Theorem 6.5) from Lemma 6.6.* The proof is by induction on the order of $G$. If the order of $G$ is 1, then the result is trivial. Now suppose that $G$ has order at least 2 and that the result holds for all graphs of lesser order. Set $k = \Delta(G) + 1$ and let $v \in V(G)$ be arbitrary. By induction, we have

$$\chi'(G - v) \leq \Delta(G - v) + 1 \leq \Delta(G) + 1 = k.$$

Furthermore, $v$ and all its neighbors have degree at most $\Delta(G) = k - 1$. Therefore, it follows by Lemma 6.6 that $\chi'(G) \leq k = \Delta(G) + 1$. ∎

Note that Lemma 6.6 appears stronger than what we need; in our proof of Vizing's theorem, there are no vertices of degree $k = \Delta(G) + 1$. This strengthening in fact simplifies the proof of the lemma, a not uncommon occurrence in inductive proofs.

*Proof of Lemma 6.6.* The proof is by induction on $k$. In the base case $k = 1$, we have a graph $G$ a vertex $v$ of $G$ such that $\chi'(G - v) \leq 1$, and further, $v$ and all its neighbors have degree at most 1, and $v$ has at most one neighbor of degree 1. This implies that $v$ lies on at most one edge of $G$, which is incident to no other edges. It follows that $\chi'(G) \leq 1$, so the lemma holds for $k = 1$.

Now assume that $k \geq 2$ and take $v$ to be a vertex of a graph $G$ satisfying the conditions of the lemma. By possibly adding vertices of degree 1 adjacent to the neighbors of $v$, we may assume that one neighbor of $v$ has degree $k$, while the others all have degree $k - 1$.

For a given coloring of the edges of $G - v$ by the colors $\{1, 2, \ldots, k\}$, let $A_i$ (for $1 \leq i \leq k$) denote the set of neighbors of $v$ that are not incident to an edge of color $i$ (thus, these are the vertices at which the color $i$ is available to be used). Choose a $k$-edge-coloring of $G - v$ in order to minimize $\sum |A_i|^2$.

First we show that we are done if $|A_i| = 1$ for some color $i$, and then we show how this property must be true. Suppose, without loss of generality, that $|A_k| = 1$. Let $w$ denote the unique member of $A_k$, so $w$ is the only neighbor of $v$ that has color $k$ available for one of its edges. Let $F$ denote the set consisting of the edge $vw$ and all edges that are colored $k$ in this $k$-edge-coloring of $G - v$.

Since $A_k = \{w\}$, every neighbor of $v$ is incident to precisely one edge of $F$. Therefore, in the graph $G - F$, at most one neighbor of $v$ has degree $k - 1$, while all its other neighbors have degree at most $k - 2$. Moreover, our $k$-edge-coloring of $G - v$ restricts to give a $(k - 1)$-edge-coloring of $(G - F) - v$. Therefore, we have by induction that $\chi'(G - F) \leq k - 1$. By assigning the color $k$ to all edges in $F$, this gives a $k$-edge-coloring of $G$, as desired.

To finish the proof, we must show how to guarantee that $|A_i| = 1$ for some color $i$. Suppose to the contrary that we have chosen a $k$-edge-coloring of $G - v$ to minimize $\sum |A_i|^2$ but that $|A_i| \neq 1$ for all $1 \leq i \leq k$. As we have assumed that one neighbor of $v$ has degree $k$ while the others have degree $k-1$, one edge incident to $v$ has one color available while the others have two colors available. Thus, the sum of available colors over all edges incident to $v$ is $2 \deg v - 1$. Counting this quantity another way, we see that

$$\sum_{i=1}^{k} |A_i| = 2 \deg v - 1 \leq 2k - 1.$$

This implies that some $A_i$ has cardinality less than 2, and since we have assumed that no $A_i$ has cardinality 1, we may assume without loss of generality that $|A_1| = 0$. Because $2 \deg v - 1$ is odd, we also see that some $A_i$ must have odd cardinality. Suppose, again without loss of generality, that $|A_2| \geq 3$.

We now show that this leads to a contradiction. Let $H$ denote the subgraph of $G - v$ formed by all edges of colors 1 and 2, and let $u \in A_2 \setminus A_1$ be arbitrary. Because $\Delta(H) \leq 2$, every component of $H$ is either a path or a cycle. Moreover, since $u \in A_2 \setminus A_1$, we know that $\deg_H u = 1$ and thus its component in $H$ is a path. By exchanging the colors 1 and 2 along this path, we produce a different edge coloring of $G - v$ that reduces $|A_1|^2 + |A_2|^2$ (specifically, by $2|A_2|$), contradicting our choice of this edge coloring and completing the proof of the lemma. ∎

## Class one and class two graphs

Vizing's theorem (Theorem 6.5) states that the chromatic index of a graph is either equal to the maximum degree of the graph or to one more than the maximum degree. If $\chi'(G) = \Delta(G)$, the graph is said to be of **class one**, and if $\chi'(G) = 1 + \Delta(G)$, the graph is said to be of **class two**. It is difficult to determine which class a given graph belongs to in general. Holyer [130] showed that it is NP-complete to determine whether a graph is $k$-edge colorable, even for cubic graphs.

Exercises 6.12 and 6.14 ask the reader to prove that even cycles and complete bipartite graphs are of class one. These are special cases of the line coloring theorem of König [149].

**Theorem 6.7 (König's line coloring theorem).** *Every bipartite graph is of class one.*

*Proof.* Our proof is by induction on the size of the graph. For the base case, note that the theorem is true for graphs without edges. Now suppose that $G$ is a bipartite graph and that the result holds for all bipartite graphs on fewer edges. Take $e = uv$ to be an arbitrary edge of $G$, so by induction, $\chi'(G - e) = \Delta(G - e)$. Let $\Delta = \Delta(G)$.

Color the edges of $G - e$ with $\Delta$ colors. Because $u$ and $v$ both have degree at most $\Delta - 1$ in $G - e$, there is some color, which we may assume without loss of generality is the color 1, that is not used on an edge incident to $u$. There is also a color not used on an edge incident to $v$. If this color is also 1, then we may color the edge $e$ by 1 and we are done. Thus, we may assume without loss of generality that the color 2 is not used on an edge incident to $v$. As we are done otherwise, we may further assume that $u$ is incident to an edge colored 2 and that $v$ is incident to an edge colored 1.

Let $H$ denote the subgraph of $G - e$ formed by all edges of colors 1 and 2. Because $\Delta(H) \leq 2$, every component of $H$ is either a path or a cycle. Moreover, since $\deg_H u = 1$, we know that $u$ is an end-vertex of a path in $H$; let $P$ denote this path. The edges of $P$ must alternate in color 2, 1, 2, ..., and since $G$ is bipartite, the only way this path could include $v$ would be if $v$ was incident to an edge colored 2, which we have assumed is not the case. Therefore, $P$ contains $u$ but not $v$. By interchanging the colors 1 and 2 along all the edges of $P$, we therefore produce a new $\Delta$-edge-coloring of $G - e$ in which neither $u$ nor $v$ is incident with an edge colored 1. We may then assign the color 1 to $e = uv$ to obtain a proper $\Delta$-edge-coloring of $G$, completing the proof. ∎

Another sufficient condition for a graph to be class one is given by a result of Fournier [96]. This result can be viewed as a strengthening of Vizing's theorem (Theorem 6.5) and follows from a very similar proof also using Lemma 6.6 (Exercise 6.20).

**Theorem 6.8 (Fournier's theorem).** *Let $D$ denote the set of vertices of maximum degree in the graph $G$. If $G[D]$ is a forest, then $G$ is of class one.*

Although it is not obvious, there are considerably more class one graphs than class two graphs, relatively speaking. It follows from Fournier's theorem that every graph with a unique vertex of maximum degree is of class one. In 1977, Erdős and Wilson [85] proved that the probability that a graph of order $n$ (chosen uniformly at random) has a unique vertex of maximum degree (and is therefore of class one) tends to 1 as $n$ approaches infinity. Thus, in the parlance of the probabilistic method, almost all graphs are of class one.

We have observed that if $G$ is a graph of size $m \geq 1$, then $\chi'(G) \geq m/\alpha'(G)$. Thus, if $G$ is of class one, its maximum degree must also be at least $m/\alpha'(G)$. Put another way, every class one graph $G$ of size $m \geq 1$ must satisfy

$$m \leq \alpha'(G) \cdot \Delta(G).$$

This establishes a sufficient condition for a graph to be of class two, originally due to Beineke and Wilson [21].

**Theorem 6.9.** *If $G$ has more than $\alpha'(G) \cdot \Delta(G)$ edges, then it is of class two.*

A graph $G$ of order $n$ and size $m$ is called **overfull** if $m > \Delta(G) \cdot \lfloor n/2 \rfloor$. Since $\alpha'(G) \leq \lfloor n/2 \rfloor$ for every graph $G$ of order $n$, if $G$ is overfull, then

$$m > \Delta(G) \cdot \lfloor n/2 \rfloor \geq \alpha'(G) \cdot \Delta(G),$$

giving the following immediate consequence of Theorem 6.9.

**Corollary 6.10.** *Every overfull graph is of class two.*

It should be mentioned that Theorem 6.9 and its corollary provide only sufficient conditions for a graph to be of class two. There are graphs with relatively few edges, such as odd cycles, that are of class two (Exercise 6.13). Chetwynd and Hilton [47] conjectured, however, a close connection between class two graphs and overfull subgraphs.

**Conjecture 6.11 (Overfull conjecture).** *Let $G$ be a graph of order $n$ with $\Delta(G) > n/3$. Then $G$ is of class two if and only if $G$ contains an overfull subgraph $H$ with $\Delta(H) = \Delta(G)$.*

Note that if $G$ does contain an overfull subgraph $H$ with $\Delta(H) = \Delta(G)$, then $G$ must be of class two (Exercise 6.21). The overfull conjecture has been established for many classes of graphs such as complete multipartite graphs, but it remains open in general and is one of the primary open questions in graph colorings.

# 6.3  Critical and perfect graphs

We return to the context of vertex coloring here, beginning with an important family of graphs. For an integer $k \geq 2$, we say that a graph $G$ is **critically $k$-chromatic** or simply **$k$-critical** if $\chi(G) = k$ but $\chi(G - v) < k$ for every vertex $v$ of $G$. By convention, we also define $K_1$ to be 1-critical. For example, the graph $K_2$ is the only 2-critical graph, while the odd cycles are the only 3-critical graphs (Exercise 6.23). It follows quickly that every $k$-critical graph is connected (Exercise 6.24).

If a graph $G$ is $k$-chromatic, then it must contain a $k$-critical induced subgraph; in particular, its $k$-chromatic induced subgraph of minimum order must be $k$-critical. We begin with an easy result.

**Theorem 6.12.** *If a graph $G$ is $k$-critical, then $\delta(G) \geq k - 1$.*

*Proof.* Let $G$ be a $k$-critical graph and suppose to the contrary that $v \in V(G)$ has $\deg v \leq k - 2$. Because $G$ is $k$-critical, there is a proper $(k - 1)$-coloring of $G - v$. Because $v$ has at most $k - 2$ neighbors, one of these $k - 1$ colors is not used on the vertices of $N(v)$, so we can extend the coloring to a proper $(k - 1)$-coloring of $G$. However, this implies that $\chi(G) \leq k - 1$, which contradicts our hypothesis that $G$ is $k$-critical. ∎

Theorem 6.12 implies that a $k$-critical graph with $n$ vertices must have at least $(k-1)n/2$ edges. In 2014, Kostochka and Yancey [153] proved a remarkable strengthening of this bound, establishing conjectures of Gallai and Ore in the process.

**Theorem 6.13.** *If $G$ is a $k$-critical graph with $n$ vertices and $m$ edges, then*

$$m \geq \frac{(k+1)(k-2)n - k(k-3)}{2(k-1)}.$$

While the proof of Theorem 6.13 is quite involved, Kostochka and Yancey [152] have given a simplified (though still sophisticated) proof in the case $k = 4$, where the result states that a 4-critical graph of order $n$ must have at least

$$m \geq \frac{5n-2}{3}$$

edges. In Section 6.4, we show that this special case has significant applications.

Our next result about the structure of $k$-critical graphs, due to Dirac [66], is more significant. Recall from Section 4.3 that a graph $G$ with at least two vertices is $k$-edge-connected if one must remove $k$ or more edges to disconnect it.

**Theorem 6.14.** *If the graph $G$ is $k$-critical for $k \geq 2$, then $G$ is $(k-1)$-edge-connected.*

*Proof.* The only 2-critical graph is $K_2$, which is 1-edge-connected, so we may assume that $k \geq 3$. Suppose that the graph $G$ is $k$-critical but is not $(k-1)$-edge-connected for some $k \geq 3$. We will produce a proper $(k-1)$-coloring of $G$, a contradiction.

Because we have assumed that $G$ is not $(k-1)$-edge-connected, there is a set of at most $k-2$ edges whose removal disconnects $G$. Thus, we can partition the vertices of $G$ into two proper subgraphs $G_1$ and $G_2$ so that the number of edges joining the vertices of $G_1$ to the vertices of $G_2$ in $G$ is at most $k-2$.

Since $G$ is $k$-critical, we have that $\chi(G_1), \chi(G_2) \leq k-1$. Choose proper $(k-1)$-colorings of both subgraphs. We keep this coloring of $G_2$ and the color classes of this coloring of $G_1$, but we permute the actual colors used in the coloring of $G_1$. To this end, let $U_1, U_2, \ldots, U_t$ denote the color classes of $G_1$ for which there is some vertex of $U_i$ that is adjacent to a vertex of $G_2$. For $1 \leq i \leq t$, let $k_i$ denote the number of edges between the vertices of $U_i$ and those of $G_2$. Thus $k_i \geq 1$ for every $1 \leq i \leq t$ and

$$k_1 + k_2 + \cdots + k_t \leq k-2.$$

We now assign new colors to the color classes of this coloring of $G_1$, starting with $U_1, U_2, \ldots, U_t$ in that order, and then proceeding to the other color classes. When we come to assign a new color to $U_i$, we will have assigned $i-1$ colors to $U_1, U_2, \ldots, U_{i-1}$, and none of these can be assigned to $U_i$. Also, since the vertices of $U_i$ have at most $k_i$ neighbors in $G_2$, there are at most $k_i$ additional

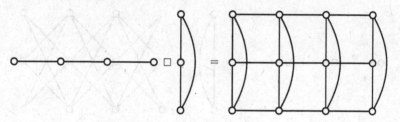

**Figure 6.5.** The cartesian product of a pair graphs

colors that we cannot assign to $U_i$. Still, the number of colors available to be assigned to $U_i$ is at least

$$(k-1) - (i-1) - k_i \geq (k-1) - (k_1 + k_2 + \cdots + k_{i-1}) - k_i$$
$$\geq (k-1) - (k_1 + k_2 + \cdots + k_t)$$
$$\geq 1.$$

This shows that we are guaranteed to be able to find a color for each of the color classes $U_1, U_2, \ldots, U_t$. The other color classes of $G_1$—those that are not adjacent to any vertices of $G_2$—may be given any remaining colors, yielding a proper $(k-1)$-coloring of $G$ and completing the proof. ∎

We conclude our discussion of critical graphs with a topic about which very little has been proved. By definition, if we delete a vertex from a $k$-critical graph, then the chromatic number decreases by 1, but what if we delete two vertices? A graph is called **$k$-double-critical** if it has chromatic number $k \geq 3$, and the deletion of any pair of adjacent vertices decreases the chromatic number by 2. Complete graphs are double-critical, and in 1966, Lovász (see Erdős [78]) asked if they are the only such graphs.

**Conjecture 6.15.** *For all $k \geq 3$, the only $k$-double-critical graph is $K_k$.*

## Graph operations

There are many ways to combine two graphs to produce a new one, and many of these interact nicely with chromatic numbers. For example, the chromatic number of a disconnected graph is the maximum of the chromatic numbers of its components, so

$$\chi(G \cup H) = \max\{\chi(G), \chi(H)\}$$

for all vertex-disjoint graphs $G$ and $H$. The reader is asked to prove the following result about the join $G \vee H$ and cartesian product $G \square H$ in Exercise 6.26.

**Theorem 6.16.** *For any pair of vertex-disjoint graphs $G$ and $H$, we have*

$$\chi(G \vee H) = \chi(G) + \chi(H) \quad and \quad \chi(G \square H) = \max\{\chi(G), \chi(H)\}.$$

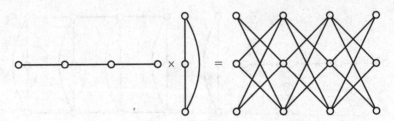

**Figure 6.6.** The tensor product of a pair graphs

There is an operation that does not behave so nicely, however. Given vertex-disjoint graphs $G$ and $H$, their **tensor product** is the graph $G \times H$ with vertex set $V(G \times H) = V(G) \times V(H)$ and in which two vertices $(u_1, v_1)$ and $(u_2, v_2)$ are adjacent if and only if $u_1 u_2 \in E(G)$ and $v_1 v_2 \in E(H)$. An example of the tensor product is shown in Figure 6.6. It may be contrasted with the cartesian product of the same graphs shown in Figure 6.5.

It is not hard to show that $\chi(G \times H) \le \min\{\chi(G), \chi(H)\}$ for all vertex-disjoint graphs $G$ and $H$ (Exercise 6.27), and Hedetniemi conjectured in his 1966 thesis [125] that equality always holds. This conjecture was widely believed to be true until Shitov [220] produced an ingenious counterexample in 2019.

## Nordhaus–Gaddum inequalities

In a brief 1956 paper, Nordhaus and Gaddum established a pair of inequalities bounding the sum and product of the chromatic numbers of a graph and its complement. When similar results are found for other graph parameters, they are now called **Nordhaus–Gaddum-type inequalities** (see Aouchiche and Hansen [8]), but these are the originals.

**Theorem 6.17 (Nordhaus–Gaddum inequalities).** *For every graph $G$ of order $n$,*

$$n \le \chi(G) \cdot \chi(\overline{G}) \le \left(\frac{n+1}{2}\right)^2 \quad and \quad 2\sqrt{n} \le \chi(G) + \chi(\overline{G}) \le n+1.$$

*Proof.* The theorem will follow if we prove that

$$\sqrt{n} \le \sqrt{\chi(G) \cdot \chi(\overline{G})} \le \frac{\chi(G) + \chi(\overline{G})}{2} \le \frac{n+1}{2}.$$

Moreover, the second inequality above follows from the inequality of arithmetic and geometric means, which states that $(x+y)/2 \ge \sqrt{xy}$ for all $x, y \ge 0$.

For the first inequality, we note that in any proper $\chi(G)$-coloring of $G$, at least one of the color classes must contain at least $n/\chi(G)$ vertices. As the vertices of such a color class form an independent set in $G$, they form a clique in $\overline{G}$. This shows that $\chi(\overline{G}) \ge n/\chi(G)$, and so $\chi(G) \cdot \chi(\overline{G}) \ge n$.

To prove the remaining inequality, $\chi(G) + \chi(\overline{G}) \leq n + 1$, we use induction on $n$. The base case $n = 1$ holds evidently, so take $G$ to be a graph on $n \geq 2$ vertices and choose an arbitrary vertex $v$ of $G$. By induction, there is some $1 \leq k \leq n$ so that $G - v$ has a proper $k$-coloring and $\overline{G} - v$ has a proper $(n - k)$-coloring. Fix such colorings of $G - v$ and $\overline{G} - v$. If $v$ can be assigned one of the $k$ colors used to color $G - v$, then by assigning $v$ a new color in $\overline{G}$, we are done. The same holds if $v$ can be assigned one of the $n - k$ colors used to color $\overline{G} - v$. If neither of these things is possible, then that means that $v$ is adjacent to at least $k$ vertices in $G$ and at least $n - k$ vertices in $\overline{G}$, but that is impossible, because $G$ only has $n$ vertices including $v$. ∎

The proof we have presented is not the original. In particular, the idea to relate the two Nordhaus–Gaddum inequalities using arithmetic and geometric means was first observed by Chartrand and Mitchem [45]. The bound $\chi(G) \cdot \chi(\overline{G}) \geq n$ had been observed before Nordhaus and Gaddum by Zykov [265]. One can also derive the bound $\chi(G) + \chi(\overline{G}) \leq n + 1$ by applying Theorem 6.2 to both $G$ and $\overline{G}$ and then performing algebraic manipulation, as observed by Bondy [28].

## Reed's $\omega$, $\Delta$, and $\chi$ conjecture

As we have seen, the two most immediate bounds on the chromatic number are

$$\omega(G) \leq \chi(G) \leq 1 + \Delta(G).$$

In 1998, Reed [196] made a conjecture that would bound $\chi(G)$ by the (rounded up) average of these two quantities.

**Conjecture 6.18 (Reed's $\omega$, $\Delta$, and $\chi$ conjecture).** *For every graph $G$,*

$$\chi(G) \leq \left\lceil \frac{\omega(G) + 1 + \Delta(G)}{2} \right\rceil.$$

This conjecture remains open even for the case in which $\omega(G) = 2$, that is, for graphs with at least one edge but no $K_3$. In this special case, Reed's conjecture is equivalent to the statement that $\chi(G) \leq \Delta(G)/2 + 2$ for every triangle-free graph $G$.

We now establish a related inequality, first found by Brigham and Dutton [35]. By Exercise 6.4, if $G$ is a graph of order $n$, then

$$\omega(G) \leq \chi(G) \leq n + 1 - \alpha(G).$$

Our next result shows that $\chi(G)$ is bounded above by the average of these two quantities.

**Theorem 6.19.** *For every graph $G$ of order $n$,*

$$\chi(G) \leq \frac{\omega(G) + n + 1 - \alpha(G)}{2}.$$

*Proof.* We prove the theorem using induction on $n$. The result is true for $K_1$, so now assume that $n \geq 2$ and that the result is true for all graphs of lesser order. Let $S \subseteq V(G)$ denote a maximum independent set, so $|S| = \alpha(G)$. If $|S| = n$, so $G$ has no edges, then $\chi(G) = 1$ and the bound holds, so we may assume that $G - S$ has at least one vertex.

Since $G - S$ has $n - \alpha(G)$ vertices, we have by induction that

$$\chi(G - S) \leq \frac{\omega(G - S) + n + 1 - \alpha(G) - \alpha(G - S)}{2}.$$

First suppose that $\omega(G - S) = \omega(G)$. In particular, this implies that every vertex of $S$ has a nonneighbor in $G - S$. Therefore, we can color $G$ by first coloring $G - S$ with $\chi(G - S)$ colors, and then assigning each vertex of $S$ the color of one of its nonneighbors in $G - S$. Thus in this case we have

$$\chi(G) \leq \chi(G - S) \leq \frac{\omega(G) + n + 1 - \alpha(G) - \alpha(G - S)}{2},$$

which is better than the bound we seek to prove.

The other case is when $\omega(G - S) < \omega(G)$. In this case, we can color $G$ by first coloring $G - S$ with $\chi(G - S)$ colors, and then adding a new color and assigning it to every vertex of $S$. This gives us that

$$\chi(G) \leq 1 + \chi(G - S) \leq \frac{\omega(G - S) + n + 3 - \alpha(G) - \alpha(G - S)}{2},$$

which is at most the bound we seek to establishing because $\omega(G - S) \leq \omega(G) - 1$ and $\alpha(G - S) \geq 1$. ■

## Perfect graphs

As we have frequently noted, $\chi(G) \geq \omega(G)$. We might ask: for which graphs does equality hold? Such graphs, however, are unlikely to have any nice characterization because if $G$ is *any* $k$-chromatic graph, then the disjoint union $G \cup K_k$ has chromatic number equal to its clique number. Instead, we ask when equality holds for a graph and all of its induced subgraphs. A graph $G$ is called **perfect** if $\chi(H) = \omega(H)$ for every induced subgraph $H$ of $G$.

For example, the complete graph $K_n$ is perfect. To show this, we must note that every induced subgraph $H$ of $K_n$ is also complete and so satisfies $\chi(H) = \omega(H)$. Similarly, every bipartite graph $G$ is perfect because if $H$ is an induced subgraph of a bipartite graph, then $\chi(H) = \omega(H) = 2$ if $H$ has any edges and $\chi(H) = \omega(H) = 1$ otherwise. The odd cycles provide an example of a family of graphs that is *not* perfect, because for $k \geq 2$, we have $\chi(C_{2k+1}) = 3$ while $\omega(C_{2k+1}) = 2$.

Perfect graphs were defined in 1963 by Berge [24], who made two conjectures that attracted significant interest. First, Berge conjectured that a graph $G$ is perfect if and only if $\overline{G}$ is perfect. This conjecture, referred to as the **perfect graph conjecture**, was proved by Lovász [158, 159] in 1972.

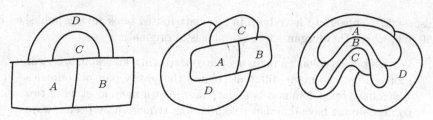

**Figure 6.7.** Renditions of the examples of map colorings in De Morgan's 1852 letter to Hamilton

**Theorem 6.20 (Perfect graph theorem).** *A graph is perfect if and only if its complement is perfect.*

Berge's second conjecture concerned the characterization of perfect graphs. In what was generally referred to as the **strong perfect graph conjecture**, Berge conjectured that a graph $G$ is perfect if and only if neither $G$ nor $\overline{G}$ contains an induced odd cycle of length 5 or more. This conjecture was finally proved by Chudnovsky, Robertson, Seymour, and Thomas [48] in 2006. The proof is long and technical, and based on a deep structural decomposition of these graphs.

**Theorem 6.21 (The strong perfect graph theorem).** *A graph $G$ is perfect if and only if neither $G$ nor $\overline{G}$ contains an induced odd cycle of length 5 or more.*

 **6.4** **Maps and planar graphs**

The first written reference to the four color problem appears to be in a letter dated October 23, 1852, from Augustus De Morgan to Sir William Rowan Hamilton. De Morgan wrote, in the language of the time:

> A student of mine asked me to day to give him a reason for a fact which I did not know was a fact — and do not yet. He says that if a figure be any how divided and the compartments differently coloured so that figures with any portion of common boundary line are differently coloured — four colours may be wanted, but no more.

The student referred to by De Morgan was Frederick Guthrie, who later attributed the problem to his brother Francis [113]. In his reply only three days later, Hamilton, the discoverer of the quaternions and the man for whom hamiltonian graphs are named, expressed disinterest:

> I am not likely to attempt your "quaternion of colours" very soon.

Undeterred by Hamilton's lack of interest, De Morgan continued to spread the problem. His first published account of the problem (brought to light by

Wilson [257]) appears to have been in an unattributed book review published in 1860 [63], where De Morgan wrote (emphasis in original):

> When a person colours a map, say the countries in a kingdom, it is clear he must have so many different colours that every pair of countries which have some common boundary *line*—not a mere meeting of two corners—must have different colours. Now, it must have been always known to map-colourers that *four* different colours are enough.

This myth that mapmakers cared about the problem appears to have been a fiction created by De Morgan. As the historian May wrote [163]:

> In the first place there is no evidence that mapmakers were or are aware of the sufficiency of four colors. A sampling of atlases in the large collection of the Library of Congress indicates no tendency to minimize the number of colors used. Maps utilizing only four colors are rare, and those that do usually require only three. Books on cartography and the history of mapmaking do not mention the four-color property, though they often discuss various other problems relating to the coloring of maps.

May concluded that "if cartographers are aware of the four-color conjecture, they have certainly kept the secret well." Nevertheless, the four color problem motivated the development of graph theory for over a century.

The question De Morgan asked was if four colors would suffice to color any map on the plane so that no two adjacent countries were colored alike. Two countries are considered to be **adjacent** if they share a common boundary line—not simply a point.

To place De Morgan's question in the context of graph theory, we say that a plane graph is **$k$-region colorable** if its regions can be colored with $k$ or fewer colors so that adjacent regions are colored differently. The **four color problem** asks if every plane graph is 4-region colorable.

A more common statement of the four color problem is that every planar graph is 4-colorable, but it is not immediately apparent that this is the same problem De Morgan popularized. To see the equivalence of these two formulations, we must introduce the concept of the dual of a plane graph.

For a given connected plane graph $G$, we construct a pseudograph $G^*$ as follows. A vertex is placed in each region of $G$, and these vertices constitute the vertex set of $G^*$. Two distinct vertices of $G^*$ are then joined by an edge for each edge common to the boundaries of the two corresponding regions of $G$. In addition, a loop is added at a vertex $v$ of $G^*$ for each bridge of $G$ that belongs to the boundary of the corresponding region. Each edge of $G^*$ is drawn so that it crosses its associated edge of $G$ but no other edge of $G$ or $G^*$ (which is always possible); hence, $G^*$ is planar.

The pseudograph $G^*$ is referred to as the **dual** of $G$. In addition to being planar, $G^*$ has the same number of edges as $G$, and can be drawn so that each region of $G^*$ contains a single vertex of $G$; indeed, $(G^*)^* \cong G$. An example is

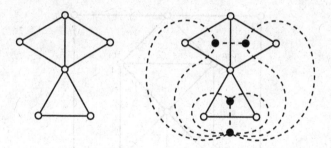

**Figure 6.8.** A plane graph and its dual

shown in Figure 6.8, with the vertices of $G^*$ represented by solid circles. The concept of a dual graph can be used to prove Theorem 6.22 (see Exercise 6.34).

**Theorem 6.22.** *Every planar graph is 4-colorable if and only if every plane graph is 4-region colorable.*

Through the efforts of De Morgan, the four color problem had been circulating for nearly twenty years when Arthur Cayley asked during a meeting of the London Mathematical Society on June 13, 1878 if the problem had been solved. Soon afterward, Cayley [40] published a paper in the *Proceedings of the Royal Geographical Society* in which he presented his views on why the problem appeared to be so difficult. From his discussion, one might very well infer the existence of planar graphs with an arbitrarily large chromatic number.

One of the most important and notorious events related to the four color problem occurred on July 17, 1879, when the magazine *Nature* carried an announcement [183] that the four color conjecture had been verified by Alfred Bray Kempe. Kempe published his proof in 1879 [141], and then he published what he thought was a simpler proof in 1880 in *Nature* [142], writing "I have succeeded in obtaining the following simple solution in which mathematical formulae are conspicuous by their absence." For approximately ten years, the four color conjecture was considered to be settled. Then in 1890, Percy John Heawood [123] discovered an error in Kempe's proof.

All was not lost, however, as Heawood was able to use Kempe's technique to prove that every planar graph is 5-colorable. This result is referred to, quite naturally, as the five color theorem. (The reader is asked to prove the **six color theorem** in Exercise 6.32.)

**Theorem 6.23 (Five color theorem).** *Every planar graph is 5-colorable.*

*Proof.* We proceed by induction on the order $n$ of the graph. Clearly, the result is true if $1 \leq n \leq 5$. Assume now that $G$ is a planar graph of order $n \geq 6$ and that every planar graph of order $n - 1$ is 5-colorable.

Recall that Theorem 5.3 states that a planar graph of order $n$ may have at most $3n - 6$ edges, and it follows immediately (Corollary 5.4) that every planar graph has a vertex of degree at most 5. Take $v$ to be a vertex of $G$ with $\deg v \leq 5$.

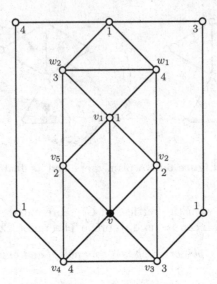

**Figure 6.9.** A graph demonstrating why the proof of the five color theorem does not prove the four color theorem

By our induction hypotheses, $G - v$ has a proper coloring with the colors 1, 2, 3, 4, and 5. If one of these colors is not used to color the neighbors of $v$, then we can assign this missing color to $v$ to see that $G$ is 5-colorable. Hence we may assume that $\deg v = 5$ and that all five colors are used to color the neighbors of $v$. Embed $G$ in the plane and label the neighbors of $v$ as $v_1$, $v_2$, $v_3$, $v_4$, and $v_5$, arranged cyclically about $v$. We may assume that each $v_i$ has the color $i$.

Let $H_{1,3}$ denote the subgraph of $G - v$ induced by the set of vertices colored 1 or 3, so $v_1, v_3 \in V(H_{1,3})$. If $v_1$ and $v_3$ belong to different components of $H_{1,3}$, then we may interchange the colors 1 and 3 in the component of $H_{1,3}$ containing $v_1$ and assign the color 1 to $v$ to obtain a proper 5-coloring of $G$.

Therefore, we may assume that $v_1$ and $v_3$ belong to the same component of $H_{1,3}$. This implies that $G - v$ contains a $v_1$-$v_3$ path $P$, every vertex of which is colored 1 or 3. This path and the vertex $v$ together constitute a cycle in $G$ that encloses either $v_2$ or both $v_4$ and $v_5$. In particular, this implies that every path from $v_2$ to $v_4$ passes through either $v$ or a vertex colored 1 or 3.

Let $H_{2,4}$ denote the subgraph of $G - v$ induced by the set of vertices colored 2 or 4. By our observation above, $v_2$ and $v_4$ must belong to different components of $H_{2,4}$. Hence we may interchange the colors 2 and 4 in the component of $H_{2,4}$ containing $v_2$ and assign the color 2 to $v$ to obtain a proper 5-coloring of $G$. ∎

It is not immediately obvious why the proof of the five color theorem we have given cannot be adapted to prove the four color theorem. Indeed, Kempe thought that it could be. Consider the graph $G$ of Figure 6.9. The graph $G - v$ has a 4-coloring, and we would like to swap neighbor colors to eliminate one color. It is impossible to remove colors 1, 3, or 4 by swapping; thus, we must swap color 2.

By examining the right side of the graph, we could switch colors 2 and 4 in the component of $H_{2,4}$ containing $v_2$. Similarly, on the left side, we could switch colors 2 and 3 in the component of $H_{2,3}$ containing $v_5$. If both swaps could be performed simultaneously, then the color 2 could be assigned to $v$. Alas, this is not possible, as both $w_1$ and $w_2$ would have color 2.

##  The four color problem becomes a theorem

In 1976, Appel and Haken [9] announced a proof of the four color theorem.

**Theorem 6.24 (Four color theorem).** *Every planar graph is 4-colorable.*

The details of the proof were contained in two papers, published in 1977 by Appel and Haken [10] and Appel, Haken, and Koch [11]. The proof uses an inductive structure like our proof of the five color theorem, but it appeals to Euler's formula (Theorem 5.1) and a procedure known as discharging to produce what they referred to as a set $\mathcal{U}$ of unavoidable configurations in any plane triangulation (comparable to a vertex of degree at most 5 in Heawood's proof). Their set $\mathcal{U}$ contained 1482 configurations. They then showed that a proper 4-coloring of a plane triangulation containing any element of $\mathcal{U}$ can be obtained from proper 4-colorings of smaller plane triangulations (comparable to expanding a proper 5-coloring of $G - v$ to such a coloring of $G$). This part of their proof relied heavily on computer calculations.

In 1997, Robertson, Sanders, Seymour, and Thomas [203] obtained a stream-lined proof of the four color theorem using the same general approach of Appel, Haken, and Koch. Their proof has fewer than 700 configurations and is based on a simpler discharging procedure, but it still rests on computer calculations. To this day, no proof of the four color theorem has been found that does not involve extensive computer calculations.

##  Hadwiger's conjecture

We have frequently noted that for every graph $G$, the clique number $\omega(G)$ is a lower bound for $\chi(G)$; that is, $\chi(G) \geq \omega(G)$. This bound may be quite poor, though; in particular, a $k$-chromatic graph need not contain $K_k$. However, Hadwiger [116] conjectured the following.

**Conjecture 6.25 (Hadwiger's conjecture).** *Every $k$-chromatic graph contains $K_k$ as a minor.*

In 1937, Wagner [249] proved that every planar graph is 4-colorable if and only if every 5-chromatic graph contains $K_5$ as a minor, that is, Wagner showed the equivalence of the four color problem and Hadwiger's conjecture for $k = 5$ (albeit six years before Hadwiger stated his conjecture). Hadwiger's conjecture

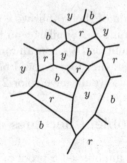

**Figure 6.10.** A 3-colored map presented by Heawood

can therefore be considered a generalization of the four color conjecture, and the four color theorem establishes Hadwiger's conjecture for $k = 5$.

Using the four color theorem, Robertson, Seymour, and Thomas [205] verified Hadwiger's Conjecture for $k = 6$. For integers $k \geq 6$, the conjecture remains open.

 **The three color problem**

The **three color problem** asks: which planar graphs are 3-colorable? This problem is almost as old as the four color problem. The first significant result about 3-coloring planar graphs, Theorem 6.26 below, was mentioned by Heawood in his 1890 paper [123] where he re-opened the four color problem and proved the five color theorem. Heawood included a proof of Theorem 6.26 in his second paper on coloring [124], published eight years later. That second paper also contained the 3-colored map shown in Figure 6.10.

**Theorem 6.26 (Heawood's theorem).** *A maximal planar graph of order 3 or more has chromatic number 3 if and only if it is eulerian.*

This result does not settle the three color problem, because it only applies to maximal planar graphs. Indeed, the three color problem is still attracting an immense amount of interest. One of the seminal results in this study is the following 1959 result of Grötzsch [110]. The proof we present is due to Kostochka and Yancey [152] and uses the lower bound on the number of edges in a 4-critical graph provided by their Theorem 6.13.

**Theorem 6.27 (Grötzsch's three color theorem).** *Every planar graph without triangles is 3-colorable.*

*Proof.* Suppose otherwise and choose $G$ to be a triangle-free plane graph of minimum order that is not 3-colorable. Let $G$ have $n$ vertices, $m$ edges, and $r$ regions, so by our assumption, all triangle-free plane graphs on fewer than $n$ vertices are 3-colorable.

First suppose that $G$ has no regions bounded by a 4-cycle. Because $G$ is a minimal counterexample, it cannot have a bridge (if it did have a bridge, we could remove it, 3-color the two components, and then obtain a 3-coloring of $G$). Therefore, every edge of $G$ lies on two regions, and in this case, every region contains at least 5 edges. By double-counting region-edge pairs, we see that $5r \leq 2m$, so $r \leq 2m/5$. Substituting this bound into Euler's formula (Theorem 5.1) shows that

$$m = n + r - 2 \leq n + \frac{2m}{5} - 2,$$

and solving this for $m$ gives

$$m \leq \frac{5n - 10}{3}.$$

This is a contradiction to Theorem 6.13. Therefore, we may assume that $G$ has at least one 4-region because we are done otherwise.

Let $u_1, u_2, u_3, u_4$ be a 4-region of $G$, in counterclockwise order. Since $G$ has no triangles, we know that $u_1u_3, u_2u_4 \notin E(G)$. Let $G'$ denote the graph obtained from $G$ by "gluing" $u_1$ and $u_3$ together, by which we mean that we remove both $u_1$ and $u_3$ from $G$ and then create a new vertex called $u'$ that is adjacent to every vertex adjacent to either $u_1$ or $u_3$, so $N_{G'}(u') = N_G(u_1) \cup N_G(u_3)$. If $G'$ is triangle-free, then the minimality of $G$ ensures that $G'$ is 3-colorable. We can then extend that proper 3-coloring of $G'$ to a proper 3-coloring of $G$ by assigning $u_1$ and $u_3$ the same color in $G$ as $u'$ has in this coloring of $G'$, and so in this case, we are done.

It follows that $G'$ must contain a triangle. Since $G$ is triangle-free, this triangle must include the vertex $u'$. This means that $G$ itself must contain a path of the form $u_1, v, w, u_3$. Furthermore, since $G$ is triangle-free, $\{v, w\}$ must be disjoint from $\{u_2, u_4\}$. Thus, we have the situation depicted below.

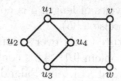

Now consider the graph $G''$ obtained from $G$ by gluing $u_2$ and $u_4$ together as before. This graph cannot contain a triangle, because any path of the form $u_2, v', w', u_4$ would have to cross the path $u_1, v, w, u_3$. Therefore, the minimality of $G$ ensures that $G''$ is 3-colorable, and as we argued with $G'$, this implies that $G$ itself is 3-colorable, completing the proof. ∎

In 1963, Grünbaum [111] proved a strengthening of Grötzsch's three color theorem that allows a graph to have three triangles. Allowing three triangles is best possible because $K_4$ is planar, not 3-colorable, and has four triangles. Grünbaum's proof of his theorem was flawed, and the first correct proof was given by Aksenov over a decade later [4]. A proof using Theorem 6.13 has been given by Borodin, Kostochka, Lidický, and Yancey [34]

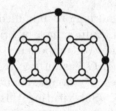

**Figure 6.11.** The graph shown has no 4-cycles and only four triangles, but cannot be 3-colored; the solid vertices must all receive different colors

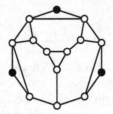

**Figure 6.12.** A counterexample to Steinberg's conjecture can be built from 12 copies of this graph

**Theorem 6.28 (Grünbaum's three color theorem).** *Every planar graph with at most three triangles is 3-colorable.*

## ⊳ Restricted cycles and Steinberg's conjecture

Instead of restricting the number of triangles as in Grötzsch's and Grünbaum's three color theorems, one might instead allow unlimited triangles but forbid cycles of other lengths. Forbidding cycles of length 4 is not enough, as demonstrated by the graph in Figure 6.11 (Exercise 6.35).

In 1976, Steinberg conjectured that every planar graph without 4-cycles or 5-cycles is 3-colorable [222, problem 9.1]. This conjecture was disproved in 2016 by Cohen-Addad, Hebdige, Král', Li, and Salgado [54] using 12 copies of the graph in Figure 6.12. This graph does not contain any 4- or 5-cycles. Additionally, it does not have a proper 3-coloring where all three solid vertices share the same color.

Even though Steinberg's conjecture is now known to be false, a large family of weaker results have been proved. In 1990, Erdős asked [222, problem 9.2] if there is an integer $k$ such that every planar graph without cycles of lengths 4 through $k$ is 3-colorable. Within a year, it was proved that $k = 11$ sufficed, then $k = 10$, then $k = 8$, and currently the best result is the following, established by Borodin, Glebov, Raspaud, and Salavatipour [33] in 2005.

**Theorem 6.29.** *Every planar graph without cycles of lengths 4 through 7 is 3-colorable.*

It is an open question if every planar graph without cycles of lengths 4, 5, or 6 is 3-colorable. We prove a slightly weaker result than Theorem 6.29, using Theorem 6.13. Our proof, which follows Hamaker and Vatter [119] and is inspired by the argument that Abbott and Zhou [1] used to prove that $k = 11$ suffices, begins by establishing the following. Note that in a graph without 4-cycles, it must be the case that no two triangles share an edge.

**Lemma 6.30.** *If $G$ is a connected plane graph of minimum degree 3 in which no two triangles share an edge, then strictly less than $2/3$ of its regions are triangles.*

*Proof.* Let $G$ be a connected plane graph with $n$ vertices, $m$ edges, and $r$ regions. Further, let $n_3$ denote the number of degree 3 vertices in $G$, let $r_3$ denote the number of its triangular regions, and let $m_3$ denote the number of edges contained in one of these triangular regions. Note that since no two triangles share an edge, $r_3 = m_3/3$. By double counting edges, since the minimum degree of $G$ is 3, we have

$$2m = \sum_{v \in V(G)} \deg v \geq 3n_3 + 4(n - n_3) = 4n - n_3,$$

so $n_3 \geq 4n - 2m$.

Now let $v$ be a vertex of degree 3 in $G$. Since no edge is contained in two triangles, at least one of the edges incident to $v$ must not be part of a triangle, and so contributes to $m - m_3$. As this edge might be incident to two vertices of degree 3, the most we can claim is that $m - m_3 \geq n_3/2$, or after rearranging, $m_3 \leq m - n_3/2$. Combining this with our inequality on $n_3$, we have

$$r_3 = \frac{m_3}{3} \leq \frac{m - n_3/2}{3} \leq \frac{m - (2n - m)}{3} \leq \frac{2m - 2n}{3}.$$

Recall that by Euler's formula (Theorem 5.1), $m - n = r - 2$. This implies that

$$r_3 \leq \frac{2m - 2n}{3} = \frac{2r - 4}{3},$$

which completes the proof. ∎

**Theorem 6.31.** *Every planar graph without cycles of lengths 4 through 8 is 3-colorable.*

*Proof.* Suppose that the result is not true and take $G$ to be a plane graph of minimal order $n$ that is not 3-colorable despite having no cycles of lengths 4 through 8. Let $m$ denote the number of edges of $G$ and $r$ denote the number of regions. As it is a minimal counterexample, $G$ must be 4-critical, so we have $m \geq 5n/3 - 2/3$ by Theorem 6.13.

Let $r_3$ denote the number of triangular regions in $G$. No two triangles of $G$ may share an edge because $G$ does not contain any 4-cycles, and $G$ is connected because it is 4-critical (see Exercise 6.24). Hence $r_3 < 2r/3$ by Lemma 6.30.

The number of non-triangular regions of $G$ is by definition $r - r_3$. The sum of the number of edges in these regions is at least $9(r - r_3)$, because each of these regions has at least nine edges. We can obtain an upper bound on this quantity by observing that the sum of the number of edges in every region of $G$ is at most $2m$ (because every edge borders at most two regions), and the sum of the number of edges in every triangular region of $G$ is precisely $3r_3$. Therefore, we have

$$9(r - r_3) \leq 2m - 3r_3.$$

Combining this with the fact that $r_3 < 2r/3$, we obtain

$$r \leq r_3 + \frac{2m - 3r_3}{9} = \frac{2m}{9} + \frac{2r_3}{3} < \frac{2m}{9} + \frac{4r}{9}.$$

This implies that $r < 2m/5$. We have $m = n + r - 2$ by Euler's formula (Theorem 5.1), so

$$m = n + r - 2 < n + \frac{2m}{5} - 2.$$

However, this implies that $m < (5n - 10)/3$, and that contradicts the fact that $m \geq (5n - 2)/3$. ∎

 Edge coloring planar graphs

The four color theorem implies a bound on the edge chromatic number of bridgeless cubic planar graphs. To establish this connection, we make use of the Klein four-group $\mathbb{Z}_2 \times \mathbb{Z}_2$. Using $\oplus$ as the binary operation for this group and denoting its elements by $(0,0)$, $(0,1)$, $(1,0)$, and $(1,1)$, we define addition by

$$(a, b) \oplus (c, d) = (a + c \bmod 2, b + d \bmod 2).$$

Note that $(a, b) \oplus (c, d) = (0,0)$ in $\mathbb{Z}_2 \times \mathbb{Z}_2$ if and only if $(a, b) = (c, d)$.

**Theorem 6.32.** *Every bridgeless, cubic planar graph is 3-edge colorable.*

*Proof.* Let $G$ be a bridgeless cubic plane graph. By the region-coloring version of the four color theorem, $G$ is 4-region colorable. Color the regions of $G$ with the elements of the Klein four-group $\mathbb{Z}_2 \times \mathbb{Z}_2$. Since $G$ has no bridges, each edge of $G$ lies on the boundary of two regions. Define the color of an edge to be the sum of the colors of the two regions it bounds. Since no two bordering regions share the same color, it follows that no edge is assigned the color $(0,0)$. Moreover, the sum of the colors of the three edges incident with any particular vertex is $(0,0)$, and it can be checked that this means that the three edges must have the three colors $(0,1)$, $(1,0)$, and $(1,1)$. Therefore, this is a 3-edge coloring of $G$. ∎

Using the terminology of Section 6.2, we can say that every bridgeless cubic planar graph is of class one.

In 1965, Vizing showed that every planar graph with maximum degree at least 8 is of class one. For every integer $k$ with $2 \leq k \leq 5$ there is a planar graph with maximum degree $k$ of class two (see Exercise 6.36). In 2001, Sanders and

Zhao [211] resolved one of the missing cases, $k = 7$, by showing that if $G$ is a planar graph with $\Delta(G) = 7$, then $G$ is of class one. Thus, only one case remains. Vizing conjectured that these graphs are also class one.

**Conjecture 6.33** (**Vizing's planar graph conjecture**). *Every planar graph with maximum degree 6 is of class one.*

## Snarks

Theorem 6.32 shows how the four color theorem implies that every bridgeless cubic planar graph is 3-edge-colorable. In fact the converse is true, too: the fact that every bridgeless cubic planar graph is 3-edge-colorable implies the four color theorem. This equivalence was observed by Tait [228] in 1880, and for this reason 3-edge-colorings of cubic graphs are often called **Tait colorings**.

Tait in fact believed that all cubic graphs were 3-edge-colorable, although he had clearly overlooked bridges, because no cubic graph with a bridge is 3-edge-colorable (Exercise 6.17). However, even when the bridgeless hypothesis is added, there are still counterexamples. One such counterexample is the Petersen graph, which was studied by Petersen [191] for precisely this reason in 1898. Almost fifty years passed until another was found. In 1975, Isaacs wrote about the search for such graphs that, "I recommend it as a pleasant diversion for any mathematician," but that anyone who so indulges will "be vividly impressed with the maddening difficulty of finding [one]". Motivated by this, in his Mathematical Games column in *Scientific American* in 1976, Gardner [100] proposed calling such graphs snarks, after the elusive fictional species from Lewis Carroll's poem, "The Hunting of the Snark", and the name has stuck.

A **snark** is therefore a non-3-edge-colorable bridgeless cubic graph, and the four color theorem is equivalent to the fact that no snark is planar. For reasons of minimality, other conditions are often imposed. For example, if a snark were to contain a triangle, then by contracting the vertices of that triangle to a single vertex, one would obtain a smaller snark. Thus, snarks are frequently also required to have girth of at least 5 and to satisfy a condition related to edge-connectivity, although these additional requirements vary.

It was fifty years from Petersen's study of the Petersen graph until Blanuša [26] found the two **Blanuša snarks** of order 18 shown on the left of Figure 6.13. The **Descartes snark** (discovered by Tutte [64] in 1948) has order 210. Szekeres [225] discovered the **Szekeres snark** of order 50 in 1973. Until 1973, these were the only known snarks. In 1975, however, Isaacs [132] described two infinite families of snarks, one of which contained all previously known snarks, while the second family was completely new. This second family contains the **flower snarks**, one example of which is shown in Figure 6.13. In addition, Isaacs found a snark that belonged to neither family. This **double-star snark** is shown on the far right of Figure 6.13.

All of these snarks appear to bear some resemblance to the Petersen graph. In fact, Tutte [241] conjectured in 1966 that this is always the case.

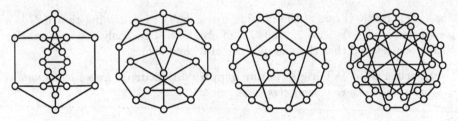

**Figure 6.13.** From left to right, the two Blanuša snarks, a flower snark, and the double-star snark

**Conjecture 6.34 (Tutte's snark conjecture).** *Every snark has the Petersen graph as a minor.*

Because any graph that contains the Petersen graph as a minor must not be planar, the snark conjecture would be a strengthening of the four color theorem. Thomas [230, p. 214] announced in 1999 that he had found a proof of the snark conjecture, together with Robertson, Sanders, and Seymour, and building on their simplified proof of the four color theorem [203]. However, while some parts of this proof have been published [71, 206–208], it remains incomplete.

# Exercises for Chapter 6

### Section 6.1. Vertex coloring

**6.1.** Prove that, for every integer $k \geq 2$, $\chi(C_{2k}) = 2$ and $\chi(C_{2k-1}) = 3$.

**6.2.** Prove that the size of every $k$-chromatic graph is at least $\binom{k}{2}$.

**6.3.** For every integer $k \geq 2$, give an example of a regular $k$-chromatic graph that is not complete.

**6.4.** Show that for every graph $G$ of order $n$,

$$\frac{n}{\alpha(G)} \leq \chi(G) \leq n + 1 - \alpha(G).$$

**6.5.** Let $G$ be a $k$-chromatic graph, where $k \geq 2$, and let $r$ be a positive integer such that $r \geq \Delta(G)$. Prove that there exists an $r$-regular $k$-chromatic graph $H$ such that $G$ is an induced subgraph of $H$.

**6.6.** Show that for every graph $G$, there is an order of its vertices so that the greedy coloring uses only $\chi(G)$ colors.

**6.7.** For every integer $k \geq 3$, give a tree and an ordering of the vertices of that tree so that the greedy coloring requires at least $k$ colors (even though trees are 2-colorable).

**6.8.** Show that for every graph $G$, there is a graph $H$ on the same vertex set that satisfies $E(H) \supseteq E(G)$ and $\chi(H) = \chi(G)$, and for which greedy coloring with any vertex ordering uses only $\chi(G)$ colors.

**6.9.** State and prove a version of Brooks' theorem (Theorem 6.4) for graphs that are not necessarily connected.

**6.10.** For the double star $T$ containing two vertices of degree 4 shown below, what upper bounds on $\chi(T)$ are given by Theorems 6.2, 6.3, and Brooks' theorem (Theorem 6.4)?

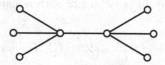

## Section 6.2. Edge coloring

**6.11.** Prove that if $G$ is a graph of size $m$, then $\chi'(G) \geq m/\alpha'(G)$.

**6.12.** Without using Kőnig's line coloring theorem (Theorem 6.7), prove that even cycles are of class one, so they have edge-chromatic number 2.

**6.13.** Prove that odd cycles are of class two, so they have edge-chromatic number 3.

**6.14.** Without using Kőnig's line coloring theorem (Theorem 6.7), prove that the complete bipartite graph $K_{k,\ell}$ is of class one, so its edge-chromatic number is $\max\{k, \ell\}$.

**6.15.** Prove that $\chi'(K_n) = n$ for every odd integer $n$.

**6.16.** Prove that every hamiltonian cubic graph is of class one.

**6.17.** Show that every cubic graph with a bridge is of class two.

**6.18.** Show that the Petersen graph is of class two.

**6.19.** Determine the edge chromatic number of the graph shown below.

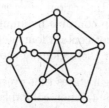

**6.20.** Prove Fournier's theorem (Theorem 6.8) from Lemma 6.6.

**6.21.** Prove the direction of the overfull conjecture stating that if the graph $G$ contains an overfull subgraph $H$ with $\Delta(H) = \Delta(G)$, then $G$ is of class two.

**6.22.** Show that every nontrivial self-complementary regular graph is of class two.

## Section 6.3.  Critical and perfect graphs

**6.23.** Prove that the odd cycles are the only 3-critical graphs.

**6.24.** Prove, without using Theorem 6.14, that every $k$-critical graph must be connected.

**6.25.** Prove that if $G$ is $k$-critical, then for every vertex $v \in V(G)$, there is a proper $k$-coloring of $G$ in which $v$ does not share its color with any other vertex.

**6.26.** Prove Theorem 6.16: for any pair of vertex-disjoint graphs $G$ and $H$, we have

$$\chi(G \vee H) = \chi(G) + \chi(H) \quad \text{and} \quad \chi(G \square H) = \max\{\chi(G), \chi(H)\}.$$

**6.27.** Prove that if $G$ and $H$ are vertex-disjoint graphs, then $\chi(G \times H) \leq \min\{\chi(G), \chi(H)\}$.

**6.28.** Prove that the graph of the octahedron (the graph $K_{2,2,2}$) is perfect.

**6.29.** We know that the odd cycles $C_{2k+1}$ are not perfect for all $k \geq 2$. Show that their complements are also not perfect by determining $\chi(\overline{C}_{2k+1})$ and $\omega(\overline{C}_{2k+1})$.

**6.30.** For each integer $n \geq 7$, give an example of a graph $G_n$ of order $n$ such that no induced subgraph of $G_n$ is an odd cycle of length at least 5 but $G_n$ is not perfect.

**6.31.** For the double star $T$ of Exercise 6.10, what upper bound on $\chi(T)$ is given by Theorem 6.19.

## Section 6.4.  Maps and planar graphs

**6.32.** Prove that every planar graph is 6-colorable. (Obviously, do not use the four or five color theorems.)

**6.33.** It was once thought that the regions of every map can be colored with four or fewer colors because no map contains five mutually adjacent regions. Show that there exist maps that do not contain four mutually adjacent regions but yet four colors are required to color the regions of these maps.

**6.34.** Prove Theorem 6.22, which states that every planar graph is 4-colorable if and only if every plane graph is 4-region colorable.

**6.35.** Prove that the graph in Figure 6.11 has chromatic number 4.

**6.36.** Show that for every integer $k$ with $2 \leq k \leq 5$, there is a planar graph of class one and a planar graph of class two, both having maximum degree $k$.

**6.37.** Is the graph shown from Exercise 6.19 a snark?

# Flows

**7**

This chapter focuses on the study of flows in digraphs. We start by examining networks, a type of digraph that is useful for modeling certain real-world problems. Networks can also be used to analyze connectedness in digraphs. We will also discuss the connection between coloring the regions of a plane graph, as explored in Chapter 6, and a certain type of flow in a digraph.

## 7.1 Networks

A **network** $N$ is a digraph $D$ with two distinguished vertices $u$ and $v$, called the **source** and **sink**, respectively, together with a function $c \colon E(D) \to \mathbb{R}$ such that $c(a) \geq 0$ for every $a \in E(D)$. The digraph $D$ is called the **underlying digraph** of $N$ and the function $c$ is called the **capacity function** of $N$. The value $c(a) = c(x, y)$ of an arc $a = (x, y)$ of $D$ is called the **capacity** of $a$. Any vertex of $N$ distinct from $u$ and $v$ is called an **intermediate vertex** of $N$.

The source $u$ of $N$ can be thought of as the location from which material is shipped and then transported through $N$, eventually reaching its destination, namely the sink $v$ of $N$. The capacity of an arc $(x, y)$ in $N$ may be thought of as the maximum amount of material that can be transported from $x$ to $y$ along $(x, y)$ per unit time. The problem then is to maximize the *flow* of material that can be transported from the source $u$ to the sink $v$ without exceeding the capacity of any arc.

A network $N$ can be represented by drawing the underlying digraph $D$ of $N$ and labeling each arc of $D$ with its capacity. A network is shown in Figure 7.1. In this network, the capacity of the arc $(x, y)$ is then $c(x, y) = 4$. While, in general, there may be more than one source from which material originates and more than one sink providing destinations of the material, it suffices to consider a network with a single source and a single sink (see Exercise 7.17).

For a vertex $x$ in a digraph $D$, the out-neighborhood $N^+(x)$ and the in-neighborhood $N^-(x)$ of $x$ are defined by

$$
\begin{aligned}
N^+(x) &= \{y \in V(D) : (x, y) \in E(D)\} \text{ and} \\
N^-(x) &= \{y \in V(D) : (y, x) \in E(D)\}.
\end{aligned}
$$

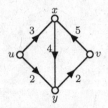

**Figure 7.1.** A network

Thus $|N^+(x)| = \operatorname{od} x$ and $|N^-(x)| = \operatorname{id} x$.

For a digraph $D$ and a function $g: E(D) \to \mathbb{R}$, it is convenient to introduce some notation. For subsets $X$ and $Y$ of $V(D)$, define the set $[X,Y]$ and the number $g(X,Y)$ by

$$[X,Y] = \{(x,y) : x \in X, y \in Y\}$$

and

$$g(X,Y) = \sum_{(x,y)\in[X,Y]} g(x,y),$$

where $g(X,Y) = 0$ if $[X,Y] = \varnothing$. For $x \in V(D)$,

$$g^+(x) = \sum_{y\in N^+(x)} g(x,y) \ \text{ and } \ g^-(x) = \sum_{y\in N^-(x)} g(y,x). \qquad (7.1)$$

More generally, for $X \subseteq V(D)$,

$$g^+(X) = \sum_{x\in X} g^+(x) \ \text{ and } \ g^-(X) = \sum_{x\in X} g^-(x).$$

A **flow** in a network $N$ with underlying digraph $D$, source $u$, sink $v$ and capacity function $c$ is a function $f: E(D) \to \mathbb{R}$ satisfying

$$0 \le f(a) \le c(a) \text{ for every arc } a \text{ of } D \qquad (7.2)$$

such that

$$f^+(x) = f^-(x) \text{ for each intermediate vertex } x \text{ of } D. \qquad (7.3)$$

If $f$ is a function on $E(D)$ defined by $f(a) = 0$ for every arc $a$ of $D$, then $f$ satisfies both (7.2) and (7.3) and so $f$ is a flow, called the **zero flow**.

The value $f(a) = f(x,y)$ of an arc $a = (x,y)$ is called the **flow along the arc** $a$ and can be interpreted as the rate at which material is transported along $a$ under the flow $f$. Condition (7.2) requires that the flow along $a$ cannot exceed the capacity of $a$. Condition (7.3) is referred to as the **conservation equation** and states that the rate at which material is transported into an intermediate vertex $x$ equals the rate at which material is transported out of $x$.

For a flow $f$ in a network $N$, the **net flow out of a vertex** $x$ is defined by

$$f^+(x) - f^-(x), \qquad (7.4)$$

while the **net flow into** $x$ is

$$f^-(x) - f^+(x). \tag{7.5}$$

By the conservation equation, Equation (7.3), it follows that for every intermediate vertex $x$ of $D$, the net flow out of $x$ equals the net flow into $x$ and the common value of (7.4) and (7.5) is 0.

If $f(a) = c(a)$ for an arc $a$ in a network $N$, then the arc $a$ is said to be **saturated** with respect to the flow $f$. On the other hand, if $f(a) < c(a)$, then the arc $a$ is **unsaturated**. An example of a flow in a network is shown in Figure 7.2. The first number associated with an arc is its capacity and for each arc of $N$, the capacity of the arc is a fixed number, while the second number is the flow along the arc. In general, many flows are possible for a given network. While the arc $(x, t)$ is saturated for the flow $f$ shown in the network in Figure 7.2 since $f(x, t) = c(x, t)$, the arc $(w, y)$ is unsaturated since $f(w, y) < c(w, y)$. In this example, the net flow out of the source $u$ is 3 and the net flow into the sink $v$ is also 3. As we will soon see, that these two numbers are equal is true in general.

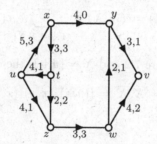

**Figure 7.2.** A flow in a network

**Theorem 7.1.** *Let $u$ and $v$ be the source and sink, respectively, of a network $N$ with underlying digraph $D$ and let $f$ be a flow defined on $N$. Then the net flow out of $u$ equals the net flow into $v$, that is,*

$$f^+(u) - f^-(u) = f^-(v) - f^+(v).$$

*Proof.* Since $f^+(V(D)) = f^-(V(D))$, it follows that

$$\sum_{x \in V(D)} f^+(x) = \sum_{x \in V(D)} f^-(x). \tag{7.6}$$

By (7.3),

$$f^+(x) = f^-(x) \text{ when } x \neq u, v. \tag{7.7}$$

By (7.6) then,

$$f^+(u) + f^+(v) = f^-(u) + f^-(v),$$

giving the desired result. ∎

## ⬡ Maximum flows

The **value of a flow** $f$ in a network $N$, denoted by val($f$), is defined as the net flow out of the source of $N$. By Theorem 7.1, val($f$) is also the net flow into the sink of $N$. For the flow $f$ defined on the network in Figure 7.2, we have val($f$) = 3.

There are certain flows of particular and obvious interest to us. A flow in a network $N$ whose value is maximum among all flows that can be defined on $N$ is called a **maximum flow**. Thus, a flow $f$ defined on $N$ is a maximum flow if val($f$) $\geq$ val($f'$) for every flow $f'$ defined on $N$. For a given network, a major goal is to find a maximum flow. For the purpose of doing this, it will be convenient to introduce another concept.

Let $N$ be a network with underlying digraph $D$, source $u$, sink $v$ and capacity function $c$. For a set $X$ of vertices in $D$, let $\overline{X} = V(D) \setminus X$. A **cut** in $N$ is a set of arcs of the form $[X, \overline{X}]$, where $u \in X$ and $v \in \overline{X}$. If $K = [X, \overline{X}]$ is a cut in $N$, then the **capacity** of $K$, denoted by cap($K$), is

$$\text{cap}(K) = c(X, \overline{X}) = \sum_{(x,y) \in [X, \overline{X}]} c(x, y).$$

For the network $N$ of Figure 7.2 and $X = \{u, x\}$, the cut

$$K = [X, \overline{X}] = \{(u, z), (x, y), (x, t)\}$$

in $N$ has capacity

$$\text{cap}(K) = c(u, z) + c(x, y) + c(x, t) = 4 + 4 + 3 = 11.$$

If $K$ is a cut in a network $N$, then any path from the source $u$ to the sink $v$ must contain at least one arc of $K$. Consequently, if all arcs of $K$ were removed from the underlying digraph $D$ of $N$, then there would be no path from $u$ to $v$. So just as a vertex-cut in a graph $G$ separates some pair of vertices in $G$ and an edge-cut in $G$ separates some pair of vertices in $G$, a cut in the underlying digraph $D$ of a network $N$ separates $u$ and $v$ in a certain sense.

Let $N$ be a network with underlying digraph $D$, source $u$ and sink $v$. For a set $X$ of vertices of $D$ with $u \in X$ and $v \in \overline{X}$ and a flow $f$ defined on $N$, the **net flow out of $X$** is $f^+(X) - f^-(X)$ and the **net flow into $X$** is $f^-(X) - f^+(X)$. It then follows (see Exercise 7.7) that

$$f^+(X) - f^-(X) = f(X, \overline{X}) - f(\overline{X}, X). \tag{7.8}$$

For the set $X = \{u, x, t\}$ in the network $N$ in Figure 7.2, $f^+(X) = 10$ and $f^-(X) = 7$. For this network then, the net flow out of $X$ is $f^+(X) - f^-(X) = 10 - 7 = 3 = \text{val}(f)$. We now show that for a network $N$ and any cut $K = [X, \overline{X}]$ in $N$, the value of any flow in $N$ is the net flow out of $X$ and that this value never exceeds the capacity of $K$.

**Theorem 7.2.** *Let* $f$ *be a flow in a network* $N$ *and let* $K = [X, \overline{X}]$ *be a cut in* $N$. *Then*

$$\text{val}(f) = f^+(X) - f^-(X) \leq \text{cap}(K).$$

*Proof.* Let $D$ be the underlying digraph of $N$, let $u$ and $v$ be the source and sink, respectively, of $N$ and let $c$ be the capacity function of $N$.

Since $f^+(x) - f^-(x) = 0$ for every $x \in X \setminus \{u\}$, it follows that

$$\sum_{x \in X} (f^+(x) - f^-(x)) = f^+(u) - f^-(u) = \text{val}(f).$$

Furthermore,

$$\sum_{x \in X} (f^+(x) - f^-(x)) = \sum_{x \in X} f^+(x) - \sum_{x \in X} f^-(x) = f^+(X) - f^-(X)$$

and so $\text{val}(f) = f^+(X) - f^-(X)$. Since $0 \leq f(a) \leq c(a)$ for every arc $a$ of $N$, it follows that

$$\begin{aligned} \text{val}(f) &= f^+(X) - f^-(X) = f(X, \overline{X}) - f(\overline{X}, X) \\ &\leq f(X, \overline{X}) \leq c(X, \overline{X}) = \text{cap}(K), \end{aligned}$$

giving the desired result. ∎

## ⊳ Minimum cuts

There are, in general, many cuts in a network $N$ and each cut has a capacity. Any cut in $N$ whose capacity is minimum among all cuts in $N$ is called a **minimum cut**. That is, a cut $K$ in $N$ is a minimum cut if $\text{cap}(K) \leq \text{cap}(K')$ for every cut $K'$ in $N$. The following two corollaries provide some important information about minimum cuts and maximum flows.

**Corollary 7.3.** *If* $f$ *is a flow in a network* $N$ *and* $K$ *is a cut in* $N$ *such that* $\text{val}(f) = \text{cap}(K)$, *then* $f$ *is a maximum flow and* $K$ *is a minimum cut in* $N$.

*Proof.* If $f^*$ is a maximum flow in $N$ and $K^*$ is a minimum cut, then $\text{val}(f^*) \leq \text{cap}(K^*)$ by Theorem 7.2. Consequently,

$$\text{val}(f) \leq \text{val}(f^*) \leq \text{cap}(K^*) \leq \text{cap}(K). \tag{7.9}$$

Since $\text{val}(f) = \text{cap}(K)$, it follows that there is equality throughout (7.9) and so $\text{val}(f) = \text{val}(f^*)$ and $\text{cap}(K^*) = \text{cap}(K)$, that is, $f$ is a maximum flow and $K$ is a minimum cut. ∎

**Corollary 7.4.** *If* $f$ *is flow in a network* $N$ *with capacity function* $c$ *and* $[X, \overline{X}]$ *is a cut in* $N$ *such that*

$$f(a) = c(a) \text{ for all } a \in [X, \overline{X}]$$

*and*

$$f(a) = 0 \quad for \ all \ a \in [\overline{X}, X],$$

*then $f$ is a maximum flow in $N$ and $[X, \overline{X}]$ is a minimum cut.*

Corollary 7.4 (see Exercise 7.8) suggests how the values of a flow $f$ should be defined on the arcs of a minimum cut in order for $f$ to be a maximum flow.

A network $N$ with source $u$ and sink $v$ is shown in Figure 7.3 together with a flow $f$ defined on $N$. As always, the first number associated with an arc $a$ is its capacity $c(a)$ and the second number is the flow $f(a)$. If $X = \{u, x, y\}$, then $K = [X, \overline{X}]$ is a cut in $N$. Since $f(a) = c(a)$ for all $a \in [X, \overline{X}]$ and $f(a) = 0$ for all $a \in [\overline{X}, X]$, it follows by Corollary 7.4 that $f$ is a maximum flow and $K$ is a minimum cut. Since $\text{cap}(K) = 4$, the value of the maximum flow $f$ is 4.

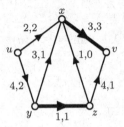

**Figure 7.3.** A cut $K = [X, \overline{X}]$ in a network, where $X = \{u, x, y\}$

# 7.2   Max-flow min-cut theorem

According to Corollary 7.3, if it should ever occur that the value of some flow $f$ in a network $N$ equals the capacity of some cut $K$ in $N$, then $f$ must be a maximum flow and $K$ is a minimum cut. In this section, we prove the converse of this result; that is, we show that the value of a maximum flow $f$ is equal to the capacity of a minimum cut $K$. In preparation for proving this, some additional terminology is useful.

For a digraph $D$, an **$x$-$y$ semipath** in $D$ is an alternating sequence

$$P: x = w_0, a_1, w_1, a_2, w_2, \ldots, w_{k-1}, a_k, w_k = y$$

of distinct vertices and arcs of $D$ beginning with $x$ and ending with $y$ such that either $a_i = (w_{i-1}, w_i)$ or $a_i = (w_i, w_{i-1})$ for each $i$ ($1 \le i \le k$). In this case, $(w_{i-1}, w_i)$ is called a **forward arc** of $P$ and $(w_i, w_{i-1})$ is a **backward arc** of $P$. Hence, when proceeding from $x$ to $y$ along the semipath $P$ in $D$, we move in the direction of a forward arc on $P$ and move opposite to the direction of a backward arc on $P$.

Let $N$ be a network with underlying digraph $D$ and capacity function $c$ and on which is defined a flow $f$. Recall that an arc $a$ is unsaturated if $f(a) < c(a)$. A semipath

$$P\colon w_0, a_1, w_1, a_2, \ldots, w_{k-1}, a_k, w_k$$

in $D$ is said to be **$f$-unsaturated** if for each $i$ $(1 \le i \le k)$,
  (i) $a_i$ is unsaturated whenever $a_i$ is a forward arc and
  (ii) $f(a_i) > 0$ whenever $a_i$ is a backward arc.
A trivial semipath in $D$ is vacuously $f$-unsaturated. If $P$ is an $f$-unsaturated $u$-$v$ semipath in $D$ where $u$ and $v$ are the source and sink, respectively, then $P$ is called an **$f$-augmenting semipath**. As we will see, given an $f$-augmenting semipath in the underlying digraph $D$ of a network $N$, it is possible to augment (alter) the values of the flow $f$ on each arc of $P$ to obtain a new flow $f'$ whose value exceeds that of $f$. For example, consider the flow $f$ in the network $N$ in Figure 7.2. Then

$$P\colon u, (t, u), t, (x, t), x, (x, y), y, (y, v), v$$

is an $f$-augmenting semipath.

Recall that the value of the flow $f$ defined on the network $N$ in Figure 7.2 is $f(u, x) + f(u, z) - f(t, u) = 3 + 1 - 1 = 3$. Let $f'$ be the function defined on $E(D)$ by redefining the flow of each arc on the $f$-augmenting semipath as follows:

$$f'(a) = \begin{cases} f(a) + 1 & \text{if } a \in \{(x, y), (y, v)\} \\ f(a) - 1 & \text{if } a \in \{(t, u), (x, t)\} \\ f(a) & \text{otherwise.} \end{cases}$$

Then $f'$ is also a flow in $N$ and $\mathrm{val}(f') = 4$. Consequently, $f$ is not a maximum flow. In fact, $f'$ is not a maximum flow either. That $f$ is not a maximum flow and that $N$ contains an $f$-augmenting semipath is not a coincidence, as Ford and Fulkerson [94] showed.

**Theorem 7.5.** *Let $N$ be a network with underlying digraph $D$. A flow $f$ in $N$ is a maximum flow if and only if there is no $f$-augmenting semipath in $D$.*

*Proof.* Let $u$ and $v$ be the source and sink, respectively, of $N$ and let $c$ be the capacity function. Suppose first that $D$ contains an $f$-augmenting semipath

$$P\colon u = w_0, a_1, w_1, a_2, w_2, \ldots, w_{k-1}, a_k, w_k = v.$$

Then, for $1 \le i \le k$,
  (a) $f(a_i) < c(a_i)$ whenever $a_i$ is a forward arc and
  (b) $f(a_i) > 0$ whenever $a_i$ is a backward arc.
If there is at least one forward arc $a_i = (w_{i-1}, w_i)$ on $P$, let $\epsilon_1$ be the minimum value of $c(a_i) - f(a_i)$ over all forward arcs $a_i$. If there is at least one backward arc $a_i = (w_i, w_{i-1})$ on $P$, let $\epsilon_2$ be the minimum value of $f(a_i)$ over all such backward arcs $a_i$. If only one of $\epsilon_1$ and $\epsilon_2$ is defined, then denote this number by $\epsilon$; otherwise, let $\epsilon = \min(\epsilon_1, \epsilon_2)$.

Define a function $f'$ on $E(D)$ by

$$f'(a) = \begin{cases} f(a) + \epsilon & \text{if } a \text{ is a forward arc on } P \\ f(a) - \epsilon & \text{if } a \text{ is a backward arc on } P \\ f(a) & \text{if } a \notin E(P). \end{cases}$$

Then $f'$ is also a flow. Since $\text{val}(f') = \text{val}(f) + \epsilon$, it follows that $f$ is not a maximum flow.

We now turn to the converse. Assume that there is no $f$-augmenting semipath in $D$. Let $X$ be the set of all vertices $x$ in $D$ for which there exists an $f$-unsaturated $u$-$x$ semipath. Then $u \in X$ and, by assumption, $v \notin X$. Thus, $K = [X, \overline{X}]$ is a cut in $N$.

Let $(y, z)$ be an arc in $K$. Since $y \in X$, there exists an $f$-unsaturated $u$-$y$ · semipath $P$ in $D$. Since $z \in \overline{X}$, there is no $f$-unsaturated $u$-$z$ semipath in $D$, which implies that $f(y, z) = c(y, z)$. Similarly, if $(w, x) \in [\overline{X}, X]$, then $f(w, x) = 0$. Since $K = [X, \overline{X}]$ is a cut such that $f(a) = c(a)$ for every arc $a \in [X, \overline{X}]$ and $f(a) = 0$ for every arc $a \in [\overline{X}, X]$, it follows by Corollary 7.4 that $f$ is a maximum flow. ∎

With the aid of Theorem 7.5, Ford and Fulkerson [94] proved a famous result in 1956 that is known as the **max-flow min-cut theorem**. Independently, and also in 1956, Elias, Feinstein and Shannon [74] discovered and proved the very same result.

**Theorem 7.6 (Max-flow min-cut theorem).** *In any network, the value of a maximum flow equals the capacity of a minimum cut.*

*Proof.* Let $f$ be a maximum flow in a network $N$ having capacity function $c$. By Theorem 7.5, there is no $f$-augmenting semipath in the underlying digraph $D$ of $N$. By the proof of Theorem 7.5, since $D$ contains no $f$-augmenting semipath, there is a minimum cut $K = [X, \overline{X}]$ such that

$$f(a) = \begin{cases} c(a) & \text{if } a \in K \\ 0 & \text{if } a \in [\overline{X}, X]. \end{cases}$$

Therefore, by (7.8),

$$\begin{aligned} \text{val}(f) &= f^+(X) - f^-(X) = f(X, \overline{X}) - f(\overline{X}, X) \\ &= c(X, \overline{X}) - 0 = \text{cap}(K), \end{aligned}$$

as desired. ∎

As an illustration of Theorem 7.5 and its proof, consider the network shown in Figure 7.4 where the labels on each arc are the capacity $c(a)$ and the flow $f(a)$, respectively.
Now

$$P: u, (u, r), r, (r, y), y, (y, v), v$$

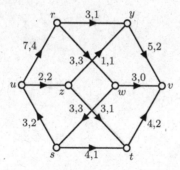

**Figure 7.4.** A flow $f$ in a network.

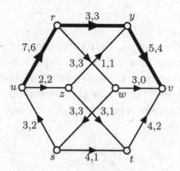

**Figure 7.5.** The $f$-augmenting semipath and the new flow $f'$ in the network

is an $f$-augmenting semipath with $\epsilon = 2$. (Thus $f$ is not a maximum flow by Theorem 7.5 since $P$ is an $f$-augmenting semipath.) The new flow $f'$ is shown in Figure 7.5.

If there is no $f'$-augmenting semipath, then by Theorem 7.5 we may conclude that $f'$ is a maximum flow. However,

$$P': u, (s, u), s, (w, s), w, (w, v), v$$

is an $f'$-augmenting semipath again with $\epsilon = 2$ and a new flow $f''$ can be obtained from $f'$ by augmenting the flow of each arc of $P'$ by $\epsilon = 2$, increasing the flow by $\epsilon = 2$ along each forward arc of $P'$ and decreasing the flow along each backward arc by $\epsilon = 2$, as shown in Figure 7.6.

Now there is no $f''$-augmenting semipath so that by Theorem 7.5 $f''$ is a maximum flow in $N$. Furthermore, following the second half of the proof of Theorem 7.5, we may construct a minimum cut. Let $X$ be the set of all vertices $x$ for which there is an $f''$-unsaturated $u$-$x$ semipath. Thus, $X = \{u, r\}$. Then, $K = [X, \overline{X}]$ is a minimum cut and as expected, $\text{cap}(K) = \text{val}(f'') = 8$.

The procedure just outlined is a somewhat simplified version of the algorithm given by Ford and Fulkerson [95] for finding a maximum flow in a network in which all the capacities are rational numbers. Ford and Fulkerson showed,

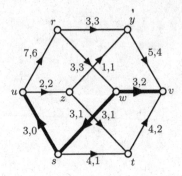

**Figure 7.6.** The $f'$-augmenting semipath and the new flow $f''$ in the network

with a rather complicated example, that their procedure might not terminate if the capacities are irrational. Even if all the capacities are rational, which is a reasonable expectation for networks encountered in discrete mathematics, their algorithm may not be efficient. Consider the network $N$ in Figure 7.7, where the labels on the arcs indicate their capacities and $C$ is any constant.

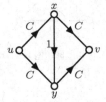

**Figure 7.7.** A network $N$

If we unwisely always choose augmenting semipaths that include the only arc with capacity equal to 1 (which is possible in the original Ford–Fulkerson Algorithm), then the maximum flow is $2C$, no matter what constant $C$ we choose, and the number of iterations of the algorithm is $2C$ if we start with the zero flow. Of course, optimally, the algorithm takes just two iterations.

A slight refinement of the Ford–Fulkerson Algorithm, first suggested by Edmonds and Karp [70], always selects a shortest augmenting semipath. This refined algorithm runs in polynomial time, even with irrational capacities.

# 7.3  Menger's theorems for digraphs

Since the vertex and edge forms of Menger's theorem, Theorems 4.18 and 4.20, both deal with sets that separate two vertices in a graph, it is perhaps not surprising that these two theorems are closely related to the max-flow min-cut theorem which deals with deleting the arcs of a cut, thereby separating the source and sink of a network.

Recall that the vertex form of Menger's theorem states that if $u$ and $v$ are nonadjacent vertices of a graph $G$, then the maximum number of internally disjoint $u$-$v$ paths in $G$ equals the minimum number of vertices that separate $u$ and $v$. That is, Menger's theorem is a "max-min" theorem. In Chapter 4, we saw that there are several forms of Menger's theorem, all of which have analogues for digraphs. In fact, the proofs given of these results in Chapter 4 apply to digraphs as well. All of these results can be proved, either directly or indirectly, using the max-flow min-cut theorem. In each case, the object is to construct an appropriate network from the given graph or digraph.

While we have defined separating sets of vertices and separating sets of edges in a graph, we need analogous terminology for digraphs. Let $D$ be a digraph and let $u$ and $v$ be two nonadjacent vertices of $D$. A set $S \subseteq V(D) \setminus \{u, v\}$ is said to be a **$u$-$v$ separating set of vertices** if every $u$-$v$ path in $P$ contains at least one vertex of $S$. A set $S$ of arcs in $D$ is a **$u$-$v$ separating set of arcs** if every $u$-$v$ path in $P$ contains at least one arc of $S$. We now present the arc form of Menger's theorem and use the max-flow min-cut theorem (Theorem 7.6) in its proof.

**Theorem 7.7 (Menger's theorem, arc form).** *For distinct vertices $u$ and $v$ in a digraph $D$, the maximum number of pairwise arc-disjoint $u$-$v$ paths in $D$ equals the minimum number of arcs in a $u$-$v$ separating set of arcs.*

*Proof.* Suppose that the maximum number of $u$-$v$ paths in a collection of pairwise arc-disjoint $u$-$v$ paths in $D$ is $k$ and the minimum number of arcs in a $u$-$v$ separating set $S$ is $\ell$. Since each of these $u$-$v$ paths contains at least one arc of $S$ and no arc in $S$ belongs to more than one such $u$-$v$ path, it follows that $k \leq \ell$. It remains to show that $\ell \leq k$.

We now construct a network $N$ with underlying digraph $D$, source $u$ and sink $v$ by defining a capacity function $c$ on $E(D)$ such that $c(a) = 1$ for each arc $a$ of $D$. By the max-flow min-cut theorem (Theorem 7.6), the value of a maximum flow in $N$ equals the capacity of a minimum cut. Let $f$ be a maximum flow in $N$ and $K$ a minimum cut. Thus, $\text{cap}(K) = \text{val}(f)$. We next show that $\ell \leq \text{cap}(K)$ and $\text{val}(f) \leq k$, from which it will follow that $\ell \leq k$ and so $\ell = k$.

Since $K$ is a minimum cut in $N$, the set $K$ is a $u$-$v$ separating set of arcs in $D$. Therefore, $\ell \leq |K| = \text{cap}(K)$.

Since $f$ is an integer-valued function defined on $E(D)$ such that $0 \leq f(a) \leq c(a)$ for every arc $a$ of $D$, it follows that either $f(a) = 0$ or $f(a) = 1$ for every arc $a$ of $D$. Let $D_1$ be the digraph obtained from $D$ by deleting all arcs $a$ from $D$ for which $f(a) = 0$. Consequently, $f(a) = 1$ for each arc $a$ of $D_1$. Since $f$ is a flow in $D$, it follows by (7.7) that

$$f^+(x) = f^-(x) \quad \text{for each } x \in V(D) \setminus \{u, v\}$$

and that
$$f^+(u) - f^-(u) = \text{val}(f) = f^-(v) - f^+(v).$$
However, for each vertex $x$ in $D$,
$$f^+(x) = \text{od}_{D_1} x \quad \text{and} \quad f^-(x) = \text{id}_{D_1} x.$$
Therefore,
$$\text{od}_{D_1} w = \text{id}_{D_1} w \quad \text{if } w \in V(D) \setminus \{u, v\}$$
and
$$\text{od}_{D_1} u - \text{id}_{D_1} u = \text{val}(f) = \text{id}_{D_1} v - \text{od}_{D_1} v.$$
Since $\text{od}_{D_1} u = \text{id}_{D_1} u + \text{val}(f)$ and $\text{id}_{D_1} v = \text{od}_{D_1} v + \text{val}(f)$, it follows by Exercise 3.20 in Chapter 3 that the digraph $D_1$ and $D$ as well contain $\text{val}(f)$ arc-disjoint $u$-$v$ paths and so $k \geq \text{val}(f)$. Therefore, $\ell = k$. ∎

With the aid of the proof of Theorem 7.7, we can determine the maximum number of arc-disjoint $u$-$v$ paths in the digraph $D$ of Figure 7.8(a). Define a capacity function $c$ on $E(D)$ such that $c(a) = 1$ for each arc $a$ of $D$. We then have a network $N$ with underlying digraph $D$, source $u$, sink $v$ and capacity function $c$. Using the proof of Theorem 7.5, we obtain a maximum flow $f$ and minimum cut $K = [X, \overline{X}]$, where $X = \{u, u_4, u_5\}$, so that $\text{val}(f) = \text{cap}(K) = 3$. See Figure 7.8(b). Hence,

$$K = \{(u, u_1), (u, u_2), (u_5, v)\}.$$

Consequently, the maximum number of arc-disjoint $u$-$v$ paths in $D$ is 3. Three such paths are

$$P: u, u_1, u_2, u_3, v, \quad P': u, u_2, v \text{ and } P'': u, u_5, v.$$

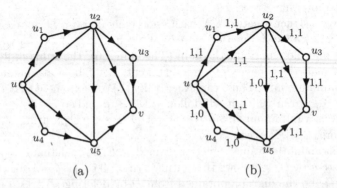

(a)                                        (b)

**Figure 7.8.** Determining the maximum number of internally disjoint $u$-$v$ paths in a digraph

Next we use the max-flow min-cut theorem (Theorem 7.6) to prove the directed vertex form of Menger's theorem.

**Theorem 7.8 (Menger's theorem, directed vertex form).** *Let $D$ be a digraph and let $u$ and $v$ be distinct vertices of $D$ such that $(u,v) \notin E(D)$. Then the maximum number of internally disjoint $u$-$v$ paths in $D$ equals the minimum number of vertices in a $u$-$v$ separating set of vertices of $D$.*

*Proof.* Suppose that the maximum number of $u$-$v$ paths in a collection of internally disjoint $u$-$v$ paths in $D$ is $k$ and the minimum number of vertices in a $u$-$v$ separating set $S$ of vertices of $D$ is $\ell$. Since each of these $u$-$v$ paths contains a vertex of $S$ and no vertex of $S$ lies on more than one such path, it follows that $k \leq \ell$. It remains to show that $\ell \leq k$.

Construct a new digraph $D'$ from $D$ by replacing each vertex $t \neq u, v$ by two new vertices $t'$ and $t''$ and the arc $(t', t'')$. Suppose that $x$ and $y$ are two vertices of $D$ such that neither $x$ nor $y$ is $u$ or $v$, that is, $x, y \notin \{u, v\}$. If $(x, y)$ is an arc of $D$, then $(x, y)$ is replaced by $(x'', y')$. If $(u, x)$ is an arc of $D$, then $(u, x)$ is replaced by $(u, x')$. If $(x, u)$ is an arc of $D$, then $(x, u)$ is replaced by $(x'', u)$. If $(x, v)$ is an arc of $D$, then $(x, v)$ is replaced by $(x'', v)$; while if $(v, x)$ is an arc of $D$, then $(v, x)$ is replaced by $(v, x')$.

Suppose that $k'$ is the maximum number of pairwise arc-disjoint $u$-$v$ paths in $D'$ and $\ell'$ is the minimum number of arcs in a $u$-$v$ separating set of arcs in $D'$. By Theorem 7.7, $\ell' = k'$. We show that $\ell \leq \ell'$ and $k' \leq k$, which will give us the desired inequality $\ell \leq k$.

Let $A$ be a $u$-$v$ separating set of arcs in $D'$ with $|A| = \ell'$. Thus, $A$ contains no arc of the form $(z, u)$ or $(v, z)$ where $z \neq u, v$. Each arc $a \in A$ is either of the form

$$a = (u, x'), \; a = (x', x''), \; a = (y'', x') \text{ or } a = (x'', v)$$

for some vertices $x, y \in V(D) \setminus \{u, v\}$. Regardless of what the arc $a$ is, denote the vertex $x$ involved in $a$ by $w_a$. Furthermore, let

$$W = \{w_a : a \in A\} \text{ so that } W \subseteq V(D) \setminus \{u, v\}$$

and $|W| \leq |A| = \ell'$. Since $A$ is a $u$-$v$ separating set of arcs in $D'$, the set $W$ is a $u$-$v$ separating set of vertices in $D$. Thus, $\ell \leq |W| \leq \ell'$ and so $\ell \leq \ell'$.

Next, let $P_1', P_2', \ldots, P_{k'}'$ be a collection of $k'$ pairwise arc-disjoint $u$-$v$ paths in $D'$. Then each path $P_i'$ $(1 \leq i \leq k)$ is of the form

$$P_i' : u, x_1', x_1'', x_2', x_2'', \ldots, x_r', x_r'', v$$

and gives rise to the path

$$P_i : u, x_1, x_2, \ldots, x_r, v.$$

From the manner in which the digraph $D'$ is constructed and the fact that $P_1', P_2', \ldots, P_{k'}'$ are pairwise arc-disjoint, it follows that $P_1, P_2, \ldots, P_{k'}$ are internally disjoint $u$-$v$ paths in $D$. Thus, $k \geq k'$. Hence, $\ell \leq \ell' = k' \leq k$ and so $\ell = k$. ∎

As an illustration of the proof of Theorem 7.8, consider the digraph $D$ of Figure 7.9. We seek the maximum number of internally disjoint $u$-$v$ paths in $D$. Although this is rather easy to do in this case, we construct the digraph $D'$ described in the proof of Theorem 7.8 and determine the maximum number of pairwise arc-disjoint $u$-$v$ paths in $D'$. This number is 2 and two such paths are

$$P_1': u, w', w'', z', z'', v \quad \text{and} \quad P_2': u, y', y'', x', x'', v.$$

As described in the proof of Theorem 7.8, this gives rise to the paths

$$P_1: u, w, z, v \quad \text{and} \quad P_2: u, y, x, v,$$

which constitute a maximum set of internally disjoint $u$-$v$ paths in $D$.

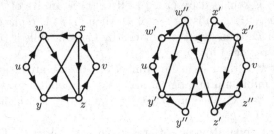

**Figure 7.9.** Illustrating the proof of Theorem 7.8

 **7.4**   **A connection to coloring**

As we will see in this section, there is a strong connection between the region colorings considered in Section 6.4 and a certain type of flow. Consider an oriented graph $D$ and a function $\phi : E(D) \to \mathbb{Z}$ for which

$$\phi^+(v) = \phi^-(v) \tag{7.10}$$

for every vertex $v$ of $D$. Such a function is called a **flow** on the oriented graph $D$. The property (7.10) of $\phi$ is called the **conservation property**, just as in Equation (7.3). Here, however, Equation (7.10) must hold for *every* vertex $v$ of $D$.

For an integer $k \geq 2$, if a flow $\phi$ on an oriented graph $D$ has the property that $|\phi(e)| < k$ for every arc $e$ of $D$, then $\phi$ is called a **$k$-flow** on $D$. Furthermore, if $0 < |\phi(e)| < k$ for every arc $e$ of $D$ (that is, $\phi(e)$ is never 0), then $\phi$ is called a **nowhere-zero $k$-flow** on $D$. Hence, a nowhere-zero $k$-flow on $D$ has the property that

$$\phi(e) \in \{\pm 1, \pm 2, \ldots, \pm(k-1)\}$$

for every arc $e$ of $D$.

If an orientation $D$ of a graph $G$ has a nowhere-zero $k$-flow $\phi$ and $D'$ is the orientation of $G$ obtained by replacing some arc $(u,v)$ with the arc $(v,u)$, then $D'$ has the nowhere-zero $k$-flow $\phi_{D'}$ defined by

$$\phi_{D'}(e) = \begin{cases} \phi(e) & \text{if } e \neq (u,v) \\ -\phi(e) & \text{if } e = (u,v) \end{cases}.$$

Thus if some orientation of $G$ has a nowhere-zero $k$-flow, then so does *every* orientation of $G$. Consequently, when we say that a graph $G$ has a nowhere-zero $k$-flow, we mean, in fact, that every orientation of $G$ has a nowhere-zero $k$-flow.

Nowhere-zero flows in planar graphs are of particular interest because of their relationship to region colorings. Let $G$ be a bridgeless plane graph, with the edges of $G$ oriented arbitrarily, and let $c$ be a $k$-region coloring of $G$. Thus for each region $R$ of $G$, the region $R$ is colored $c(R)$, where $c(R) \in \{1, 2, \ldots, k\}$. For each oriented edge $e = (u,v)$ of $G$, define $\phi(e)$ to be $c(R_1) - c(R_2)$, where $R_1$ is the region to the right $e = (u,v)$ as we travel along $e$ from $u$ to $v$ and $R_2$ is the region to the left of $e$. In Figure 7.10, an example of this construction is given for the orientation $D$ of plane graph $G$ whose regions have been colored with colors 1, 2 and 3.

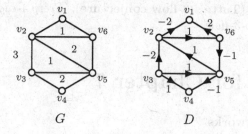

**Figure 7.10.** Flow construction

It is straightforward to verify that the integer-valued function $\phi$ defined above on the bridgeless plane graph $G$ is a nowhere-zero $k$-flow (see Exercise 7.20). Thus if a bridgeless plane graph $G$ is $k$-region colorable, then $G$ has a nowhere-zero $k$-flow. Therefore, another consequence of the four color theorem can be given; this is a flow analogue of Theorem 6.32.

**Corollary 7.9.** *Every bridgeless planar graph has a nowhere-zero 4-flow.*

## ▷ Tutte's flow conjectures

In general, no graph with a bridge has a nowhere-zero $k$-flow for any positive integer $k$ (see Exercise 7.21). For bridgeless graphs, Tutte [238] conjectured in 1954

that there exists an integer $k$ so that every bridgeless graph has a nowhere-zero
$k$-flow. This conjecture was proved independently in the 1970s by Kilpatrick [144]
and Jaeger [134], who showed that every bridgeless graph has a nowhere-zero
8-flow. A few years later, their result was improved by Seymour [219].

**Theorem 7.10.** *Every bridgeless graph has a nowhere-zero 6-flow.*

Tutte, however, had conjectured that $k = 5$ suffices, and this remains open.

**Conjecture 7.11 (Tutte's 5-flow conjecture).** *Every bridgeless graph has a*
*nowhere-zero 5-flow.*

Over a decade after making his 5-flow conjecture, Tutte [241] conjectured that
the Petersen graph is essentially the only obstruction to a nowhere-zero 4-flow.

**Conjecture 7.12 (Tutte's 4-flow conjecture).** *Every bridgeless graph without*
*a Petersen minor has a nowhere-zero 4-flow.*

It can be shown that a cubic graph has a nowhere-zero 4-flow if an only if
it is 3-edge-colorable. Thus in the special case of cubic graphs, Tutte's 4-flow
conjecture is equivalent to his snark conjecture (Conjecture 6.34), which itself
would be a strengthening of the four color theorem.

In 1972 [222, footnote 20], Tutte also made a conjecture about 3-flows.

**Conjecture 7.13 (Tutte's 3-flow conjecture).** *Every 4-edge-connected graph*
*has a nowhere-zero 3-flow.*

# Exercises for Chapter 7

### Section 7.1.  Networks

**7.1.** Let $N$ be the network with source $u$ and sink $v$ shown in Figure 7.11, where
each arc is labeled with its capacity.
  (a) Show that no flow can have a value exceeding 9.
  (b) Give an example of a flow $f$ on $N$ such that $\mathrm{val}(f) = 9$.

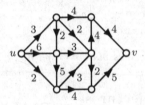

**Figure 7.11.** The network $N$ in Exercise 7.1

**7.2.** For the network $N$ shown in Figure 7.12 with source $u$ and sink $v$, each arc has unlimited capacity. A flow $f$ in the network is indicated by the labels on the arcs.

   (a) Determine the missing flows $a$, $b$ and $c$.

   (b) Determine val($f$).

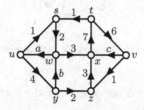

**Figure 7.12.** The network $N$ in Exercise 7.2

**7.3.** Assume that the network $N$ shown in Figure 7.13 has unlimited capacities. Give an example of a flow $f$ on $N$ where the flow along each arc is a positive integer and where the maximum of the flows along the arcs is as small as possible.

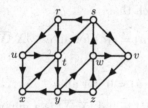

**Figure 7.13.** The network $N$ in Exercise 7.3

**7.4.** Let $N$ be a network with underlying digraph $D$, source $u$, sink $v$, capacity function $c$ and let $f$ be a flow on $N$. Suppose that $t \in V(D) \setminus \{u, v\}$ such that $\mathrm{id}\, t = 0$ and that $N'$ is obtained from $N$ by deleting $t$. Define a function $f'$ on $E(D - t)$ by $f'(x, y) = f(x, y)$ for all arcs $(x, y) \in E(D - t)$. Show that $f'$ is a flow on $N'$ and that val($f$) = val($f'$). Show that the same conclusion holds if $\mathrm{od}\, t = 0$.

**7.5.** A network $N$ with source $u$ and sink $v$ is shown in Figure 7.14, where each arc is labeled with its capacity. Describe a flow on $N$ where the flow along each arc is a positive integer.

   (a) Determine the net flow out of each vertex.

   (b) What is the maximum value of a flow $f$ in the network $N$?

   (c) Give an example of a minimum cut $[X, \overline{X}]$ and determine $f(X, \overline{X}) - f(\overline{X}, X)$.

**7.6.** Let $u$ and $v$ be two vertices of a digraph $D$ and let $A$ be a set of arcs of $D$ such that every $u$-$v$ path in $D$ contains at least one arc of $A$.

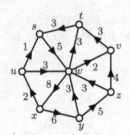

**Figure 7.14.** The network $N$ in Exercises 7.5 and 7.10

(a) Show that there exists a set of arcs of the form $[X, \overline{X}]$ where $u \in X$ and $v \in \overline{X}$ and $[X, \overline{X}] \subseteq A$.

(b) Show that $[X, \overline{X}]$ may be a proper subset of $A$.

**7.7.** Let $N$ be a network with underlying digraph $D$, source $u$ and sink $v$. For a set $X$ of vertices of $D$ with $u \in X$ and $v \in \overline{X}$ and a flow $f$ defined on $N$, prove that $f^+(X) - f^-(X) = f(X, \overline{X}) - f(\overline{X}, X)$.

**7.8.** Prove Corollary 7.4: if $f$ is flow in a network $N$ with capacity function $c$ and $[X, \overline{X}]$ is a cut in $N$ such that

$$f(a) = c(a) \text{ for all } a \in [X, \overline{X}]$$

and

$$f(a) = 0 \text{ for all } a \in [\overline{X}, X],$$

then $f$ is a maximum flow in $N$ and $[X, \overline{X}]$ is a minimum cut.

**7.9.** Show that the converse of Corollary 7.4 is true.

**7.10.** Let $N$ be the network $N$ of Figure 7.14 in Exercise 7.5.

(a) Show that $N$ has a flow $f$ other than the zero flow with $\mathrm{val}(f) = 0$.

(b) Discuss a sufficient condition for a network to have a flow $f$ other than the zero flow with $\mathrm{val}(f) = 0$.

## Section 7.2.  Max-flow min-cut theorem

**7.11.** Find a maximum flow $f$ and a minimum cut $K$ in the network $N$ in Figure 7.15.

**7.12.** Find a maximum flow $f$ and a minimum cut $K$ in the network $N$ in Figure 7.16.

**7.13.** Let $N$ be a network with capacity function $c$ and suppose that $[X, \overline{X}]$ is a minimum cut in $N$. Prove or disprove:

(a) If $f_1$ and $f_2$ are flows in $N$ that agree on $[X, \overline{X}]$ and $[\overline{X}, X]$, then $f_1$ and $f_2$ are maximum flows in $N$.

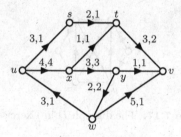

**Figure 7.15.** The network $N$ in Exercise 7.11

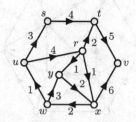

**Figure 7.16.** The network $N$ in Exercise 7.12

(b) If $f_1$ and $f_2$ are maximum flows in $N$, then $f_1$ and $f_2$ agree on $[X, \overline{X}]$ and $[\overline{X}, X]$.

**7.14.** Define a **generalized network** $N$ to be a digraph $D$ with two distinguished vertices $u$ and $v$ called the *source* and *sink*, respectively, together with two nonnegative integer-valued functions $c_1$ and $c_2$ on $E(D)$. A *flow* in $N$ is a real-valued function $f$ on $E(D)$ that satisfies (7.3) (that is, $f^+(x) = f^-(x)$ for each intermediate vertex $x$ of $D$) as well as

$$c_1(a) \le f(a) \le c_2(a) \text{ for every arc } a \text{ of } D. \qquad (7.11)$$

Give an example of a nontrivial generalized network $N$ that has no (legal) flow.

### Section 7.3. Menger's theorems for digraphs

**7.15.** Use the proof of Theorem 8.8 to find the maximum number of internally disjoint $u$-$v$ paths in the digraph $D$ shown in Figure 7.17.

**7.16.** Find the maximum number of internally disjoint $u$-$v$ paths in the digraph $D$ shown in Figure 7.18.

**7.17.** Define a **multisource/multisink network** $N$ to consist of a digraph $D$, two nonempty subsets $S$ and $T$ of vertices and a nonnegative real-valued function $c$ defined on $E(D)$. Then $D$ is called the *underlying digraph* of $N$, the vertices in $S$ are called the *sources* of $N$, the vertices in $T$ are called the *sinks* of $N$, and $c$ is called the *capacity function* of $N$. A *flow* in $N$ is a real-valued function $f$ on $E(D)$ satisfying (7.2) (that is, $0 \le f(a) \le c(a)$ for every arc $a$ of $D$) and

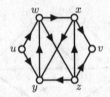

**Figure 7.17.** The digraph $D$ in Exercise 7.15

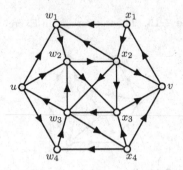

**Figure 7.18.** The digraph $D$ in Exercise 7.16

$$f^+(x) = f^-(x) \text{ for each } x \in V(D) \setminus S \setminus T.$$

Using the natural definition for maximum flow and minimum cut in a multi-source/multisink network, explain how the problem of determining a maximum flow can be reduced to the case of networks with a single source and sink.

**7.18.** For a given graph $G$, prove the edge form of Menger's theorem (Theorem 4.20) by applying Theorem 7.7 to the symmetric digraph $D$ whose underlying graph is $G$.

**7.19.** For a graph $G$, prove the vertex form of Menger's theorem (Theorem 4.18) for a graph $G$ by applying Theorem 7.8 to the symmetric digraph $D$ whose underlying graph is $G$.

## Section 7.4. A connection to coloring

**7.20.** Show that if a bridgeless plane graph is $k$-region colorable, then $G$ has a nowhere-zero $k$-flow.

**7.21.** Prove that no graph with a bridge has a nowhere-zero $k$-flow for any positive integer $k$.

**7.22.** Show that the Petersen graph does not have a nowhere-zero 4-flow.

**7.23.** Show that the Petersen graph has a nowhere-zero 5-flow.

# Factors and covers

**8**

In this chapter we study subsets of either vertices or edges with special properties. Some of these subsets deal with the idea of independence (in which every two elements in the set are nonadjacent). Such a set of edges is called a matching in a graph. Two other fundamental concepts are covers and domination which are also discussed in this chapter. Additional problems we consider deal with collections of subgraphs of a given graph $G$ such that each edge belongs to exactly one subgraph in the collection. Collections of subgraphs with this property are often divided into two categories, called decompositions and factorizations, depending on whether the subgraphs are also required to be spanning.

##  8.1  Matchings and 1-factors

Recall that a set of edges in a graph $G$ is **independent** if no two edges in the set are adjacent and that the **edge independence number** $\alpha'(G)$ is the maximum number of edges in an independent set of edges. If $\{e_1, e_2, \ldots, e_k\}$ is a set of independent edges in a graph $G$ where $e_i = u_i v_i$ for $1 \leq i \leq k$, then the vertices $u_1, u_2, \ldots, u_k$ are **matched** to the vertices $v_1, v_2, \ldots, v_k$, and thus we call an independent set of edges a **matching**.

A matching of maximum size in a graph $G$ is a **maximum matching**, and thus $\alpha'(G)$ is the number of edges in a maximum matching. In fact, $\alpha'(G)$ is sometimes referred to as the **matching number** of $G$. (We will see that the edge independence number has a connection with other parameters in Section 8.2.) In the graph $G$ of Figure 8.1, the set $M_1 = \{e_1, e_4\}$ is a matching that is not a maximum matching, while $M_2 = \{e_1, e_3, e_5\}$ and $M_3 = \{e_1, e_3, e_6\}$ are maximum matchings in $G$. Hence, $\alpha'(G) = 3$.

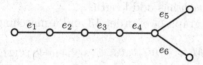

**Figure 8.1.** Matchings and maximum matchings

If $M$ is a matching in a graph $G$ with the property that every vertex of $G$ is incident with an edge of $M$, then $M$ is a perfect matching. Clearly, if $G$ has a perfect matching $M$, then $G$ has even order and the edge-induced subgraph $G[M]$ is a 1-regular spanning subgraph of $G$. Since the graph $G$ of Figure 8.1 has odd order, it cannot have a perfect matching.

If $M$ is a matching in a graph $G$, then every vertex of $G$ is incident with at most one edge of $M$. A vertex that is incident with no edges of $M$ is referred to as an **$M$-unmatched vertex** or simply an **unmatched vertex** if the matching $M$ is clear. The following theorem will prove to be useful.

**Theorem 8.1.** *Let $M_1$ and $M_2$ be matchings in a graph $G$. Then each component of the spanning subgraph $H$ of $G$ with $E(H) = (M_1 \setminus M_2) \cup (M_2 \setminus M_1)$ is one of the following:*

   *(i)  an isolated vertex,*

   *(ii)  an even cycle whose edges are alternately in $M_1$ and in $M_2$,*

   *(iii)  a nontrivial path whose edges are alternately in $M_1$ and in $M_2$ and such that each end-vertex of the path is either $M_1$-unmatched or $M_2$-unmatched but not both.*

*Proof.* First, we note that $\Delta(H) \leq 2$, for if $H$ contains a vertex $v$ such that $\deg_H v \geq 3$, then $v$ is incident with at least two edges in the same matching. Since $\Delta(H) \leq 2$, every component of $H$ is a path (possibly trivial) or a cycle. Since no two edges in a matching are adjacent, the edges of each cycle and path in $H$ are alternately in $M_1$ and in $M_2$. Thus each cycle in $H$ is even.

Suppose that $e = uv$ is an edge of $H$ and $u$ is the end-vertex of a path $P$ that is a component of $H$. The proof will be complete once we have shown that $u$ is $M_1$-unmatched or $M_2$-unmatched but not both. Since $e \in E(H)$, it follows that $e \in M_1 \setminus M_2$ or $e \in M_2 \setminus M_1$. If $e \in M_1 \setminus M_2$, then $u$ is not $M_1$-unmatched. We show that $u$ is $M_2$-unmatched. If this is not the case, then there is an edge $f$ in $M_2$ (thus $f \neq e$) such that $f$ is incident to $u$. Since $e$ and $f$ are adjacent, $f \notin M_1$. Thus, $f \in M_2 \setminus M_1 \subseteq E(H)$. This, however, is impossible since $u$ is the end-vertex of $P$. Therefore, $u$ is $M_2$-unmatched. Similarly, if $e \in M_2 \setminus M_1$, then $u$ is $M_1$-unmatched. ∎

In order to present a characterization of maximum matchings in a graph, we introduce two new terms. Let $M$ be a matching in a graph $G$. An **$M$-alternating path** of $G$ is a path whose edges are alternately in $M$ and not in $M$. An **$M$-augmenting path** is an $M$-alternating path $P$ both of whose end-vertices are unmatched, that is, the first and last edges of $P$ do not belong to $M$. Necessarily, every $M$-augmenting path has odd length.

Figure 8.2 shows a graph $G$ of order 13 and a matching

$$M = \{v_2v_4, v_3v_6, v_5v_8, v_7v_{10}, v_{11}v_{12}\}$$

of size 5, whose edges are indicated by bold lines. The path $P\colon v_1, v_2, v_4, v_7, v_{10}, v_9$ is then an $M$-augmenting $v_1$-$v_9$ path in $G$.

The following result is due to Berge [23].

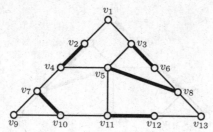

**Figure 8.2.** A matching $M$ in a graph $G$ and an $M$-augmenting path
$P\colon v_1, v_2, v_4, v_7, v_{10}, v_9$

**Theorem 8.2.** *A matching $M$ in a graph $G$ is a maximum matching if and only if $G$ contains no $M$-augmenting path.*

*Proof.* Let $M$ be a maximum matching in a graph $G$ and assume, to the contrary, that $G$ contains an $M$-augmenting path $P$. Let $M'$ denote the edges of $P$ belonging to $M$, and let $M'' = E(P) \setminus M'$. Since $|M''| = |M'| + 1$, the set $(M \setminus M') \cup M''$ is a matching having cardinality exceeding that of $M$, producing a contradiction.

For the converse, suppose that $M_1$ is a matching of a graph $G$ that is not a maximum matching. Let $M_2$ be a maximum matching in $G$ and let $H$ be the spanning subgraph of $G$ with

$$E(H) = (M_1 \setminus M_2) \cup (M_2 \setminus M_1).$$

Since $|M_2| > |M_1|$, it follows by Theorem 8.1 that there is a component $P$ of $H$ that is a path whose edges are alternately in $M_1$ and in $M_2$ such that $P$ contains more edges belonging to $M_2$ than to $M_1$. Necessarily, each end-vertex of $P$ is $M_1$-unmatched, which implies that $P$ is an $M_1$-augmenting path. ∎

Since the path $P\colon v_1, v_2, v_4, v_7, v_{10}, v_9$ in the graph $G$ of Figure 8.2 is an $M$-augmenting path for the matching

$$M = \{v_2v_4, v_3v_6, v_5v_8, v_7v_{10}, v_{11}v_{12}\},$$

it follows by Theorem 8.2 that $M$ is not a maximum matching in $G$. Following the proof of Theorem 8.2, we let $M' = E(P) \cap M$ and $M'' = E(P) \setminus M'$. Then $(M \setminus M') \cup M''$ is a matching whose size exceeds that of $M$. In fact, $(M \setminus M') \cup M''$ is a maximum matching in the graph $G$ of Figure 8.2 (see Figure 8.3).

Although matchings are of interest in all graphs, they are of particular interest in bipartite graphs. Let $G$ be a bipartite graph with partite sets $U$ and $W$, where $|U| \leq |W|$. Necessarily, for any matching of $k$ edges, $k \leq |U|$. If $|U| = k$, then $U$ is said to be **matched** to a subset of $W$.

## Hall's theorem

For a bipartite graph $G$ with partite sets $U$ and $W$ and for $S \subseteq U$, let $N(S)$ be the set of all vertices in $W$ having a neighbor in $S$. The condition that

$$|N(S)| \geq |S| \quad \text{for all } S \subseteq U$$

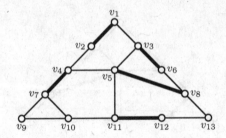

**Figure 8.3.** A maximum matching in a graph $G$

is referred to as **Hall's condition**. This condition is named for Philip Hall (1904–1982). The following 1935 theorem of Hall [118] shows that this condition is necessary and sufficient for one partite set of a bipartite graph to be matched to a subset of the other.

**Theorem 8.3 (Hall's theorem).** *Let $G$ be a bipartite graph with partite sets $U$ and $W$. Then $U$ can be matched to a subset of $W$ if and only if Hall's condition is satisfied.*

*Proof.* Suppose that $U$ can be matched to a subset of $W$ under a matching $M^*$. Then every nonempty subset $S$ of $U$ can be matched under $M^*$ to some subset of $W$, implying that $|N(S)| \geq |S|$; so, $U$ satisfies Hall's condition.

To verify the converse, suppose that there exists a bipartite graph $G$ for which Hall's condition is satisfied, but $U$ cannot be matched to a subset of $W$. Let $M$ be a maximum matching in $G$. By assumption, there is a vertex $u$ in $U$ that is unmatched. Let $X$ be the set of all vertices of $G$ that are connected to $u$ by an $M$-alternating path. Since $M$ is a maximum matching, it follows by Theorem 8.2 that $u$ is the only unmatched vertex in $X$.

Let $U_1 = X \cap U$ and let $W_1 = X \cap W$. Since no vertex of $X \setminus \{u\}$ is unmatched, it follows that $U_1 \setminus \{u\}$ is matched under $M$ to $W_1$. Therefore, $|W_1| = |U_1| - 1$ and $W_1 \subseteq N(U_1)$. Furthermore, for every $w \in N(U_1)$, the graph $G$ contains an $M$-alternating $u$-$w$ path and so $N(U_1) \subseteq W_1$. Thus, $N(U_1) = W_1$ and

$$|N(U_1)| = |W_1| = |U_1| - 1 < |U_1|.$$

This, however, contradicts the fact that Hall's condition is satisfied. ∎

If $G$ is a bipartite graph with partite sets $U$ and $W$ where $|U| \leq |W|$ and Hall's condition is satisfied, then by Theorem 8.3, $G$ contains a matching of size $|U|$, which is a maximum matching. If $|U| = |W|$, then such a matching is a perfect matching in $G$. The following result is another consequence of Theorem 8.3 (see Exercise 8.4).

**Theorem 8.4.** *Every $r$-regular bipartite graph $(r \geq 1)$ has a perfect matching.*

A collection $S_1, S_2, \ldots, S_n$, $n \geq 1$, of finite nonempty sets is said to have a **system of distinct representatives** if there exists a set $\{s_1, s_2, \ldots, s_n\}$ of distinct elements such that $s_i \in S_i$ for $1 \leq i \leq n$.

**Theorem 8.5.** *A collection* $\{S_1, S_2, \ldots, S_n\}$ *of finite nonempty sets has a system of distinct representatives if and only if for each integer* $k$ *with* $1 \leq k \leq n$, *the union of any* $k$ *of these sets contains at least* $k$ *elements.*

*Proof.* Assume first that $\{S_1, S_2, \ldots, S_n\}$ has a system of distinct representatives. Then, for each integer $k$ with $1 \leq k \leq n$, the union of any $k$ of these sets contains at least $k$ elements.

For the converse, suppose that $\{S_1, S_2, \ldots, S_n\}$ is a collection of $n$ sets such that for each integer $k$ with $1 \leq k \leq n$, the union of any $k$ of these sets contains at least $k$ elements. We now consider the bipartite graph $G$ with partite sets

$$U = \{S_1, S_2, \ldots, S_n\} \quad \text{and} \quad W = S_1 \cup S_2 \cup \cdots \cup S_n$$

such that a vertex $S_i$ $(1 \leq i \leq n)$ in $U$ is adjacent to a vertex $w$ in $W$ if $w \in S_i$. Let $X$ be any subset of $U$ with $|X| = k$, where $1 \leq k \leq n$. Since the union of any $k$ sets in $U$ contains at least $k$ elements, it follows that $|N(X)| \geq |X|$. Thus, $G$ satisfies Hall's condition. By Theorem 8.3, $G$ contains a matching from $U$ to a subset of $W$. This matching pairs off the sets $S_1, S_2, \ldots, S_n$ with $n$ distinct elements in $S_1 \cup S_2 \cup \ldots \cup S_n$, producing a system of distinct representatives for $\{S_1, S_2, \ldots, S_n\}$. ∎

Theorem 8.5 is also due to Hall. Indeed, it was through systems of distinct representatives that Hall proved the theorem that bears his name.

## 1-Factors

We have already noted that if $M$ is a perfect matching in a graph $G$, then $G[M]$ is a 1-regular spanning subgraph of $G$. Any spanning subgraph of a graph $G$ is referred to as a **factor**. A $k$-regular factor is called a **$k$-factor**. Thus, a 1-factor of a graph $G$ is the subgraph induced by a perfect matching in $G$. Therefore, all theorems stated earlier that concern perfect matchings can be restated in terms of 1-factors.

Clearly, if a graph $G$ has a 1-factor, then $G$ has even order. It turns out that for a graph $G$ to contain a 1-factor, this depends not only on the order of $G$ being even but also on the orders of components of certain subgraphs of $G$. We refer to a component of a graph as **odd** or **even** according to whether its order is odd or even.

While the graph $G$ of Figure 8.4 has even order, it does not contain a 1-factor. Let $S = \{s_1, s_2, s_3\}$. Since each of the components $G_1, G_2, \ldots, G_5$ of $G - S$ is odd, it follows that if $G$ has a 1-factor $F$, then some edge of $F$ must join a vertex of $S$ and a vertex of $G_i$ for each $i$ $(1 \leq i \leq 5)$. However, since there are more odd components in $G - S$ than vertices in $S$, this is impossible.

The explanation as to why the graph $G$ in Figure 8.4 does not have a 1-factor is, in fact, the key reason as to why any graph does not have a 1-factor. Tutte [239] characterized graphs containing a 1-factor and the proof we present here of his

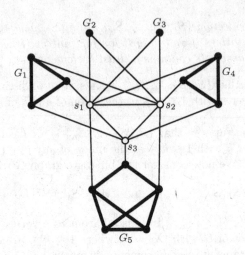

**Figure 8.4.** A graph that contains no 1-factor

result is due to Anderson [6]. The number of components in a graph $G$ is denoted by $k(G)$, while the odd components is denoted by $k_o(G)$.

**Theorem 8.6 (Tutte's theorem).** *A nontrivial graph $G$ contains a 1-factor if and only if $k_o(G - S) \leq |S|$ for every proper subset $S$ of $V(G)$.*

*Proof.* First, suppose that $G$ contains a 1-factor $F$. Let $S$ be a proper subset of $V(G)$. If $G - S$ has no odd components, then $k_o(G - S) = 0$ and $k_o(G - S) \leq |S|$. Thus, we may assume that $k_o(G - S) = k \geq 1$. Let $G_1, G_2, \ldots, G_k$ be the odd components of $G - S$. (There may be some even components of $G - S$ as well.) For each odd component $G_i$ of $G - S$, there is at least one edge of $F$ joining a vertex of $G_i$ and a vertex of $S$. Thus, $k_o(G - S) \leq |S|$.

We now verify the converse. Let $G$ be a graph such that $k_o(G - S) \leq |S|$ for every proper subset $S$ of $V(G)$. In particular, $k_o(G - \varnothing) \leq |\varnothing| = 0$, implying that every component of $G$ is even and so $G$ itself has even order. We now show that $G$ has a 1-factor by employing induction on the (even) order of $G$. Since $K_2$ is the only graph of order 2 having no odd components and $K_2$ has a 1-factor, the base case of the induction is verified.

For a given even integer $n \geq 4$, assume that all graphs $H$ of even order less than $n$ and satisfying $k_o(H - S) \leq |S|$ for every proper subset $S$ of $V(H)$ contain a 1-factor. Now, let $G$ be a graph of order $n$ satisfying $k_o(G - S) \leq |S|$ for every proper subset $S$ of $V(G)$. As we saw above, every component of $G$ has even order. We show that $G$ has a 1-factor.

For a vertex $v$ of $G$ that is not a cut-vertex and $R = \{v\}$, it follows that $k_o(G - R) = |R| = 1$. Hence, there are nonempty proper subsets $T$ of $V(G)$ for which $k_o(G - T) = |T|$. Among all such sets $T$, let $S$ be one of maximum cardinality. Suppose that $k_o(G - S) = |S| = k \geq 1$ and let $G_1, G_2, \ldots, G_k$ be the odd components of $G - S$.

We claim that $k(G - S) = k$, that is, $G_1, G_2, \ldots, G_k$ are the *only* components of $G - S$. Assume, to the contrary, that $G - S$ has an even component $G_0$. Let $v_0$ be a vertex of $G_0$ that is not a cut-vertex of $G_0$. Let $S_0 = S \cup \{v_0\}$. Since $G_0$ has even order,

$$k_o(G - S_0) \geq k_o(G - S) + 1 = k + 1.$$

On the other hand, $k_o(G - S_0) \leq |S_0| = k + 1$. Therefore,

$$k_o(G - S_0) = |S_0| = k + 1,$$

which is impossible. Thus, as claimed, $G_1, G_2, \ldots, G_k$ are the only components of $G - S$.

For each integer $i$ $(1 \leq i \leq k)$, let $S_i$ denote the set of those vertices in $S$ adjacent to at least one vertex of $G_i$. Since $G$ has only even components, each set $S_i$ is nonempty.

We claim, for each integer $\ell$ with $1 \leq \ell \leq k$, that the union of any $\ell$ of the sets $S_1, S_2, \ldots, S_k$ contains at least $\ell$ vertices. Assume, to the contrary, that this is not the case. Then there is an integer $j$ such that the union $S'$ of $j$ of the sets $S_1, S_2, \ldots, S_k$ has fewer than $j$ elements. Suppose that $S_1, S_2, \ldots, S_j$ have this property. Thus,

$$S' = S_1 \cup S_2 \cup \cdots \cup S_j \quad \text{and} \quad |S'| < j.$$

Then $G_1, G_2, \ldots, G_j$ are at least some of the components of $G - S'$ and so

$$k_o(G - S') \geq j > |S'|,$$

which contradicts the hypothesis. Thus, as claimed, for each integer $\ell$ with $1 \leq \ell \leq k$, the union of any $\ell$ of the sets $S_1, S_2, \ldots, S_k$ contains at least $\ell$ vertices.

By Theorem 8.5, there is a set $\{v_1, v_2, \ldots, v_k\}$ of $k$ distinct vertices of $S$ such that $v_i \in S_i$ for $1 \leq i \leq k$. Since every component $G_i$ of $G - S$ contains a vertex $u_i$ such that $u_i v_i$ is an edge of $G$, it follows that $\{u_i v_i : 1 \leq i \leq k\}$ is a matching of $G$.

We now show that for each nontrivial component $G_i$ of $G - S$ $(1 \leq i \leq k)$, the graph $G_i - u_i$ contains a 1-factor. Let $W$ be a proper subset of $V(G_i - u_i)$. We claim that

$$k_o(G_i - u_i - W) \leq |W|.$$

Assume, to the contrary, that $k_o(G_i - u_i - W) > |W|$. Since $G_i$ has odd order, $G_i - u_i$ has even order and so $k_o(G_i - u_i - W)$ and $|W|$ are both even or both odd. Hence,

$$k_o(G_i - u_i - W) \geq |W| + 2.$$

Let $X = S \cup W \cup \{u_i\}$. Then

$$
\begin{aligned}
|X| &= |S| + |W| + 1 = |S| + (|W| + 2) - 1 \\
&\leq k_o(G - S) + k_o(G_i - u_i - W) - 1 \\
&= k_o(G - X) \leq |X|,
\end{aligned}
$$

which implies that $k_o(G - X) = |X|$ and contradicts the defining property of $S$. Thus, as claimed, $k_o(G_i - u_i - W) \le |W|$.

Therefore, by the induction hypothesis, for each nontrivial component $G_i$ of $G - S$ ($1 \le i \le k$), the graph $G_i - u_i$ has a 1-factor. The collection of 1-factors of $G_i - u_i$ for all nontrivial graphs $G_i$ of $G - S$ together with the edges in $\{u_iv_i : 1 \le i \le k\}$ produce a 1-factor of $G$. ∎

##  Petersen's theorem

Clearly, every 1-regular graph contains a 1-factor, while the only 2-regular graphs containing a 1-factor are those that are the union of even cycles. Determining which cubic graphs contain a 1-factor is considerably more complex. One of the best known theorems in this area is due to Petersen [190].

The relatively short proof of Petersen's theorem given below is not that of Petersen but makes use of Tutte's theorem, which appeared decades after Petersen's result.

**Theorem 8.7 (Petersen's theorem).** *Every bridgeless cubic graph contains a 1-factor.*

*Proof.* Let $G$ be a bridgeless cubic graph and let $S$ be a proper subset of $V(G)$ with $|S| = k$. We show that $k_o(G - S) \le |S|$. This is true if $G - S$ has no odd components; so we assume that $G - S$ has $\ell \ge 1$ odd components, say $G_1, G_2, \ldots, G_\ell$.

Let $E_i$ ($1 \le i \le \ell$) denote the set of edges joining the vertices of $G_i$ and the vertices of $S$. Since $G$ is cubic, every vertex of $G_i$ has degree 3 in $G$. Since the sum of the degrees in $G$ of the vertices of $G_i$ is odd and the sum of the degrees in $G_i$ of the vertices of $G_i$ is even, it follows that $|E_i|$ is odd. Since $G$ is bridgeless, $|E_i| \ne 1$ and so $|E_i| \ge 3$ for $1 \le i \le \ell$. This implies that there are at least $3\ell$ edges joining the vertices of $G - S$ and the vertices of $S$. Since $|S| = k$, at most $3k$ edges join the vertices of $G - S$ and the vertices of $S$. Thus,

$$3k_o(G - S) = 3\ell \le 3k = 3|S|$$

and so $k_o(G - S) \le |S|$. By Tutte's theorem (Theorem 8.6), $G$ has a 1-factor. ∎

Indeed, Petersen showed that a cubic graph with at most two bridges contains a 1-factor (see Exercise 8.12). This result cannot be extended further, however, since the graph $G$ of Figure 8.5 is cubic and contains three bridges but no 1-factor since $k_o(G - v) = 3 > 1 = |\{v\}|$.

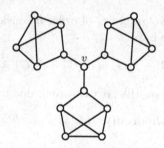

**Figure 8.5.** A cubic graph with no 1-factor

 # 8.2 Independence and covers

We now explore not only the edge independence number in more detail, but also the vertex independence number and two related parameters. A vertex and an edge are said to **cover each other** in a graph $G$ if they are incident. A **vertex cover** is a set of vertices that covers all the edges of $G$. An **edge cover** in a graph $G$ without isolated vertices is a set of edges that covers all the vertices of $G$. The minimum cardinality of a vertex cover in a graph $G$ is called the **vertex covering number** and is denoted by $\beta(G)$. As expected, the **edge covering number** $\beta'(G)$ of a graph $G$ (without isolated vertices) is the minimum cardinality of an edge cover.

For $s \le t$, we have $\beta(K_{s,t}) = s$ and $\beta'(K_{s,t}) = t$. Recall that the (vertex) independence number $\alpha(G)$ of $G$ is the maximum cardinality among the independent sets of vertices in $G$. We have already seen that $\alpha(K_{s,t}) = t$. An independent set $S$ of vertices in a graph $G$ is **maximal independent** if $S$ is not a proper subset of any other independent set of vertices in $G$. Therefore, there are two maximal independent sets of vertices in $K_{s,t}$, one with $s$ vertices and the other with $t$ vertices, namely the two partite sets of $K_{s,t}$. Furthermore, $\alpha'(K_{s,t}) = s$. As another illustration of these four parameters, we note that for $n \ge 2$, $\alpha(K_n) = 1$, $\alpha'(K_n) = \lfloor n/2 \rfloor$, $\beta(K_n) = n - 1$ and $\beta'(K_n) = \lceil n/2 \rceil$. The independence and covering concepts are summarized below.

| $\alpha(G)$ | vertex independence number | maximum number of vertices, no two of which are adjacent |
|---|---|---|
| $\alpha'(G)$ | edge independence number | maximum number of edges, no two of which are adjacent |
| $\beta(G)$ | vertex covering number | minimum number of vertices that cover the edges of $G$ |
| $\beta'(G)$ | edge covering number | minimum number of edges that cover the vertices of $G$ |

Observe that for the two graphs $G$ of order $n$ considered above, namely $K_{s,t}$ with $n = s + t$ and $K_n$, we have

$$\alpha(G) + \beta(G) = \alpha'(G) + \beta'(G) = n.$$

These two examples illustrate the next theorem, due to Gallai [99].

**Theorem 8.8 (Gallai's theorem).** *If $G$ is a graph of order $n$ with no isolated vertices, then*

$$\alpha(G) + \beta(G) = n \tag{8.1}$$

*and*

$$\alpha'(G) + \beta'(G) = n. \tag{8.2}$$

*Proof.* Let $G$ be a graph of order $n$ with no isolated vertices. We begin with (8.1). Let $U$ be an independent set of vertices with $|U| = \alpha(G)$. Then the set $V(G) \setminus U$ is a vertex cover in $G$. Therefore, $\beta(G) \le n - \alpha(G)$. If, however, $W$ is a set of $\beta(G)$ vertices that covers all edges of $G$, then $V(G) \setminus W$ is an independent set; thus $\alpha(G) \ge n - \beta(G)$. This proves (8.1).

Next, we verify (8.2). Let $E_1$ be a maximum independent set of edges and so $|E_1| = \alpha'(G)$. Then $E_1$ covers $2\alpha'(G)$ vertices of $G$. The remaining $n - 2\alpha'(G)$ vertices can be covered by $n - 2\alpha'(G)$ edges not in $E_1$. Thus, $\beta'(G) \le \alpha'(G) + (n - 2\alpha'(G)) = n - \alpha'(G)$ and so $\alpha'(G) + \beta'(G) \le n$.

Let $E'$ be an edge cover in $G$ with $|E'| = \beta'(G)$. The minimality of $E'$ implies that each component of $G[E']$ is a tree. Select from each component of $G[E']$ one edge, denoting the resulting set of edges by $E''$. Then $E''$ is independent and so $|E''| \le \alpha'(G)$. If $G[E']$ is a forest with $k$ components, then, by Corollary 1.13, the size of $G[E']$ is $n - k$. Thus,

$$\alpha'(G) + \beta'(G) \ge |E''| + |E'| = k + (n - k) = n,$$

completing the proof of (8.2) and the theorem. ∎

An elementary relationship involving the independence and covering numbers is given next.

**Theorem 8.9.** *If $G$ is a graph having no isolated vertices, then*

$$\beta(G) \ge \alpha'(G) \quad and \quad \beta'(G) \ge \alpha(G).$$

*Proof.* Let $S$ be a vertex cover in $G$ and let $X$ be an independent set of edges with $|X| = \alpha'(G)$. For each edge $e$ of $X$, there is a vertex $v_e$ in $S$ that is incident with $e$. Furthermore, for every two distinct edges $e$ and $f$ of $X$, the vertices $v_e$ and $v_f$ are distinct. Thus, $|S| \ge |X| = \alpha'(G)$, which implies that $\beta(G) \ge \alpha'(G)$.

For a graph $G$ of order $n$, it therefore follows by Gallai's theorem (Theorem 8.8) that

$$n - \alpha(G) = \beta(G) \ge \alpha'(G) = n - \beta'(G)$$

and so $\beta'(G) \ge \alpha(G)$. ∎

While $\beta(G) \geq \alpha'(G)$ for every graph $G$, equality does not hold in general. If, however, $G$ is bipartite, then $\beta(G) = \alpha'(G)$. This result was discovered independently by Kőnig [150] and Egerváry [72].

**Theorem 8.10 (Kőnig–Egerváry theorem).** *If $G$ is a bipartite graph, then*

$$\beta(G) = \alpha'(G).$$

*Proof.* Since $\beta(G) \geq \alpha'(G)$, it suffices to show that $\beta(G) \leq \alpha'(G)$. Let $U$ and $W$ be the partite sets of $G$ and let $M$ be a maximum matching. Then $\alpha'(G) = |M|$. Denote by $A$ the set of all $M$-unmatched vertices in $U$. (If $A = \varnothing$, then the proof is complete.) Observe that $|M| = |U| - |A|$. Let $S$ be the set of all vertices of $G$ that are connected to some vertex in $A$ by an $M$-alternating path. Define $U' = S \cap U$ and $W' = S \cap W$.

As in the proof of Hall's theorem 8.3, we have that $U' \setminus A$ is matched to $W'$ and that $N(U') = W'$. Since $U' \setminus A$ is matched to $W'$, it follows that $|U'| - |W'| = |A|$.

Observe that $C = (U \setminus U') \cup W'$ is a vertex cover in $G$; for otherwise, there is an edge $vw$ in $G$ such that $v \in U'$ and $w \notin W'$. Furthermore,

$$|C| = |U| - |U'| + |W'| = |U| - |A| = |M|.$$

Therefore, $\beta(G) \leq |C| = |M| = \alpha'(G)$ and the proof is complete. ∎

# 8.3 Domination

A vertex $v$ in a graph $G$ is said to **dominate** itself and each of its neighbors, that is, $v$ dominates the vertices in its closed neighborhood $N[v]$. A set $S$ of vertices is a **dominating set** if every vertex of $G$ is dominated by at least one vertex of $S$. Equivalently, a set $S$ of vertices of $G$ is a dominating set if every vertex in $V(G) \setminus S$ is adjacent to at least one vertex in $S$. The minimum cardinality among the dominating sets of $G$ is called the **domination number** and is denoted by $\gamma(G)$. A dominating set of cardinality $\gamma(G)$ is then referred to as a **minimum dominating set**. Although $\gamma(G)$ is the same notation that is used for the genus of a graph $G$, the notation in both instances is common, and we will never use the terms domination number and the genus of a graph in the same discussion.

The sets $S_1 = \{v_1, v_2, y_1, y_2\}$ and $S_2 = \{w_1, w_2, x\}$ are both dominating sets for the graph $G$ of Figure 8.6. Since the dominating set $S_2$ consists of three vertices, $\gamma(G) \leq 3$. A vertex of degree 4 dominates five vertices. Because $\Delta(G) = 4$ and $G$ has order 11, two vertices of $G$ can dominate at most ten vertices of $G$ and so $\gamma(G) \geq 3$. Thus, $\gamma(G) = 3$.

Dominating sets appear to have their origins in the game of chess, where the goal is to cover or dominate the squares of a chessboard by certain chess pieces. In 1862 Carl Friedrich de Jaenisch [61] considered the problem of determining

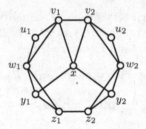

**Figure 8.6.** A graph with two minimum dominating sets

the minimum number of queens (which can move either horizontally, vertically, or diagonally over any number of unoccupied squares) that can be placed on a chessboard such that every square is either occupied by a queen or can be occupied by one of the queens in a single move. The minimum number of such queens is 5, and one possible placement of five such queens is shown in Figure 8.7.

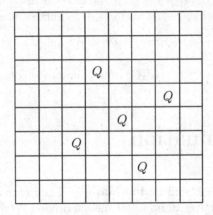

**Figure 8.7.** The minimum number of queens dominating the squares of a chessboard

Two queens on a chessboard are **attacking** if the square occupied by one of the queens can be reached by the other queen in a single move; otherwise, they are **nonattacking queens**. Note that the queens in Figure 8.7 are nonattacking, and hence the minimum number of nonattacking queens such that every square of the chessboard can be reached by one of the queens is also 5.

The connection between the chessboard problem described above and dominating sets in graphs is immediate. The 64 squares of a chessboard are the vertices of a graph $G$ and two vertices (squares) are adjacent in $G$ if each of the two squares can be reached by a queen on the other square in a single move. The graph $G$ is referred to as the **queen's graph**. Then the minimum number of queens that dominate all the squares of a chessboard is $\gamma(G)$. The minimum number of nonattacking queens that dominate all the squares of a chessboard is the minimum cardinality of a dominating set in $G$ that is also independent.

A **minimal dominating set** in a graph $G$ is a dominating set that contains no dominating set as a proper subset. A minimal dominating set of minimum cardinality is, of course, a minimum dominating set and consists of $\gamma(G)$ vertices. For the graph $G$ of Figure 8.6, the set $S_1 = \{v_1, v_2, y_1, y_2\}$ is a minimal dominating set that is not a minimum dominating set. Minimal dominating sets were characterized by Ore [185, p. 206].

**Theorem 8.11.** *A dominating set $S$ of a graph $G$ is a minimal dominating set if and only if every vertex $v$ in $S$ satisfies at least one of the following two properties:*
  *(i) there exists a vertex $w$ in $V(G) \setminus S$ such that $N(w) \cap S = \{v\}$;*
 *(ii) $v$ is adjacent to no vertex of $S$.*

*Proof.* Let $S$ be a dominating set of a graph $G$. First, observe that if each vertex $v$ in $S$ has at least one of the properties (i) and (ii), then $S \setminus \{v\}$ is not a dominating set of $G$. Consequently, $S$ is a minimal dominating set of $G$.

Conversely, assume that $S$ is a minimal dominating set of $G$. Then certainly for each $v \in S$, the set $S \setminus \{v\}$ is not a dominating set of $G$. Hence, there is a vertex $w$ in $V(G) \setminus (S \setminus \{v\})$ that is adjacent to no vertex of $S \setminus \{v\}$. If $w = v$, then $v$ is adjacent to no vertex of $S$ and (ii) holds. Suppose then that $w \neq v$. Since $S$ is a dominating set of $G$ and $w \notin S$, the vertex $w$ is adjacent to at least one vertex of $S$. However, $w$ is adjacent to no vertex of $S \setminus \{v\}$. Consequently, $N(w) \cap S = \{v\}$ and so (i) holds. ∎

Theorem 8.11 can be reworded as follows: A dominating set $S$ of a graph $G$ is a minimal dominating set of $G$ if and only if for every vertex $v$ in $S$ either (1) $v$ dominates some vertex of $V(G) \setminus S$ that is not dominated by any other vertex of $S$ or (2) no other vertex of $S$ dominates $v$.

The following result of Ore [185, p. 207] gives a property of the complementary set of a minimal dominating set in a graph without isolated vertices.

**Theorem 8.12.** *If $S$ is a minimal dominating set of a graph $G$ without isolated vertices, then $V(G) \setminus S$ is a dominating set of $G$.*

*Proof.* Let $v \in S$. Then $v$ has at least one of the two properties (i) and (ii) described in the statement of Theorem 8.11. Suppose first that there exists a vertex $w$ in $V(G) \setminus S$ such that $N(w) \cap S = \{v\}$. Hence, $v$ is adjacent to some vertex in $V(G) \setminus S$. Suppose next that $v$ is adjacent to no vertex in $S$. Then $v$ is an isolated vertex of the subgraph $G[S]$. Since $v$ is not isolated in $G$, the vertex $v$ is adjacent to some vertex of $V(G) \setminus S$. Thus, $V(G) \setminus S$ is a dominating set of $G$. ∎

For graphs $G$ without isolated vertices, we now have an upper bound for $\gamma(G)$ in terms of the order of $G$.

**Corollary 8.13.** *If $G$ is a graph of order $n$ without isolated vertices, then*

$$\gamma(G) \leq n/2.$$

*Proof.* Let $S$ be a minimal dominating set of $G$. By Theorem 8.12, $V(G) \setminus S$ is a dominating set of $G$. Thus, $\gamma(G) \leq \min\{|S|, |V(G) \setminus S|\} \leq n/2$. ∎

Nearly all connected graphs attaining the bound in Corollary 8.13 can be produced with the aid of the following operation. The **corona** $\mathrm{cor}(H)$ of a graph $H$ is that graph obtained from $H$ by adding a pendant edge to each vertex of $H$. Let $G = \mathrm{cor}(H)$, where $G$ has order $n$. Then $G$ has no isolated vertices and $\gamma(G) = n/2$. Indeed, Payan and Xuong [189] showed that every component of a graph $G$ of order $n$ without isolated vertices having $\gamma(G) = n/2$ is either $C_4$ or the corona of some (connected) graph.

Arnautov [13] and, independently, Payan [188] obtained an upper bound for the domination number that is an improvement to that given in Corollary 8.13 when $\delta(G) \geq 5$.

**Theorem 8.14.** *For any graph $G$ of order $n$ with minimum degree $\delta = \delta(G) \geq 2$,*

$$\gamma(G) \leq \frac{n(1 + \ln(\delta + 1))}{\delta + 1}.$$

The domination number of a graph without isolated vertices is also bounded above by all of the independence and covering numbers.

**Theorem 8.15.** *If $G$ is a graph without isolated vertices, then*

$$\gamma(G) \leq \min\{\alpha(G), \alpha'(G), \beta(G), \beta'(G)\}.$$

*Proof.* Since every vertex cover of a graph without isolated vertices is a dominating set, as is every maximum independent set of vertices, $\gamma(G) \leq \beta(G)$ and $\gamma(G) \leq \alpha(G)$.

Let $X$ be an edge cover of cardinality $\beta'(G)$. Then every vertex of $G$ is incident with at least one edge in $X$. Let $S$ be a set of vertices, obtained by selecting an incident vertex with each edge in $X$. Then $S$ is a dominating set of vertices and $\gamma(G) \leq |S| \leq |X| = \beta'(G)$.

Next, let $M$ be a maximum matching in $G$. We construct a set $S$ of vertices consisting of one vertex incident with an edge of $M$ for each edge of $M$. Let $uv \in M$. The vertices $u$ and $v$ cannot be adjacent to distinct unmatched vertices $x$ and $y$, respectively; for otherwise, $(x, u, v, y)$ is an $M$-augmenting path in $G$, contradicting Theorem 8.2. If $u$ is adjacent to a unmatched vertex, place $u$ in $S$; otherwise, place $v$ in $S$. This is done for each edge of $M$. Thus, $S$ is a dominating set of $G$ and $\gamma(G) \leq |S| = |M| = \alpha'(G)$. ∎

We close this section with an unsolved problem, namely a conjecture due to Vizing [245].

**Conjecture 8.16 (Vizing's conjecture).** *For every two graphs $G$ and $H$,*

$$\gamma(G \,\square\, H) \geq \gamma(G) \cdot \gamma(H).$$

# 8.4 Factorizations and decompositions

Recall that a factor of a graph $G$ is a spanning subgraph of $G$. A graph $G$ is said to be **factorable** into the factors $F_1, F_2, \ldots, F_t$ if these factors are (pairwise) edge-disjoint and $\cup_{i=1}^t E(F_i) = E(G)$. If $G$ is factored into $F_1, F_2, \ldots, F_t$, then $\mathcal{F} = \{F_1, F_2, \ldots, F_t\}$ is called a **factorization** of $G$.

If there exists a factorization $\mathcal{F}$ of a graph $G$ such that each factor in $\mathcal{F}$ is a $k$-factor (for a fixed positive integer $k$), then $G$ is **$k$-factorable**. If $G$ is a $k$-factorable graph, then necessarily $G$ is $r$-regular for some integer $r$ that is a multiple of $k$.

If a graph $G$ is factorable into $F_1, F_2, \ldots, F_t$, where each $F_i \cong F$ for some graph $F$, then we say that $G$ is **$F$-factorable** and that $G$ has an **isomorphic factorization** into (copies of) the factor $F$. Certainly, if a graph $G$ is $F$-factorable, then the size of $F$ divides the size of $G$. A graph $G$ of order $n = 2k$ is therefore 1-factorable if and only if $G$ is $kK_2$-factorable.

## 1-Factorizations

One problem in this area that has received a great deal of attention is whether a graph is 1-factorable. Of course, only $r$-regular graphs of even order can be 1-factorable, and such a 1-factorable graph would also have edge chromatic number $r$ and be of class one (see Chapter 6). Trivially, every 1-regular graph is 1-factorable. Since a 2-regular graph contains a 1-factor if and only if every component is an even cycle, it is precisely these 2-regular graphs that are 1-factorable. The situation for $r$-regular graphs, $r \geq 3$, in general, or even 3-regular graphs in particular, is considerably more complicated. By Petersen's theorem (Theorem 8.7), every bridgeless cubic graph contains a 1-factor. Consequently, every bridgeless cubic graph can be factored into a 1-factor and a 2-factor. Not every bridgeless cubic graph is 1-factorable, however. As Petersen himself observed [191], the graph shown in Figure 8.8, the Petersen graph that we first encountered in Chapter 1, is not 1-factorable (see Exercise 8.35).

The next two results describe two well-known classes of 1-factorable graphs, the first of which is due to Kőnig [149] who showed that $r$-regular bipartite graphs are 1-factorable. This result implies that $r$-regular bipartite graphs are also of class one (see Theorem 6.7).

**Theorem 8.17.** *Every $r$-regular bipartite graph, $r \geq 1$, is 1-factorable.*

*Proof.* We proceed by induction on $r$. The result is obvious if $r = 1$. Assume that every $(r-1)$-regular bipartite graph is 1-factorable where $r - 1 \geq 1$ and

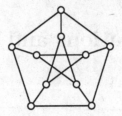

**Figure 8.8.** The Petersen graph: a bridgeless cubic graph that is not 1-factorable

let $G$ be an $r$-regular bipartite graph. By Theorem 8.4, $G$ contains a 1-factor $F_1$. Then $G - E(F_1)$ is an $(r-1)$-regular bipartite graph. By the induction hypothesis, $G - E(F_1)$ can be factored into $r-1$ 1-factors, say $F_2, F_3, \ldots, F_r$. Then $\{F_1, F_2, \ldots, F_r\}$ is a 1-factorization of $G$. ∎

As a consequence of Theorem 8.17, the 4-regular bipartite graph $K_{4,4}$ can be factored into four 1-factors. If $U = \{A, K, Q, J\}$ and $W = \{\spadesuit, \heartsuit, \diamondsuit, \clubsuit\}$ are the two partite sets of $K_{4,4}$, then one possible 1-factorization of $K_{4,4}$ is shown in Figure 8.9.

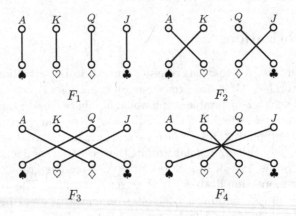

**Figure 8.9.** A 1-factorization of $K_{4,4}$

The 1-factorization of $K_{4,4}$ shown in Figure 8.9 gives rise to a partition of the edge set of $K_{4,4}$ into four perfect matchings $F_1, F_2, F_3, F_4$ whose edges can be listed as follows:

| $F_1$ | $A$ ♠ | $K$ ♡ | $Q$ ◇ | $J$ ♣ |
|---|---|---|---|---|
| $F_2$ | $J$ ◇ | $Q$ ♣ | $K$ ♠ | $A$ ♡ |
| $F_3$ | $K$ ♣ | $A$ ◇ | $J$ ♡ | $Q$ ♠ |
| $F_4$ | $Q$ ♡ | $J$ ♠ | $A$ ♣ | $K$ ◇ |

This, of course, is very suggestive of the 16 playing cards of the four denominations ace, king, queen, jack and the four suits spades, hearts, diamonds, clubs.

In the table above, each denomination and each suit appears exactly once in each row and in each column. This card problem dates back to the year 1723. Thus, the elements of the set $U$ as well as the elements of $W$ can be arranged in a $4 \times 4$ table so that every element appears exactly once in each row and column. More generally, when the elements of an $n$-element set are arranged in an $n \times n$ table so that every element appears exactly once in each row and column, a **Latin square** results. So the table above gives rise to two Latin squares, one whose entries are the denominations and one whose entries are the suits. For example, if $\Gamma$ is a (finite) group of order $n$ with $\Gamma = \{g_1, g_2, \ldots, g_n\}$, say, and an $n \times n$ table is constructed where the entry in row $i$, column $j$ of the table is the element $g_i g_j$ of $\Gamma$, then a Latin square results. The topic of Latin squares has a long history and is an area of much interest in the subject of combinatorics.

We now describe a second class of 1-factorable graphs, namely complete graphs of even order.

**Theorem 8.18.** *For each positive integer $k$, the complete graph $K_{2k}$ is 1-factorable.*

*Proof.* The result is obvious for $k = 1$, so we assume that $k \geq 2$. Denote the vertex set of $K_{2k}$ by $\{v_\infty, v_0, v_1, \ldots, v_{2k-2}\}$ and arrange the vertices $v_0, v_2, \ldots, v_{2k-2}$ cyclically in a regular $(2k-1)$-gon, placing $v_\infty$ in the center. Let $F_0$ be the 1-factor consisting of the edges $v_\infty v_0$, $v_1 v_{2k-2}$, $v_2 v_{2k-3}$, $\ldots$, $v_{k-1} v_k$. For $i = 1, 2, \ldots, 2k-2$, let $F_i$ be the 1-factor consisting of the edges $v_\infty v_i$, $v_{1+i} v_{2k-2+i}$, $v_{2+i} v_{2k-3+i}$, $\ldots$, $v_{k-1+i} v_{k+i}$ where all arithmetic is taken modulo $2k - 1$. Then $\{F_0, F_1, F_2, \cdots, F_{2k-2}\}$ is a 1-factorization of $K_{2k}$. ∎

The construction described in the proof of Theorem 8.18 is illustrated in Figure 8.10 for the graph $K_6$.

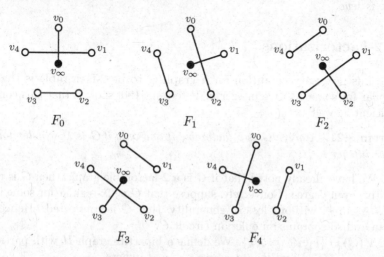

**Figure 8.10.** A 1-factorization of $K_6$

The 1-factorization obtained in the proof of Theorem 8.18 has the following nice property. In the 1-factorization of $K_{2k}$, the 1-factor $F_0$ consists of the edge $v_\infty v_0$ and all edges perpendicular to $v_0 v_\infty$, namely, $v_1 v_{2k-2}, v_2 v_{2k-3}, \ldots, v_{k-1} v_k$. If the $k$ edges of $F_1$ are rotated clockwise through an angle of $2\pi/(2k-1)$ radians, then the 1-factor $F_2$ is obtained. In general, if the edges of $F_0$ are rotated clockwise through an angle of $2\pi j/(2k-1)$ radians, where $0 \le j \le 2k-2$, then the 1-factor $F_j$ is produced. Such a factorization is called **1-rotational**.

We saw in Dirac's theorem (Corollary 3.9) that if $G$ is a graph of order $n \ge 3$ such that $\deg v \ge n/2$ for every vertex $v$ of $G$, then $G$ is hamiltonian. Consequently, if $G$ is an $r$-regular graph of even order $n \ge 4$ such that $r \ge n/2$, then $G$ contains a hamiltonian cycle $C$. Since $C$ is an even cycle, $C$ can be factored into two 1-factors. If there exists a 1-factorization of $G - E(C)$, then $G$ is 1-factorable. This is certainly the case if $r = 3$. For which values of $r$ and $n$ this can be done is not known. There is a conjecture, however, dealing with this topic that is believed to have originated in 1986 by Chetwynd and Hilton [47].

**Conjecture 8.19 (1-Factorization conjecture).** *If $G$ is an $r$-regular graph of even order $n$ such that (1) $r \ge n/2$ if $n \equiv 2 \pmod 4$ or (2) $r \ge (n-2)/2$ if $n \equiv 0 \pmod 4$, then $G$ is 1-factorable.*

The bound on $r$ in the 1-factorization theorem cannot be improved. For example, suppose that $n \equiv 2 \pmod 4$. Then $n = 2k$ for some odd integer $k \ge 3$. The graph $G = 2K_k$ is a $(k-1)$-regular graph of order $n$, where $k - 1 = \frac{n}{2} - 1$. Since $G$ does not have a 1-factor, $G$ is certainly not 1-factorable.

Employing a lengthy proof, Csaba, Kühn, Lo, Osthus and Treglown [58] showed that the 1-Factorization Conjecture is true if $n$ is sufficiently large.

**Theorem 8.20.** *For sufficiently large even integers $n$, the 1-Factorization Conjecture is true.*

## 2-Factorizations

An obvious necessary condition for a graph $G$ to be 2-factorable is that $G$ is $2k$-regular for some positive integer $k$. Petersen [190] showed that this condition is sufficient as well.

**Theorem 8.21.** *A graph $G$ is 2-factorable if and only if $G$ is $2k$-regular for some positive integer $k$.*

*Proof.* We have already noted that if $G$ is a 2-factorable graph, then $G$ is regular of positive even degree. Conversely, suppose that $G$ is $2k$-regular for some integer $k \ge 1$. Assume, without loss of generality, that $G$ is connected. Hence, $G$ is eulerian and so contains an eulerian circuit $C$.

Let $V(G) = \{v_1, v_2, \ldots, v_n\}$. We define a bipartite graph $H$ with partite sets $U = \{u_1, u_2, \ldots, u_n\}$ and $W = \{w_1, w_2, \ldots, w_n\}$, where

$$E(H) = \{u_i w_j : v_j \text{ immediately follows } v_i \text{ on } C\}.$$

The graph $H$ is $k$-regular and so, by Theorem 8.17, $H$ is 1-factorable. Hence, $\{F_1, F_2, \cdots, F_k\}$ is a 1-factorization of $H$.

Corresponding to each 1-factor $F_\ell$ of $H$ is a permutation $\alpha_\ell$ on the set $\{1, 2, \ldots, n\}$, defined by $\alpha_\ell(i) = j$ if $u_i w_j \in E(F_\ell)$. Let $\alpha_\ell$ be expressed as a product of disjoint permutation cycles. There is no permutation cycle of length 1 in this product; for if $(i)$ were a permutation cycle, then this would imply that $\alpha_\ell(i) = i$. However, this further implies that $u_i w_i \in E(F_\ell)$ and that $v_i v_i \in E(C)$, which is impossible. Also, there is no permutation cycle of length 2 in this product; for if $(i\ j)$ were a permutation cycle, then $\alpha_\ell(i) = j$ and $\alpha_\ell(j) = i$. This would indicate that $u_i w_j, u_j w_i \in E(F_\ell)$ and that $v_j$ both immediately follows and precedes $v_i$ on $C$, contradicting the fact that no edge is repeated on a circuit. Thus, the length of every permutation cycle in $\alpha_\ell$ is at least 3.

Each permutation cycle in $\alpha_\ell$ therefore gives rise to a cycle in $G$, and the product of the disjoint permutation cycles in $\alpha_\ell$ produces a collection of mutually disjoint cycles in $G$ containing all vertices of $G$, that is, a 2-factor in $G$. Since the 1-factors in $H$ are mutually edge-disjoint, the resulting 2-factors in $G$ are mutually edge-disjoint. Hence, $G$ is 2-factorable. ∎

It is an immediate consequence of Theorem 8.21 that there exists a factorization of every regular graph $G$ of positive even degree in which every factor is a union of cycles. This brings up the problem of whether there exists a factorization of $G$ in which every factor is a single cycle, that is, a hamiltonian cycle. A **hamiltonian factorization** of a graph $G$ is a factorization in which every factor is a hamiltonian cycle and a graph possessing such a factorization is **hamiltonian factorable**. Certainly, every hamiltonian factorable graph is a 2-connected regular graph of positive even degree. The converse of this statement is not true, however, as the graph $H$ of Figure 8.11 shows. Any hamiltonian cycle of $H$ necessarily contains both edges $uv$ and $xy$ and so $H$ does not contain two edge-disjoint hamiltonian cycles.

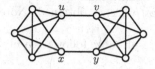

**Figure 8.11.** A 2-factorable graph that is not hamiltonian factorable

Complete graphs of odd order are not only 2-factorable, they are hamiltonian factorable. The following result and construction are credited to Walecki (see [5]).

**Theorem 8.22.** *For every positive integer $k$, the graph $K_{2k+1}$ is hamiltonian factorable.*

*Proof.* Since the result is clear for $k = 1$, we may assume that $k \geq 2$. Let $V(K_{2k+1}) = \{v_\infty, v_0, v_1, \ldots, v_{2k-1}\}$. Arrange the vertices $v_0, v_1, \ldots, v_{2k-1}$ cyclically in a regular $2k$-gon and place $v_\infty$ in some convenient position. Consider the

2-factor $F_0$ consisting of the edges of the cycle

$$v_\infty, v_0, v_1, v_{2k-1}, v_2, v_{2k-2}, \ldots, v_{k-1}, v_{k+1}, v_k, v_\infty.$$

(See $F_0$ in Figure 8.12 for the case $k = 3$.) For $i = 1, 2, \ldots, k - 1$, let $F_i$ consist of the edges of the cycle

$$v_\infty, v_i, v_{1+i}, v_{2k-1+i}, v_{2+i}, v_{2k-2+i}, \ldots, v_{k-1+i}, v_{k+1+i}, v_{k+i}, v_\infty,$$

where all arithmetic is modulo $2k$. Since each $F_i$ is a hamiltonian cycle for $0 \le i \le k - 1$, it follows that $\{F_0, F_1, F_2, \ldots, F_{k-1}\}$ is a hamiltonian factorization of $K_{2k+1}$. ∎

The hamiltonian factorization described in the proof of Theorem 8.22 is illustrated in Figure 8.12 for the complete graph $K_7$.

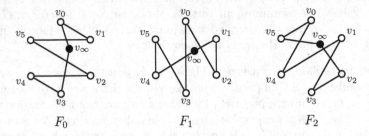

**Figure 8.12.** A hamiltonian factorization of $K_7$

Note that if we place vertex $v_\infty$ in the center of a regular $2k$-gon and rotate the edges of the hamiltonian cycle $F_0$ clockwise through an angle of $2\pi/2k = \pi/k$ radians, then the hamiltonian cycle $F_1$ is produced. Indeed, if we rotate the edges of $F_0$ clockwise through an angle of $\pi j/k$ radians for any integer $j$ with $1 \le j \le k - 1$, then the hamiltonian cycle $F_j$ is produced and the desired hamiltonian factorization of $K_{2k+1}$ is obtained. Note that this factorization is also 1-rotational.

Another factorization result now follows readily from Theorem 8.22 (see Exercise 8.44).

**Corollary 8.23.** *For each positive integer $k$, the complete graph $K_{2k}$ can be factored into $k$ hamiltonian paths.*

Using the construction employed in the proof of Theorem 8.22, we can obtain yet another factorization result (see Exercise 8.46).

**Theorem 8.24.** *For each positive integer $k$, the graph $K_{2k}$ can be factored into $k - 1$ hamiltonian cycles and a 1-factor.*

From Theorems 8.22 and 8.24, we then have the following corollary.

**Corollary 8.25.** *Every complete graph of order at least 2 can be factored into hamiltonian cycles and at most one 1-factor.*

Nash-Williams [181] conjectured that a factorization of the type stated in Corollary 8.25 not only exists for every complete graph but for every $r$-regular graph of order $n$ if $r$ is sufficiently large relative to $n$.

**Conjecture 8.26 (Hamiltonian factorization conjecture).** *If $G$ is an $r$-regular graph of order $n$ such that $r \geq n/2$, then $G$ can be factored into hamiltonian cycles and at most one 1-factor.*

Csaba, Kühn, Lo, Osthus and Treglown [58] showed that the hamiltonian factorization conjecture holds if $n$ is sufficiently large.

**Theorem 8.27.** *For every sufficiently large integer $n$, the Hamiltonian Factorization Conjecture is true.*

# Decompositions

A **decomposition** $\mathcal{D}$ of a graph $G$ is a collection $\{H_1, H_2, \ldots, H_t\}$ of nonempty subgraphs such that $H_i = G[E_i]$ for some (nonempty) subset $E_i$ of $E(G)$, where $\{E_1, E_2, \ldots, E_t\}$ is a partition of $E(G)$. Thus, no subgraph $H_i$ in a decomposition of $G$ contains isolated vertices. If $\mathcal{D}$ is a decomposition of $G$, then we say $G$ is decomposed into the subgraphs $H_1, H_2, \ldots, H_t$. Indeed, if $\mathcal{D}$ is a decomposition of a graph $G$ where each subgraph $H_i$ is a spanning subgraph of $G$, then $\{H_1, H_2, \ldots, H_t\}$ is a factorization of $G$. On the other hand, every factorization of a nonempty graph $G$ also gives rise to a decomposition of $G$.

If $\mathcal{D} = \{H_1, H_2, \ldots, H_t\}$ is a decomposition of a graph $G$ such that $H_i \cong H$ for some graph $H$ for each $i$ ($1 \leq i \leq t$), then $\mathcal{D}$ is an **$H$-decomposition** of $G$. If there exists an $H$-decomposition of a graph $G$, then $G$ is said to be **$H$-decomposable**. The graph $G = K_{2,2,2}$ (the graph of the octahedron) is $H$-decomposable for the graph $H$ shown in Figure 8.13. An $H$-decomposition of $G$ is also shown in Figure 8.13.

If $G$ is an $H$-decomposable graph for some graph $H$, then certainly $H$ is a subgraph of $G$ and the size of $H$ divides the size of $G$. Although this last condition is necessary, it is not sufficient. For example, the graph $K_{1,4}$ is a subgraph of the graph $G$ of Figure 8.13 and the size 4 of $K_{1,4}$ divides the size 12 of $G$, but yet $G$ is not $K_{1,4}$-decomposable (see Exercise 8.52).

The basic problem here is whether a given graph $G$ is $H$-decomposable for some subgraph $H$ such that the size of $H$ divides the size of $G$. Of course, every nonempty graph is $K_2$-decomposable. For a connected graph $G$ to be $P_3$-decomposable, the size of $G$ must be even. It turns out that this condition is also sufficient as we will see next.

**Theorem 8.28.** *A nontrivial connected graph $G$ is $P_3$-decomposable if and only if $G$ has even size.*

*Proof.* We have already noted that if $G$ is $P_3$-decomposable, then $G$ has even size. For the converse, assume that $G$ has even size. Suppose first that $G$ is

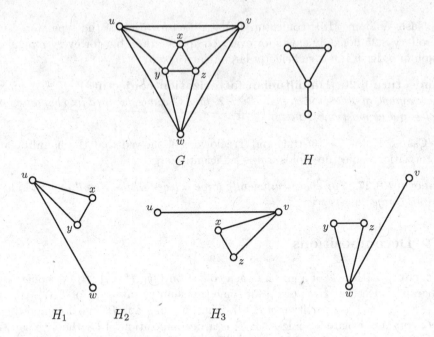

**Figure 8.13.** An $H$-decomposable graph

eulerian, where the edges of $G$ are encountered in the order $e_1, e_2, \ldots, e_m$. Then each of the sets $\{e_1, e_2\}$, $\{e_3, e_4\}$, $\ldots$,$\{e_{m-1}, e_m\}$ induces a copy of $P_3$; so $G$ is $P_3$-decomposable. Otherwise, $G$ has $2k$ odd vertices for some $k \geq 1$. By Theorem 3.3, $E(G)$ can be partitioned into subsets $E_1, E_2, \ldots, E_k$, where for each $i$, $G[E_i]$ is an open trail $T_i$ of even length connecting odd vertices of $G$. (That is, $G$ can be decomposed into $k$ open trails of even length connecting odd vertices.) Then, as with the eulerian circuit above, the edges of each trail $T_i$ can be paired off so that each pair of consecutive edges on each trail induce a copy of $P_3$. Thus, $G$ is $P_3$-decomposable. ∎

As we have noted, the vast majority of factorization and decomposition conjectures and results deal with factoring or decomposing complete graphs into a specific graph or graphs. A problem of particular interest concerns determining those complete graphs that are $K_3$-decomposable or, equivalently, $C_3$-decomposable.

A **Steiner triple system of order** $n$ consists of a set $S$ of $n$ elements and a collection $T$ of 3-element subsets of $S$, called **triples**, such that every pair of elements of $S$ belongs to exactly one triple in $T$. For example,

$$S = \{0, 1, \ldots, 6\} \text{ and } T = \{013, 124, 235, 346, 450, 561, 602\}$$

is a Steiner triple system of order 7. A Steiner triple system of order $n$ corresponds to a $K_3$-decomposition of $K_n$. In particular, the Steiner triple system of order 7 mentioned above results in the $K_3$-decomposition of $K_7$ shown in Figure 8.14.

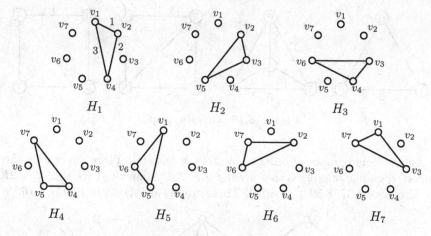

**Figure 8.14.** A $K_3$-decomposition of $K_7$

Kirkman [146] gave necessary and sufficient conditions for the existence of Steiner triple systems.

**Theorem 8.29.** *There exists a Steiner triple system of order $n \geq 3$ if and only if $n \equiv 1 \pmod{6}$ or $n \equiv 3 \pmod{6}$.*

While the necessity for the existence of a Steiner triple system of order $n$ stated in Theorem 8.29 is quite straightforward (see Exercise 8.53), the sufficiency requires the construction of appropriate Steiner triple systems.

# 8.5  Labelings of graphs

A **graph labeling** is an assignment of labels (typically positive integers or nonnegative integers) to the vertices or edges (or both) of a graph, often satisfying some prescribed requirements. A labeling that has drawn a lot of attention is Rosa's $\beta$-valuation [209], which was later called a graceful labeling by Golomb [104]. This terminology has become standard.

A nonempty graph $G$ of size $m$ is **graceful** if it is possible to label the vertices of $G$ with distinct elements from the set $\{0, 1, \ldots, m\}$ in such a way that the induced edge labeling, which assigns the integer $|i - j|$ to the edge joining vertices labeled $i$ and $j$, results in the $m$ edges of $G$ being labeled $1, 2, \ldots, m$. Such a labeling is called a **graceful labeling**. Thus, a graceful graph is a graph that admits a graceful labeling.

All of the graphs $K_3$, $K_4$, $K_4 - e$ and $C_4$ are graceful, as is illustrated in Figure 8.15. Here, the vertex labels are placed within the vertices and the induced edge labels are placed near the relevant edges.

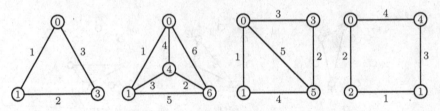

**Figure 8.15.** Graceful graphs

All connected graphs of order at most 4 are graceful. There are exactly three connected graphs of order 5 that are not graceful, namely $C_5$, $K_5$ and $K_1 \vee 2K_2$ (see Exercises 8.57, 8.59, and 8.70). These graphs are shown in Figure 8.16.

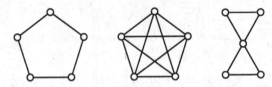

**Figure 8.16.** The three connected graphs of order 5 that are not graceful

For each graceful graph $H$ of size $m$, the complete graph $K_{2m+1}$ is $H$-decomposable. This observation is also due to Rosa [209].

**Theorem 8.30.** *If $H$ is a graceful graph of size $m$, then the complete graph $K_{2m+1}$ can be decomposed into copies of $H$.*

*Proof.* Since $H$ is graceful, there is a graceful labeling of $H$, that is, the vertices of $H$ can be labeled from a subset of $\{0, 1, \ldots, m\}$ so that the induced edge labels are $1, 2, \ldots, m$. Let $V(K_{2m+1}) = \{v_0, v_1, \ldots, v_{2m}\}$ where the vertices of $K_{2m+1}$ are arranged cyclically in a regular $(2m+1)$-gon, denoting the resulting $(2m+1)$-cycle by $C$. A vertex labeled $i$ for each integer $i$ $(0 \le i \le m)$ in $H$ is placed at $v_i$ in $K_{2m+1}$ and this is done for each vertex of $H$. Every edge of $H$ is drawn as a straight line segment in $K_{2m+1}$, denoting the resulting copy of $H$ in $K_{2m+1}$ by $H_1$. Hence, $V(H_1) \subseteq \{v_0, v_1, \ldots, v_m\}$.

Each edge $v_s v_t$ of $K_{2m+1}$ $(0 \le s, t \le 2m)$ is labeled $d_C(v_s, v_t)$, where then $1 \le d_C(v_s, v_t) \le m$. Consequently, $K_{2m+1}$ contains exactly $2m+1$ edges labeled $i$ for each $i$ $(1 \le i \le m)$ and $H_1$ contains exactly one edge labeled $i$ $(1 \le i \le m)$. Whenever an edge of $H_1$ is rotated through an angle (clockwise, say) of $2\pi k/(2m+1)$ radians, where $1 \le k \le 2m$, an edge of the same label is obtained. Denote the subgraph obtained by rotating $H_1$ through a clockwise angle of $2\pi k/(2m+1)$ radians by $H_{k+1}$. Then $H_{k+1} \cong H$ and a decomposition of $K_{2m+1}$ into $2m+1$ copies of $H$ results. ∎

While many graphs are known to be nongraceful, none of these is a tree. Indeed, it has been conjectured by Kotzig that every nontrivial tree is graceful.

**Conjecture 8.31 (Graceful tree conjecture).** *Every nontrivial tree is graceful.*

Actually, Kotzig conjectured that every tree has a more general type of labeling, called a $\rho$-labeling by Rosa [209]. A **$\rho$-labeling** of a graph $G$ with $m$ edges is a labeling of the vertices of $G$ with distinct elements from the set $\{0, 1, \ldots, 2m\}$ in such a way that the induced edge labeling, which assigns the minimum of $|i - j|$ and $2m + 1 - |i - j|$ to the edge joining vertices labeled $i$ and $j$, results in the $m$ edges of $G$ being labeled $1, 2, \ldots, m$. If the Graceful Tree Conjecture is true and $T$ is a tree of size $m$, then by Theorem 8.30, $K_{2m+1}$ is $T$-decomposable. Regardless of whether every tree is graceful, Ringel [199] made the following conjecture concerning decompositions of complete graphs into trees.

**Conjecture 8.32** (**Ringel's conjecture**). *For every tree $T$ of size $m$, the complete graph $K_{2m+1}$ is $T$-decomposable.*

Clearly, the truth of Kotzig's conjecture implies the truth of Ringel's conjecture. Recently, using probabilistic methods and $\rho$-labelings of complete graphs, Montgomery, Pokrovskiy, and Sudakov [177] showed that Ringel's conjecture is true for large $m$.

**Theorem 8.33.** *For every sufficiently large integer $m$, Ringel's conjecture is true.*

We close this section with the following necessary condition for an eulerian graph to be graceful, due to Rosa [209].

**Theorem 8.34.** *If $G$ is a graceful eulerian graph of size $m$, then*

$$m \equiv 0 \ (\mathrm{mod}\ 4) \ or \ m \equiv 3 \ (\mathrm{mod}\ 4).$$

*Proof.* Let $C \colon v_0, v_1, \ldots, v_{m-1}, v_m = v_0$ be an eulerian circuit of $G$, and let a graceful labeling of $G$ be given that assigns the integer $a_i$ ($0 \le a_i \le m$) to $v_i$ for $0 \le i \le m$, where, of course, $a_i = a_j$ if $v_i = v_j$. Thus, the label of the edge $v_{i-1}v_i$ is $|a_i - a_{i-1}|$. Observe that

$$|a_i - a_{i-1}| \equiv (a_i - a_{i-1}) \ (\mathrm{mod}\ 2)$$

for $1 \le i \le m$. Thus, the sum of the labels of the edges of $G$ is

$$\sum_{i=1}^{m} |a_i - a_{i-1}| \equiv \sum_{i=1}^{m} (a_i - a_{i-1}) \equiv 0 \ (\mathrm{mod}\ 2),$$

that is, the sum of the edge labels is even. However, the sum of the edge labels is

$$\sum_{i=1}^{m} i = \frac{m(m+1)}{2};$$

so, $m(m+1)/2$ is even. Consequently, $4 \mid m(m+1)$, which implies that $4 \mid m$ or $4 \mid (m+1)$ so that $m \equiv 0 \ (\mathrm{mod}\ 4)$ or $m \equiv 3 \ (\mathrm{mod}\ 4)$. ∎

# Exercises for Chapter 8

## Section 8.1. Matchings and 1-factors

**8.1.** Show that every tree has at most one perfect matching.

**8.2.** Determine the maximum size $m$ of a graph of order $n$ having a maximum matching of $k$ edges if (a) $n = 2k$, (b) $n = 2k + 2$.

**8.3.** Use Menger's theorem 4.18 to prove the following implication in Hall's theorem 8.3: *Let $G$ be a bipartite graph with partite sets $U$ and $W$. If Hall's condition is satisfied, then $U$ can be matched to a subset of $W$.*

**8.4.** Prove Theorem 8.4.

**8.5.** Let $G$ be a bipartite graph with partite sets $U$ and $W$ such that $|U| = |W| = r \geq 1$. Suppose that $U = \{u_1, u_2, \ldots, u_r\}$ and $W = \{w_1, w_2, \ldots, w_r\}$. Two vertices $u_i$ and $w_j$ are adjacent in $G$ if and only if $i + j \geq r + 1$. Use Hall's theorem 8.3 to show that $G$ has a perfect matching.

**8.6.** Let $G$ be a bipartite graph of size $m$ with partite sets $U$ and $W$ where $|U| = |W| = k$.
  (a) Prove that if $m \geq k^2 - k + 1$, then $G$ has a perfect matching.
  (b) Show that for $m = k^2 - k$, there exists a bipartite graph $H$ with partite sets $U$ and $W$ such that $|U| = |W| = k$ but $H$ does not have a perfect matching.

**8.7.** Let $G$ be a bipartite graph with partite sets $U$ and $W$ where $|U| \leq |W|$. The **deficiency** $\operatorname{def}(S)$ of a set $S \subseteq U$ is defined as $\max\{|A| - |N(A)|\}$, where the maximum is taken over all nonempty subsets $A$ of $S$. Show that

$$\alpha'(G) = \min\{|U|, |U| - \operatorname{def}(U)\}.$$

**8.8.** Let $G$ be a regular maximal planar graph.
  (a) Show that there is only one regular maximal planar graph $G$ whose order $n \in \{5, 6, \ldots, 11\}$.
  (b) For $G$ as in (a), show that $\overline{G}$ has a perfect matching $M$.

**8.9.** A matching $M'$ in a graph $G$ is a **maximal matching** if there exists no matching $M$ of $G$ such that $M'$ is a proper subset of $M$. Let $G$ be a nonempty graph. Prove or disprove the following.
  (a) For every two nonadjacent edges $e$ and $f$, there is a maximal matching of $G$ containing $e$ and $f$.
  (b) For every two adjacent edges $e$ and $f$, there is a maximum matching of $G$ containing one of $e$ and $f$.

**8.10.** Give an example of an infinite class of graphs $G$ such that a matching $M$ in $G$ is maximal if and only if $M$ is maximum.

**8.11.** Prove that if $M$ is a matching of a graph $G$ such that if $|M| < \frac{1}{2}\alpha'(G)$, then $M$ is not a maximal matching of $G$.

**8.12.** Prove that every cubic graph with at most two bridges contains a 1-factor.

**8.13.** Use Tutte's theorem (Theorem 8.6) to show that the graph $G$ shown in Figure 8.17 does not have a 1-factor.

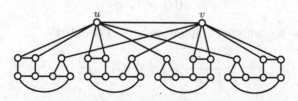

**Figure 8.17.** The graph $G$ in Exercise 8.13

**8.14.** Petersen's theorem (Theorem 8.7) states that if $G$ is a bridgeless cubic graph, then $G$ has a 1-factor.
   (a) Show that Petersen's theorem can be extended somewhat by proving that if $G$ is a bridgeless graph having exactly one vertex of degree 7 and all others of degree 3, then $G$ has a 1-factor.
   (b) Show that the result in (b) cannot be extended any further by giving an example of a bridgeless graph $G$ having exactly two vertices of degree 7 and all others of degree 3 but $G$ has no 1-factor.

**8.15.** Prove that if a connected cubic graph has a 1-factor and a bridge, then the 1-factor must contain the bridge.

**8.16.** Let $G$ be a graph every vertex of which has odd degree and let $\{V_1, V_2\}$ be a partition of $V(G)$, where $[V_1, V_2]$ is the set of edges joining $V_1$ and $V_2$. Prove that $|V_1|$ and $|[V_1, V_2]|$ are of the same parity.

**8.17.** Prove that every $(2k+1)$-regular, $2k$-edge-connected graph, $k \geq 1$, contains a 1-factor.

**8.18.** Prove that if $G$ is an $r$-regular, $(r-2)$-edge-connected graph ($r \geq 3$) of even order containing at most $r-1$ distinct edge-cuts of cardinality $r-2$, then $G$ has a 1-factor.

**8.19.** A graph $G$ is **factor-critical** if $G - v$ contains a 1-factor for every vertex $v$ of $G$. Prove that a graph $G$ of order $n$ is factor-critical if and only if $n$ is odd and $k_o(G - S) \leq |S|$ for every nonempty proper subset $S$ of $V(G)$.

## Section 8.2. Independence and covers

**8.20.** Show that a graph $G$ is bipartite if and only if $\alpha(H) \geq \frac{1}{2}|V(H)|$ for every subgraph $H$ of $G$.

**8.21.** Prove that if $G$ is a bipartite graph without isolated vertices, then

$$\alpha(G) = \beta'(G).$$

**8.22.** Let $G$ be a connected graph of order $n$. Prove that

$$\left\lceil \frac{n}{1 + \Delta(G)} \right\rceil \leq \alpha'(G) \leq \left\lfloor \frac{n}{2} \right\rfloor$$

and show that these bounds are sharp.

**8.23.** Let $G$ be a graph of order $n$ without isolated vertices. Prove that

$$\left\lceil \frac{n}{2} \right\rceil \leq \beta'(G) \leq \left\lfloor \frac{n \cdot \Delta(G)}{1 + \Delta(G)} \right\rfloor$$

and show that these bounds are sharp.

**8.24.** Characterize those nonempty graphs with the property that every two distinct maximal independent sets of vertices are disjoint.

**8.25.** Prove or disprove: A graph $G$ without isolated vertices has a perfect matching if and only if $\alpha'(G) = \beta'(G)$.

## Section 8.3. Domination

**8.26.** Determine the domination numbers of the 3-cube $Q_3$ and the 4-cube $Q_4$.

**8.27.** Determine and verify a formula for $\gamma(C_n)$.

**8.28.** Determine and verify a formula for $\gamma(P_n)$.

**8.29.** State and prove a characterization of those graphs $G$ with $\gamma(G) = 1$.

**8.30.** Prove that if $G$ is a graph of order $n$, then $\lceil \frac{n}{1+\Delta(G)} \rceil \leq \gamma(G) \leq n - \Delta(G)$.

**8.31.** Prove that if the diameter of a connected graph $G$ of order $n$ is at least 3, then $\gamma(\overline{G}) = 2$.

**8.32.** Use Theorem 8.15 to give an alternative proof of Corollary 8.13.

**8.33.** Show that if $G$ is a graph of order $n \geq 2$, then $3 \leq \gamma(G) + \gamma(\overline{G}) \leq n + 1$.

**8.34.** Prove or disprove: Every nontrivial connected graph has two disjoint minimal dominating sets.

## Section 8.4. Factorizations and decompositions

**8.35.** Show that the Petersen graph is not 1-factorable.

**8.36.** Give an example of a connected graph $G$ of composite size having the property that whenever $F$ is a factor of $G$ and the size of $F$ divides the size of $G$, then $G$ is $F$-factorable.

**8.37.** Consider the graph $Q_n$ for $n \geq 1$.
 (a) Prove that $Q_n$ is 1-factorable for all $n \geq 1$.
 (b) Prove that $Q_n$ is $k$-factorable if and only if $k \mid n$.

**8.38.** Prove Kőnig's line coloring theorem 6.7 using the fact that every $r$-regular bipartite graph is 1-factorable (Theorem 8.17).

**8.39.** It was shown (following Theorem 8.20) that the bound on $r$ in the 1-factorization theorem cannot be improved when $n \equiv 2 \pmod 4$ by observing that the graph $G = 2K_{n/2}$ contains no 1-factor and is therefore not 1-factorable. Show that if $n \geq 10$ and $n \equiv 2 \pmod 4$, then a 2-connected $r$-regular graph $G$ of order $n$, where $r = \frac{n}{2} - 1$, need not be 1-factorable.

**8.40.** Show for even integers $n \geq 8$ with $n \equiv 0 \pmod 4$ that the bound on $r$ in the 1-factorization theorem cannot be improved.

**8.41.** Use the proof of Theorem 8.21 to give a 2-factorization of the 4-regular graph $C_8^2$ (the square of the 8-cycle $C_8$).

**8.42.** Use the proof of Theorem 8.22 to produce a hamiltonian factorization of $K_9$.

**8.43.** Use Theorems 8.18 and 8.22 to prove the following: Let $r$ and $n$ be integers with $0 \leq r \leq n - 1$. Then there exists an $r$-regular graph of order $n$ if and only if $r$ and $n$ are not both odd.

**8.44.** Prove Corollary 8.23: for each positive integer $k$, the complete graph $K_{2k}$ can be factored into $k$ hamiltonian paths.

**8.45.** Show that $K_{2k+1}$ cannot be factored into hamiltonian paths.

**8.46.** Give a constructive proof of Theorem 8.24: for each positive integer $k$, the graph $K_{2k}$ can be factored into $k - 1$ hamiltonian cycles and a 1-factor.

**8.47.** Show for every positive integer $a$ and odd positive integer $b$, that there exists a factorization of $K_{2a+b+1}$ into $a$ hamiltonian cycles and $b$ 1-factors.

**8.48.** Show for every positive integer $k$ that the complete graph $K_{6k+4}$ is 3-factorable, where each 3-factor is hamiltonian.

**8.49.** By Theorem 8.21 (and Exercise 8.41) the 4-regular graph $C_8^2$ is 2-factorable. Show that $C_8^2$ is, in fact, hamiltonian factorable.

**8.50.** By Dirac's theorem (Corollary 3.10), every 4-regular graph of order 8 is hamiltonian. Suppose that a 4-regular graph $G$ of order 8 has a hamiltonian cycle $C$ with the property that $H = G - E(C) = C_3 + C_5$. Show that $G$ is hamiltonian factorable.

**8.51.** Show that there is a $P_5$-decomposition of the graph of the octahedron (see Figure 8.13).

**8.52.** Show that the graph of the octahedron (see Figure 8.13) is not $K_{1,4}$-decomposable.

**8.53.** Prove that if there exists a Steiner triple system of order $n \geq 3$, then either $n \equiv 1$ (mod 6) or $n \equiv 3$ (mod 6).

**8.54.** Find a $C_k$-decomposition of $K_9$ for all possible values of $k$.

**8.55.** Give a cycle decomposition of $K_7$ such that the decomposition contains a maximum number of cycles of distinct lengths.

**8.56.** Give an example of a cycle decomposition of $K_7$ such that each cycle in the decomposition is one of two lengths and the number of cycles of each length is the same.

### Section 8.5.  Labelings of graphs

**8.57.** Show that the graph of order 5 in Figure 8.16 obtained by identifying a vertex in two triangles is not graceful.

**8.58.** Determine whether the graph of order 6 obtained by identifying a vertex in a triangle and a vertex in a 4-cycle is graceful.

**8.59.** Prove that the complete graph $K_n$ ($n \geq 2$) is graceful if and only if $n \leq 4$.

**8.60.** Prove that every complete bipartite graph is graceful.

**8.61.** Use the fact that $K_3$ is graceful to find a $K_3$-decomposition of $K_7$.

**8.62.** Find a noncomplete regular connected $K_3$-decomposable graph.

**8.63.** Find an $F$-decomposition of $K_{12}$ where $F = 2P_2 \cup 2P_3$.

**8.64.** Find a $P_6$-decomposition of $K_{10}$.

**8.65.** For each integer $k \geq 1$, show that
   (a) $K_{2k+1}$ is $K_{1,k}$-decomposable.
   (b) $K_{2k}$ is $K_{1,k}$-decomposable.

**8.66.** Let $m$ be an even integer and let $G$ be a graceful graph of size $m$. Show that $K_{3m+1}$ can be decomposed into $3m + 1$ copies of $G$ and an $m$-regular graph of order $3m + 1$.

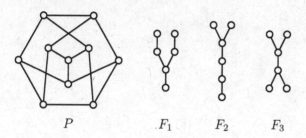

**Figure 8.18.** The Petersen graph and graphs $F_1$, $F_2$ and $F_3$ in Exercise 8.67

**8.67.** Let $P$ be the Petersen graph shown in Figure 8.18.
  (a) For each $F_i$, $i = 1, 2, 3$, shown in Figure 8.18, find an $F_i$-decomposition of $P$.
  (b) Does there exist a decomposition of the Petersen graph into $\{F_1, F_2, F_3\}$?

**8.68.** Find all subgraphs $F$ of size 3 such that the Petersen graph $P$ is $F$-decomposable.

**8.69.** Determine graceful labelings of $C_{15}$ and $C_{16}$.

**8.70.** Prove that the cycle $C_n$ is graceful if and only if $n \equiv 0 \pmod 4$ or $n \equiv 3 \pmod 4$.

**8.71.** By Exercise 8.70, the cycle $C_6$ is not graceful. Find a $\rho$-labeling of $C_6$ and use it to show that there is a $C_6$-decomposition of $K_{13}$.

**8.72.** By Exercise 8.70, the graph $G = C_9$ is not graceful.
  (a) Find a $\rho$-labeling of $G$.
  (b) What information about $G$ can be obtained from (a)?
  (c) From (a) and (b), what more general information can be obtained about a graph with a $\rho$-labeling?

**8.73.** Determine graceful labelings of $P_6$, $P_7$, $P_9$ and $P_{10}$.

**8.74.** Show that the following classes of trees are graceful:
  (a) double stars,
  (b) caterpillars.

**8.75.** Show that no disconnected forest is graceful.

**8.76.** Give an example of a disconnected graph without isolated vertices having order $n$ and size $m$ with $m \geq n - 1$ that is not graceful.

**8.77.** Prove for every integer $k \geq 2$ that there exists a graceful graph having chromatic number $k$.

**8.78.** Let $G$ be a graceful graph of order $n$ and size $m$ such that there is a graceful labeling $f$ of $G$ that assigns the labels $a_1, a_2, \ldots, a_n$ to the vertices of $G$. Show that the complementary labeling that replaces the vertex label $a_i$ by $m - a_i$ is also a graceful labeling of $G$.

# Extremal graph theory  9

Extremal graph theory is an elegant and deeply developed area of graph theory. In an **extremal problem**, one is interested in maximizing (or minimizing) some parameter of a graph (such as its number of edges) subject to some constraint. We have seen extremal problems in earlier chapters when we determined the minimum number of edges in a graph of fixed order that guarantees that it is connected, that it contains a cycle, or that it is hamiltonian. In this chapter, we consider further extremal problems, focusing on guaranteeing the existence of cliques and cycles. We go on to consider Ramsey theory, cages, and Moore graphs.

## 9.1  Avoiding a complete graph

We begin with one of the most quintessential extremal problems: determining the maximum number of edges in a graph of order $n$ that does not contain a clique of given order. The first such result was obtained by Mantel [161] in 1907.

**Theorem 9.1** (Mantel's theorem). *The maximum possible number of edges in a triangle-free graph of order $n$ is $\lfloor n^2/4 \rfloor$.*

*Proof.* Let $G = (V, E)$ be a triangle-free graph of order $n$. Since $G$ does not contain a $K_3$, every neighborhood $N(v)$ must be an independent set. Let $A \subseteq V$ be an independent set of maximum cardinality, so $\deg v \leq |A|$ for all $v \in V$. Every edge of $G$ must be incident to at least one vertex in $V \setminus A$, so we have

$$|E| \leq \sum_{v \in V \setminus A} \deg v \leq \sum_{v \in V \setminus A} |A| \leq |A| \cdot |V \setminus A|.$$

Setting $x = |A|$ and ignoring for a moment that $|A|$ must be an integer, we seek to maximize the quadratic $x(n - x) = nx - x^2$. This maximum occurs at $x = n/2$, giving $|E| \leq n^2/4$. Because $|E|$ must itself be an integer, we must actually have $|E| \leq \lfloor n^2/4 \rfloor$, proving the result. ∎

We can achieve the bound in Theorem 9.1 with a bipartite graph whose parts are nearly as equal as possible, namely $K_{\lceil n/2 \rceil, \lfloor n/2 \rfloor}$. Exercise 9.4 asks the reader to strengthen our proof by showing that these are the only graphs that attain the bound, that is, the only extremal graphs for this problem. Mantel's original proof is outlined in Exercise 9.5.

## Turán's theorem

A more general result was obtained in 1941 by Turán [234], and it is this result that is often credited as the beginning of extremal graph theory. Instead of a triangle-free graph, suppose we now want to find a graph with as many edges as possible that does not contain an $(r + 1)$-clique (for $r \geq 1$).

One family of graphs that do not contain $(r + 1)$-cliques consists of the complete $r$-partite graphs, $K_{n_1, n_2, \ldots, n_r}$ for positive integers $n_1, n_2, \ldots, n_r$. Furthermore, among the $r$-partite graphs, the graphs with the most edges are those in which the part cardinalities are as close to each other as possible: if $n_i \geq n_j + 2$, then by moving a vertex from the part of cardinality $n_i$ to the part of cardinality $n_j$, we would gain $n_i - n_j - 1 \geq 1$ edges.

In fact, for any integers $n \geq r \geq 1$, there is a unique partition $(n_1, n_2, \ldots, n_r)$ of $n$ with parts as equal as possible, or in other words, satisfying

$$n = n_1 + n_2 + \cdots + n_r, \quad n_1 \geq n_2 \geq \cdots \geq n_r \geq 1, \quad \text{and} \quad n_1 - n_r \leq 1.$$

The **Turán graph** $T(n, r)$ is defined as the complete $r$-partite graph $K_{n_1, n_2, \ldots, n_r}$ with these part sizes.

For example, for $n = 11$ and $r = 3$, the partition is $(n_1, n_2, n_3) = (4, 4, 3)$ and the Turán graph is $T(11, 3) = K_{4,4,3}$, while for $n = 12$ and $r = 8$, the partition is $(n_1, n_2, \ldots, n_8) = (2, 2, 2, 2, 1, 1, 1, 1)$ and the Turán graph is $T(12, 8) = K_{2,2,2,2,1,1,1,1}$. When $r$ evenly divides $n$, $T(n, r)$ consists of $r$ parts each of cardinality $n/r$. Otherwise, $T(n, r)$ is made up of parts of cardinality $\lceil n/r \rceil$ and $\lfloor n/r \rfloor$, the number of each determined by the remainder when $n$ is divided by $r$.

Our analysis above shows that, among all $r$-partite graphs of order $n$, the Turán graph $T(n, r)$ is the unique graph with maximum number of edges. Turán [234] proved that this remains true even when we consider arbitrary graphs avoiding $(r + 1)$-cliques. As with many famous theorems, there are a multitude of proofs of Turán's theorem; Aigner and Ziegler present five of them in *Proofs from the Book* [3]. The proof presented below is due to Erdős [79]. Turán's original proof is outlined in Exercises 9.7–9.9.

**Theorem 9.2** (Turán's theorem). *For all integers $n \geq r \geq 1$, among all graphs of order $n$ that do not contain an $(r + 1)$-clique, there exists precisely one with the maximum number of edges, namely $T(n, r)$.*

*Proof.* We proceed by induction on $r$. Since the only graph of order $n$ that does not contain a 2-clique is $T(n, 1) = \overline{K}_n$, the result holds for $r = 1$. Assume now that $r \geq 2$ and that the result holds for $r - 1$.

Let $G = (V, E)$ be a graph of order $n$ and maximum number of edges that does not contain an $(r + 1)$-clique and let $u \in V$ denote a vertex of maximum

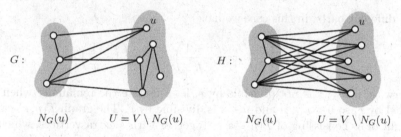

**Figure 9.1.** The operation performed in the proof of Turán's theorem

degree in $G$, so $\deg_G(u) = \Delta(G)$. Further set $S = V \setminus N_G(u)$, so $V = N_G(u) \cup S$. Note that since all of the vertices in $N_G(u)$ are adjacent to $u$, $G[N_G(u)]$ cannot contain an $r$-clique. We now have the situation shown on the left of Figure 9.1.

Form a new graph $H$ as shown on the right of Figure 9.1 by removing all edges amongst the vertices of $S$ and then making every vertex of $S$ adjacent to every vertex of $N_G(u)$. Thus $H$ consists of the join of $G[N_G(u)]$ with an independent set of order $n - \Delta(G)$.

Because $G[N_G(u)]$ does not contain an $r$-clique, it follows that $H$ does not contain an $(r+1)$-clique, and thus $H$ has at most as many edges as $G$ by our choice of $G$. In fact, we claim that $\deg_H(v) \geq \deg_G(v)$ for all $v \in V$. If $v \in N_G(u)$, then this follows because $N_H(v) \supseteq N_G(v)$ for these vertices, whereas if $v \in S$, then this follows because

$$\deg_H(v) = \deg_H(u) = \deg_G(u) = \Delta(G) \geq \deg_G(v).$$

Since we have already observed that $H$ has at most as many edges as $G$, we must have $\deg_H(v) = \deg_G(v)$ for all vertices $v \in V$. In particular, this implies that $N_H(v) = N_G(v)$ for all $v \in N_G(u)$, which shows that all of the vertices in $N_G(u)$ were already adjacent to all of the vertices of $S$. That in turn implies that $G[S]$ was already an independent set, and thus $G$ is the join of $G[N_G(u)]$ with $S$.

Since $G[N_G(u)]$ must have the maximum number of edges for its order subject to not containing an $r$-clique, our induction hypothesis implies that $G[N_G(u)]$ is a Turán graph, and is consequently $(r-1)$-partite. It follows that $G$ is itself $r$-partite. We have already observed that among all $r$-partite graphs of order $n$, the Turán graph $T(n,r)$ is the unique graph with the maximum number of edges. Thus we must have $G = T(n,r)$, as we sought to show. ∎

## Counting edges in the Turán graph

It remains to determine the number of edges in the Turán graph $T(n,r)$, which we denote by $t(n,r)$. If $r$ evenly divides $n$, then $T(n,r)$ consists of $r$ parts each of cardinality $n/r$. Since $T(n,r)$ contains edges between every pair of vertices that lie in different parts, in this case we have

$$t(n,r) = \binom{r}{2}\left(\frac{n}{r}\right)^2 = \left(\frac{r-1}{r}\right)\cdot\frac{n^2}{2} = \left(1-\frac{1}{r}\right)\cdot\frac{n^2}{2}.$$

Otherwise, if $n \geq r$ is not divisible by $r$, let $s$ denote the remainder when $n$ is divided by $r$, so $0 < s < n$ and $n - s$ is divisible by $r$. The graph $T(n,r)$ can be thought of as consisting of $T(n - s, r)$ together with $s$ extra vertices, which are distributed so that each part of $T(n - s, r)$ gets at most one. Each of these extra vertices is adjacent to all of the vertices of $T(n - s, r)$ that are not in its part together with all of the other extra vertices. Therefore,

$$t(n,r) = t(n-s,r) + s\cdot(r-1)\frac{n-s}{r} + \binom{s}{2}.$$

As we already have a formula for $t(n - s, r)$, we can simplify the above to

$$t(n,r) = \left(1-\frac{1}{r}\right)\frac{n^2-s^2}{2} + \binom{s}{2}.$$

Exercise 9.2 asks the reader to verify that $t(n,r)$ is always at most the formula in the case where $r$ divides $n$, and thus we have the following immediate consequence.

**Corollary 9.3.** *Suppose that $n \geq r \geq 1$. If a graph of order $n$ has*

$$\left(1-\frac{1}{r}\right)\cdot\frac{n^2}{2} + 1$$

*or more edges, then it contains an $(r + 1)$-clique.*

# 9.2  Containing cycles and trees

Turán's theorem tells us the minimum number of edges in a graph of order $n$ that guarantees an $(r + 1)$-clique. We now look at the minimum number of edges required to guarantee the existence of other subgraphs.

 ## Graphs with vertex-disjoint cycles

We already know that the minimum number of edges in a graph of order $n \geq 3$ that guarantees a cycle is $n$, as such a graph can no longer be a forest. Erdős and Pósa [82] established bounds on the number of edges in a graph to guarantee the existence of $k \geq 2$ vertex-disjoint cycles, which they referred to as a "question of

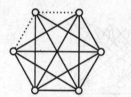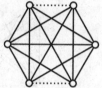

**Figure 9.2.** The two graphs of order 6 with 13 edges, with dotted lines denoting non-edges

turánian type". We consider the case $k = 2$ below. In [77], Erdős wrote that the proof of this special case was due entirely to Pósa, and it is essentially his proof that we present. Notably, Pósa was thirteen years old when he wrote [82] with Erdős.

**Theorem 9.4.** *Every graph of order $n \geq 6$ and at least $3n - 5$ edges contains two vertex-disjoint cycles.*

*Proof.* It suffices to prove the result for graphs of order $n$ with precisely $3n - 5$ edges, and to do this we use induction on $n$. As the complete graph $K_6$ has 15 edges, every graph of order 6 with 13 edges can be obtained by removing two edges from $K_6$. There are two ways to do this, shown in Figure 9.2, and it can be seen by inspection that both graphs contain two vertex-disjoint triangles. This proves the result for the base case $n = 6$.

Now assume that for a given integer $n \geq 7$, all graphs on fewer than $n$ vertices satisfy the theorem, and let $G$ be a graph of order $n$ and with $3n - 5$ edges. Using the first theorem of graph theory (Theorem 1.1), we have

$$\sum_{v \in V(G)} \deg v = 2|E(G)| = 6n - 10,$$

so there is some vertex $u \in V(G)$ with $\deg u \leq 5$. We consider three cases, depending on the degree of $u$.

First, suppose that $\deg u \leq 3$. In this case, the graph $G - u$ has order $n - 1$ and at least $3(n - 1) - 5$ edges. Thus, $G - u$ contains two vertex-disjoint cycles by induction, and we are done.

Next, suppose that $\deg u = 4$ and let $N(u) = \{v_1, v_2, v_3, v_4\}$. Here, we need two subcases, depending on whether $G[N[u]]$ is complete. If $G[N[u]] \neq K_5$, then there are at least two vertices of $N(u)$ that are not adjacent. Without loss of generality, we may assume that these vertices are $v_1$ and $v_2$. The graph $G - u + v_1 v_2$ has order $n - 1$ and $3(n - 1) - 5$ edges, so it contains two vertex-disjoint cycles by induction. At least one of these cycles, does not contain the vertex $v_1$, and thus it does not contain the edge $v_1 v_2$ and so is a cycle of $G$. Label these cycles as $C_1$ and $C_2$ so that $C_1$ is a cycle of $G$. If $C_2$ also does not contain the edge $v_1 v_2$, then it is also a cycle of $G$ and we are done. Otherwise, we may replace the edge $v_1 v_2$ on $C_2$ with the edges $v_1 u$ and $u v_2$ to obtain a second cycle of $G$ that is vertex-disjoint from $C_1$.

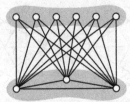

**Figure 9.3.** A graph of order 9 with 21 edges that contains neither two vertex-disjoint cycles nor a subgraph of minimum degree 4

We must also consider the case where $\deg u = 4$ and $G[N[u]] = K_5$. If there is any vertex $w \in V(G) \setminus N[u]$ that is adjacent to two or more vertices of $N(u)$, then we can find two vertex-disjoint triangles: one with $w$ and two vertices of $N[u]$, and the other with $u$ and the other two vertices of $N[u]$. Hence, we may assume that no vertex of $V(G) \setminus N[u]$ is adjacent to more than one vertex of $N(u)$. Thus the graph $G - u - v_1 - v_2$ has $n - 3$ vertices and at least $(3n - 5) - (n - 5) - 9 = 2n - 9$ edges. Because $n \geq 6$, we have $2n - 9 \geq n - 3$, and thus $G - u - v_1 - v_2$ contains at least one cycle. As that cycle and the triangle formed by $u$, $v_1$, and $v_2$ are vertex-disjoint, we are also done in this case.

Finally, suppose that $\deg u = 5$, and let $N(u) = \{v_1, v_2, v_3, v_4, v_5\}$. The subgraph $G[N[u]]$ has order 6, and thus we are done by the base case of our induction if it has 13 or more edges. Therefore, we may assume that $G[N[u]]$ has at most 12 edges. Since $\deg u = 5$, this means that $G[N(u)]$ contains at most 7 edges. This implies that there must be some neighbor $v_1$ of $u$ that is not adjacent to two other neighbors $v_2$ and $v_3$ of $u$. The graph $G - u + v_1v_2 + v_1v_3$ then has order $n - 1$ and $3(n-1) - 5$ edges, and so it must contain two vertex-disjoint cycles by induction. At least one of these two cycles must not contain the vertex $v_1$, and thus it is a cycle of $G$. Label these cycles $C_1$ and $C_2$ so that $C_1$ is a cycle of $G$.

If $C_2$ contains neither $v_1v_2$ nor $v_1v_3$, then it is also a cycle of $G$, and we are done. If $C_2$ contains precisely one of these two edges, then we may replace that edge with a path of length 2 through $u$, and we are also done. Finally, if $C_2$ contains both $v_2v_1$ and $v_1v_3$, then we may replace both of those edges with the edges $v_2u$ and $uv_3$. ∎

To see that the bound $3n - 5$ presented in Theorem 9.4 is sharp, observe that the complete 4-partite graph $K_{n-3,1,1,1}$ has order $n$ and $3n - 6$ edges (in the case of $n = 9$, this graph is shown in Figure 9.3). Every cycle in $K_{n-3,1,1,1}$ contains at least two of the three vertices of degree $n - 1$. Thus, no two cycles of $K_{n-3,1,1,1}$ are vertex-disjoint.

## Graphs with edge-disjoint cycles

If we only ask for two cycles that do not share an edge, then $n + 4$ edges suffice, as shown by the following result. This result also appeared in the paper of Erdős and Pósa [82], and as with our previous result, Erdős [77] later credited it entirely to Pósa.

**Theorem 9.5.** *Every graph of order $n \geq 5$ with at least $n + 4$ edges contains two edge-disjoint cycles.*

*Proof.* It suffices to prove the result for graphs of order $n$ and with precisely $n + 4$ edges, and to do this we use induction on $n$. If $G$ is a graph of order 5 with 9 edges, then $G$ is formed by removing a single edge from $K_5$, and this graph has two edge-disjoint cycles. Now assume that $G$ has order $n \geq 6$ and $n + 4$ edges and that the result holds for all graphs of smaller order.

We may assume that $G$ does not contain a 3-cycle or a 4-cycle, because otherwise we could remove those edges to obtain a graph of order $n$ and at least $n$ edges, which must itself contain a cycle. We now consider cases depending on the minimum degree of $G$.

If $G$ has a vertex $u$ of degree 1 then $G - u$ contains two edge-disjoint cycles by induction, and we are done. Now suppose that $G$ has a vertex $u$ of degree 2 and that $N(u) = \{v_1, v_2\}$. Because $G$ does not contain a 3-cycle, $G$ does not contain the edge $v_1 v_2$. The graph $G - u + v_1 v_2$ thus has order $n - 1$ and $n + 3$ edges, and so by induction it contains two edge-disjoint cycles. At least one of these cycles does not contain the edge $v_1 v_2$, and thus is a cycle of $G$. Label these cycles $C_1$ and $C_2$ so that $C_1$ is a cycle of $G$. If $C_2$ does not contain $v_1 v_2$, then it is also a cycle of $G$ and we are done. Otherwise, we may replace the edge $v_1 v_2$ in $C_2$ with the edges $v_1 u$ and $u v_2$ to obtain a second cycle of $G$ that is edge-disjoint from $C_1$.

Now suppose that $G$ has minimum degree at least 3. By counting edges (using the first theorem of graph theory, Theorem 1.1), we obtain the inequality

$$2n + 8 = 2|E(G)| = \sum_{v \in V(G)} \deg v \geq 3n.$$

This implies that $n \leq 8$. Let $u$ be a vertex of such a graph and label three of its neighbors as $v_1$, $v_2$, and $v_3$. These three vertices form an independent set because $G$ does not contain a 3-cycle. Thus each of these vertices must have two additional neighbors. These additional neighbors must all be distinct, because $G$ does not contain a 4-cycle. However, this implies that $n \geq 10$, which is a contradiction. ∎

The bound in Theorem 9.5 is not sharp for $n = 5$ because $n + 3 = 8$ edges suffice to guarantee two edge-disjoint cycles in this case (see Exercise 9.11), but it is sharp for $n \geq 6$, as shown in Exercise 9.13.

 ## Subgraphs of given minimum degree

We now turn to the problem of determining the number of edges that a graph must have to guarantee that it contains a subgraph of specified minimum degree.

**Theorem 9.6.** *Every graph of order $n \geq k + 1$ with at least*

$$(k - 1)n - \binom{k}{2} + 1$$

*edges contains a subgraph of minimum degree $k$.*

*Proof.* We proceed by induction on $n$, treating $k$ as fixed. In the base case where $n = k + 1$, the bound in the theorem is

$$(k - 1)n - \binom{k}{2} + 1 = (n - 2)n - \binom{n - 1}{2} + 1 = \binom{n}{2}.$$

Thus the only graph satisfying the hypotheses is $K_n = K_{k+1}$. Since $K_{k+1}$ is itself a graph of minimum degree $k$, the base case holds.

Now assume that the theorem holds for graphs of order $n - 1$, and let $G$ be a graph of order $n$ and with the number of edges specified by the theorem. If $G$ does not have minimum degree $k$ itself, then it contains a vertex $v$ with $\deg v \leq k - 1$. Therefore, the order of $G - v$ is $n - 1$, and $G - v$ has at least

$$\left( (k - 1)n - \binom{k}{2} + 1 \right) - (k - 1) = (k - 1)(n - 1) - \binom{k}{2} + 1$$

edges. It follows by the induction hypothesis that $G - v$, and thus $G$ as well, contains a subgraph of minimum degree $k$. ∎

The bound given in Theorem 9.6 cannot be improved, because the complete $k$-partite graph $K_{n-(k-1),1,1,\ldots,1}$ of order $n$ has

$$(k - 1)(n - (k - 1)) + \binom{k - 1}{2} = (k - 1)n - \binom{k}{2}$$

edges, but does not contain a subgraph of minimum degree $k$. For example, in the case $n = 9$ and $k = 4$, the graph $K_{6,1,1,1}$ shown in Figure 9.3 has no subgraph of minimum degree 4.

By Theorem 1.20, if $G$ is a graph of minimum degree at least $k$, then $G$ contains every tree with $k$ edges as a subgraph. Combining this result with Theorem 9.6 gives us the following corollary.

**Corollary 9.7.** *Every graph of order $n \geq k$ and with at least*

$$(k - 1)n - \binom{k}{2} + 1$$

*edges contains every tree with $k$ edges as a subgraph.*

The bound in Corollary 9.7 can certainly be improved (see Exercise 9.15). A 1963 conjecture of Erdős and Sós, first posed in [76], states that every graph of order $n \geq k$ with more than $(k - 1)n/2$ edges contains every tree with $k$ edges as a subgraph.

# 9.3 Ramsey theory

Ramsey theory is one of the most well-known and widely studied areas within extremal graph theory. This field is named after Frank Plumpton Ramsey (1903–1930), a British mathematician who also had interests in literature, philosophy, economics, and politics. Ramsey's most famous result, which appeared as a minor lemma in his paper, "On a Problem of Formal Logic" [194], was published the year he died.

 ## Classical Ramsey numbers

A **red-blue edge coloring** of $K_n$ is one in which every edge is colored red or blue. We do not insist that these colorings be proper, so adjacent edges are allowed to receive the same color. For positive integers $k$ and $\ell$, the **Ramsey number $R(k, \ell)$** is the least positive integer $n$ such that every red-blue edge coloring of $K_n$ contains a clique of order $k$ whose edges are all colored red (a **red $K_k$**) or a clique of order $\ell$ whose edges are all colored blue (a **blue $K_\ell$**). A specialization of Ramsey's original result [194] shows that $R(k, \ell)$ exists; we state this result here but prove something stronger in Theorem 9.10.

**Theorem 9.8 (Ramsey's theorem).** *For every two integers $k, \ell \geq 1$, the Ramsey number $R(k, \ell)$ exists.*

From the definition it is apparent that $R(k, \ell)$ is symmetric in $k$ and $\ell$, so $R(k, \ell) = R(\ell, k)$. It is also straightforward to show (Exercise 9.24) that $R(k, \ell)$ exists if either of $k$ or $\ell$ does not exceed 2 and that

$$R(1, \ell) = 1 \quad \text{and} \quad R(2, \ell) = \ell.$$

The degree of difficulty in determining the values of other Ramsey numbers increases sharply as $k$ and $\ell$ increase, and no general formulas like those above are known.

 ## The Ramsey number $R(3, 3)$

Since we have $R(1, \ell) = 1$ and $R(2, \ell) = \ell$ for all $\ell$, the first interesting Ramsey number is $R(3, 3)$.

**Theorem 9.9.** *We have $R(3, 3) = 6$.*

*Proof.* Let $V(K_5) = \{v_1, v_2, v_3, v_4, v_5\}$ and define a red-blue edge coloring of $K_5$ by coloring each edge of the 5-cycle $v_1, v_2, v_3, v_4, v_5, v_1$ red and the remaining

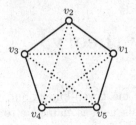

**Figure 9.4.** A depiction of a red-blue edge coloring of $K_5$ with no monochromatic $K_3$

edges blue, as indicated by Figure 9.4. Since this coloring contains neither a red $K_3$ nor a blue $K_3$, it follows that $R(3,3) \geq 6$.

To show that $R(3,3) \leq 6$, consider an arbitrary vertex $v_1$ of an arbitrary red-blue edge coloring of $K_6$. Since $v_1$ is incident with five edges, it follows that at least three of these five edges are colored the same. Suppose without loss of generality that the edges $v_1v_2$, $v_1v_3$, and $v_1v_4$ are all colored red. If any of the edges $v_2v_3$, $v_2v_4$, and $v_3v_4$ is also colored red, then we have a red $K_3$. Otherwise, all three of these edges are colored blue, so we have a blue $K_3$. This proves that $R(3,3) \leq 6$. ∎

The Ramsey number $R(k, \ell)$ can be defined without regard to red-blue edge colorings. Given a red-blue edge coloring of $K_n$, we can delete the blue edges and ignore the colors of the red edges to obtain a graph on $n$ vertices. If follows that $R(k, \ell)$ is the smallest integer $n$ such that every graph of order $n$ contains either a clique of order $k$ or an independent set of order $\ell$.

Theorem 9.9 can be phrased as a question about a party: what is the minimum number of people who need to be at a gathering to ensure that there are either three mutual acquaintances or three mutual strangers? This situation can be represented by a graph with $n$ vertices, where an edge indicates that two people are acquaintances. By Theorem 9.9, any gathering of six people will contain three mutual acquaintances (literally a clique of order 3) or three mutual strangers.

## Upper bounds on $R(k, \ell)$

The first upper bound on $R(k, \ell)$ was obtained in 1935 by Erdős and Szekeres, in a paper where they rediscovered Ramsey's theorem [84]. We establish their bound below, along with a strengthening in the case where both $R(k-1, \ell)$ and $R(k, \ell-1)$ are even that is due to Greenwood and Gleason [107].

**Theorem 9.10.** *For every two integers* $k, \ell \geq 2$, *the Ramsey number* $R(k, \ell)$ *exists and satisfies*

$$R(k, \ell) \leq R(k-1, \ell) + R(k, \ell-1).$$

*Furthermore, this inequality is strict if* $R(k-1, \ell)$ *and* $R(k, \ell-1)$ *are both even.*

*Proof.* We proceed by induction on $k + \ell$. We have already noted that $R(1, \ell) = 1$ and $R(2, \ell) = \ell$ for all $\ell \geq 1$, from which the result follows whenever $k$ or $\ell$ is at most 2. Now assume that $k, \ell \geq 3$, so $k + \ell \geq 6$.

Let $G = (V, E)$ be a graph of order $R(k - 1, \ell) + R(k, \ell - 1)$. We would like to show that $G$ has either a clique of order $k$ or an independent set of order $\ell$. Let $v$ be a vertex of $G$. There are two cases, depending on the degree of $v$.

If $\deg v \geq R(k - 1, \ell)$, then the induced subgraph $G[N(v)]$ contains either a clique of order $k - 1$ or an independent set of order $\ell$ by induction. If it contains a clique of order $k - 1$, then together with the vertex $v$ we find a clique of order $k$ in $G$. Of course, if $G[N(v)]$ contains an independent set of order $\ell$, then $G$ does as well.

Otherwise, we must have $|V \setminus N[v]| \geq R(k, \ell - 1)$, so the induced subgraph $G[V \setminus N[v]]$ contains either a clique of order $k$ or an independent set of order $\ell - 1$. If it contains a clique of order $k$, then so does $G$. If $G[V \setminus N[v]]$ contains an independent set of order $\ell - 1$, then together with the vertex $v$, we find an independent set of order $\ell$ in $G$.

Thus $R(k, \ell)$ exists and is bounded above by $R(k - 1, \ell) + R(k, \ell - 1)$. Now suppose that $R(k - 1, \ell)$ and $R(k, \ell - 1)$ are both even, and let $G = (V, E)$ be a graph of order $R(k-1, \ell) + R(k, \ell-1) - 1$. Because $G$ has odd order, and because every graph has an even number of vertices of odd degree (Corollary 1.2), $G$ must have some vertex $v$ of even degree. If $\deg v \geq R(k - 1, \ell)$, then we proceed as above to see that $G$ contains a clique of order $k$ or an independent set of order $\ell$. If $\deg v < R(k - 1, \ell)$, then since $\deg v$ and $R(k - 1, \ell)$ are both even, we must have $\deg v \leq R(k - 1, \ell) - 2$, so $|V \setminus N[v]| \geq R(k, \ell - 1)$, and we also proceed as above. ■

The bound established in Theorem 9.10 is similar to the recurrence for binomial coefficients used to construct Pascal's triangle,

$$\binom{n + 1}{k + 1} = \binom{n}{k + 1} + \binom{n}{k},$$

and this similarity leads quickly to the concrete upper bound of our next result.

**Corollary 9.11.** *For every two positive integers $k$ and $\ell$,*

$$R(k, \ell) \leq \binom{k + \ell - 2}{k - 1} = \binom{k + \ell - 2}{\ell - 1}.$$

*Proof.* We again proceed by induction on $k + \ell$. In the case where $k = 1$ or $\ell = 1$, we have that $R(1, \ell) = R(k, 1) = 1$, and also that

$$\binom{1 + \ell - 2}{1 - 1} = \binom{\ell - 1}{0} = 1 = \binom{k - 1}{k - 1} = \binom{k + 1 - 2}{k - 1}.$$

Thus, we may assume that $k$ and $\ell$ are both at least 2 and that the corollary holds for all smaller values of $k + \ell$. Theorem 9.10 and our induction hypothesis therefore imply that

$$R(k, \ell) \leq R(k - 1, \ell) + R(k, \ell - 1)$$
$$\leq \binom{k + \ell - 3}{k - 2} + \binom{k + \ell - 3}{k - 1} = \binom{k + \ell - 2}{k - 1},$$

completing the proof. ∎

The bound for $R(k, \ell)$ given in Corollary 9.11 is the true value if one of $k$ or $\ell$ is 1 or 2, or if $k = \ell = 3$. When $k = 3$ and $\ell$ is any positive integer, this result implies that

$$R(3, \ell) \leq \binom{\ell + 1}{2} = \frac{\ell^2 + \ell}{2}.$$

We improve this bound slightly below.

**Theorem 9.12.** *For every positive integer $\ell$, we have*

$$R(3, \ell) \leq \frac{\ell^2 + 3}{2}.$$

*Proof.* We proceed by induction on $\ell$. For $\ell = 1$, we have $R(3, 1) = 1 \leq 2 = (1^2 + 3)/2$. Now assume that $\ell \geq 2$ and that

$$R(3, \ell - 1) \leq \frac{(\ell - 1)^2 + 3}{2}.$$

By Theorem 9.10 and our induction hypothesis, we have

$$R(3, \ell) \leq R(2, \ell) + R(3, \ell - 1) \leq \ell + \frac{(\ell - 1)^2 + 3}{2} = \frac{\ell^2 + 4}{2}.$$

Since $R(3, \ell)$ must be an integer, it suffices to show that we never have equality above. Suppose to the contrary that we did have equality. Then $(\ell^2 + 4)/2$ must be an integer (because $R(3, \ell)$ is an integer), so $\ell = R(2, \ell)$ must be even. Thus in fact $(\ell^2 + 4)/2$ must be even. We would also have to have $R(3, \ell - 1) = ((\ell - 1)^2 + 3)/2$, and this quantity would have to be even. However, in that case the bound in Theorem 9.10 is strict, which yields a contradiction. ∎

Theorem 9.12 shows that $R(3, 4) \leq 9$ and $R(3, 5) \leq 14$. Both of these bounds can be achieved, as shown in Figure 9.5, so the bound is tight in these cases. However, the bound of Theorem 9.12 implies that $R(3, 6) \leq 19$, while the actual value is 18, as first proved in 1964 by Kéry [143].

## Diagonal Ramsey numbers

Corollary 9.11 gives an upper bound on the **diagonal Ramsey numbers** $R(k, k)$, namely

$$R(k, k) \leq \binom{2k - 2}{k - 1}.$$

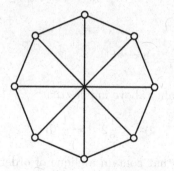 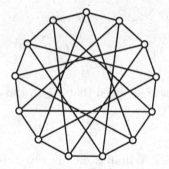

**Figure 9.5.** Graphs showing that $R(3,4) \geq 9$ and $R(3,5) \geq 14$

In theory, one could obtain lower bounds on Ramsey numbers simply by constructing large graphs without cliques or independent sets of order $k$. In practice, this has proved to be extraordinarily difficult, outside of a few small cases. In 1947, Erdős [75] presented an ingenious nonconstructive counting argument that establishes a lower bound on $R(k, k)$. A slightly weaker version of this argument is presented below, but the stronger result can be derived from the same argument with more careful analysis of the inequalities.

**Theorem 9.13.** *For every integer* $k \geq 2$,

$$R(k, k) > \lfloor 2^{k/2} \rfloor.$$

*Proof.* Our previous results show that the result is true for $k = 2$ and $k = 3$, so assume that $k \geq 4$ and let $n = \lfloor 2^{k/2} \rfloor$. We seek to show that there is a graph $G$ of order $n$ such that neither $G$ nor its complement contains $K_k$ as a subgraph. Let $\mathcal{G}$ denote the set of all graphs of order $n$ on the vertices $V = \{v_1, v_2, \ldots, v_n\}$. Two graphs in $\mathcal{G}$ are the same if they have precisely the same edge sets. There are $\binom{n}{2}$ possible edges, so $|\mathcal{G}| = 2^{\binom{n}{2}}$.

For each subset $S \subseteq V$ of order $k$, consider those graphs $G \in \mathcal{G}$ for which $G[S]$ is a complete graph. The $\binom{k}{2}$ edges joining each pair of vertices in $S$ must belong to each such graph, so the number of these graphs is $2^{\binom{n}{2} - \binom{k}{2}}$.

Letting $N$ denote the number of graphs in $\mathcal{G}$ that contain a clique of order $k$, we see that

$$N \leq \sum_{\substack{S \subseteq V, \\ |S| = k}} 2^{\binom{n}{2} - \binom{k}{2}} = \binom{n}{k} 2^{\binom{n}{2} - \binom{k}{2}}.$$

So long as $n \geq k$, we have

$$\binom{n}{k} = \frac{n(n-1)\cdots(n-k+1)}{k!} \leq \frac{n^k}{k!},$$

and this inequality is strict in our case because $n > k$. We also have, for all $k \geq 1$, that $k! \geq 2^{k-1}$. Using these two bounds together with our assumption that

$n = \lfloor 2^{k/2} \rfloor \leq 2^{k/2}$, we obtain

$$\binom{n}{k} < \frac{n^k}{k!} \leq \frac{2^{\frac{k^2}{2}}}{2^{k-1}} = 2^{\frac{k^2}{2}-k+1}.$$

As we have assumed that $k \geq 4$, the inequality above implies that

$$N < 2^{\frac{k^2}{2}-k+1} \cdot 2^{\binom{n}{2}-\binom{k}{2}} = 2^{1-\frac{k}{2}} \cdot 2^{\binom{n}{2}} \leq \frac{1}{2} 2^{\binom{n}{2}} = \frac{1}{2} |\mathcal{G}|.$$

There are $N$ graphs on the vertex set $V$ that contain a clique of order $k$ and also $N$ graphs on this vertex set that contain an independent set of order $k$ (the complements of the graphs containing a clique of order $k$). Therefore, the number of graphs that contain either a clique of order $k$ or an independent set of order $k$ is bounded above by $2N < |\mathcal{G}|$. This means that there must be at least one graph in $\mathcal{G}$ that contains neither a clique of order $k$ nor an independent set of order $k$, proving that $R(k,k) > n = \lfloor 2^{k/2} \rfloor$. ∎

The proof of Theorem 9.13 is a thinly veiled application of the probabilistic method, and we revisit the result from that perspective in the next chapter.

For all positive integers $n$, we have

$$\binom{2n}{n} = \frac{(2n)!}{(n!)^2} = \frac{(2n)(2n-1)}{n^2} \cdot \frac{(2n-2)(2n-3)}{(n-1)^2} \cdots \frac{2 \cdot 1}{1^2} < 4^n.$$

Thus the upper bound provided by Corollary 9.11 shows that $R(k,k) < 4^{k-1}$, while the lower bound of Theorem 9.13 shows that $R(k,k) > 2^{k/2}$. Erdős [80] offered \$100 for a proof of his conjecture that the **exponential growth rate**

$$c = \lim_{k \to \infty} R(k,k)^{1/k}$$

exists, and \$10,000 for a disproof. Assuming that $c$ exists, the results we have proved show that $\sqrt{2} \leq c \leq 4$, and it is remarkable that neither of these two bounds has ever been improved upon. Erdős offered an additional \$250 for the determination of $c$, and suggested in [80] that it might equal 2.

## Known values of $R(k,\ell)$

The few nontrivial classical Ramsey numbers that are known are shown in Figure 9.6. We have already mentioned that Theorem 9.12 and the constructions of Figure 9.5 give us the values of $R(3,4)$ and $R(3,5)$. These two results were first obtained by Greenwood and Gleason and appeared in 1955 [107]. The proof that $R(3,3) = 6$ had been a problem in the 1953 Putnam competition [38, problem I.2], and this appears to have been the sole motivation for Greenwood and Gleason, as their work makes no reference to Ramsey's paper [194] and also rediscovers both Theorem 9.10 and Corollary 9.11. Greenwood and Gleason also obtained

| s \ t | 1 | 2 | 3 | 4 | 5 | 6 | 7 | 8 | 9 | 10 |
|---|---|---|---|---|---|---|---|---|---|---|
| 1 | 1 | 1 | 1 | 1 | 1 | 1 | 1 | 1 | 1 | 1 |
| 2 | 1 | 2 | 3 | 4 | 5 | 6 | 7 | 8 | 9 | 10 |
| 3 | 1 | 3 | 6 | 9 | 14 | 18 | 23 | 28 | 36 | 40–42 |
| 4 | 1 | 4 | 9 | 18 | 25 | 36–41 | | | | |
| 5 | 1 | 5 | 14 | 25 | 43–48 | | | | | |

**Figure 9.6.** Known values and bounds for the Ramsey numbers $R(k, \ell)$

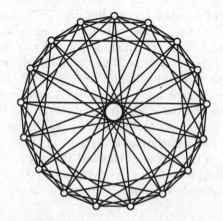

**Figure 9.7.** The Paley graph of order 17, showing that $R(4, 4) \geq 18$

the value $R(4, 4) = 18$; the upper bound follows immediately by Theorem 9.10 because $R(4, 4) \leq 2R(4, 3) = 18$, while the lower bound is obtained by considering the **Paley graph** of order 17 shown in Figure 9.7.

The fact that $R(3, 6) = 18$ was first proved in 1964 by Kéry [143]. Kalbfleisch gave an independent proof that $R(3, 6) = 18$ and provided a construction showing that $R(3, 7) \geq 23$ in his 1966 thesis [137], which seems to have included the first written use of the term "Ramsey number." A matching upper bound was given by Graver and Yackel in 1968 [106], thus establishing that $R(3, 7) = 23$.

All of the computations of Ramsey numbers mentioned so far were done by hand, but all subsequent computations have required the extensive use of computers. In 1982, Grinstead and Roberts [109] proved $28 \leq R(3, 8) \leq 29$ and $R(3, 9) = 36$. As in the $R(3, 7)$ case, the lower bound in the $R(3, 9)$ case had already been established in Kalbfleisch's thesis [137]. In 1992, McKay and Min [166] established that $R(3, 8) = 28$. This is the most recently discovered Ramsey number of the form $R(3, \ell)$.

This brings us to $R(4, 5)$. In this case, the inequality in Theorem 9.10 is strict, so we have

$$R(4, 5) < R(3, 5) + R(4, 4) = 14 + 18 = 32,$$

as first observed by Greenwood and Gleason [107]. Kalbfleisch [136] established the first nontrivial lower bound in 1965: $R(4,5) \geq 25$, which would ultimately be shown to be the true value. In his thesis [137], Kalbfleisch also proved the first nontrivial upper bound, $R(4,5) \leq 30$. This was then slowly lowered over time. First Walker [250] brought the bound down to 29 in 1968, and then to 28 three years later [251]. Twenty years passed until McKay and Radziszowski [168] brought the upper bound down to 27, and then later in unpublished work to 26, and then finally in 1995, they [169] completed the proof that $R(4,5) = 25$.

In 1989, Exoo [89] constructed a red-blue edge coloring of $K_{42}$ having no monochromatic copies of $K_5$, thus showing that $R(5,5) \geq 43$, and this construction has never been improved upon. The best upper bound on $R(5,5)$ at the time of Exoo's construction was 55. That upper bound was subsequently lowered to 53 and then to 52. Then $R(4,5) = 25$ was established, and via Theorem 9.10, that immediately gave the better bound $R(5,5) \leq 50$. In 1997, McKay and Radziszowski [170] lowered the upper bound to 49, and then in 2018, Angeltveit and McKay [7] lowered it to 48 where it stands today. In both [170] and [7], it is conjectured that $R(5,5) = 43$ and that there are precisely 656 red-blue edge colorings of $K_{42}$ with no monochromatic $K_5$.

For more bounds on the unknown classical Ramsey numbers, and for generalized Ramsey numbers, we refer to the dynamic survey of Radziszowski [193] maintained at the *Electronic Journal of Combinatorics*.

## ⬡ Multicolored Ramsey numbers

The **multicolored Ramsey number** $R(n_1, n_2, \ldots, n_k)$ is defined as the least positive integer $n$ such that if every edge of $K_n$ is colored with one of the $k$ colors 1, 2, ..., $k$, then for some value of $i$, $K_n$ contains a subgraph $K_{n_i}$ all of whose edges have color $i$. Ramsey's original theorem from [194] implies that these numbers also exist.

**Theorem 9.14.** *For every $k \geq 1$ positive integers $n_1, n_2, \ldots, n_k$, the Ramsey number $R(n_1, n_2, \ldots, n_k)$ exists.*

When $k = 2$, these are just the classical Ramsey numbers, and cases where we have $n_i \leq 2$ for some index $i$ reduce to smaller problems. Thus the multicolored Ramsey numbers first get interesting with $R(3, 3, 3)$. The value of this Ramsey number was established in the same paper of Greenwood and Gleason [107] that established the values of $R(3, 4)$ and $R(3, 5)$.

**Theorem 9.15.** *We have $R(3, 3, 3) = 17$.*

*Proof.* First we show that $R(3, 3, 3) \leq 17$. Consider any coloring of the edges of $K_{17}$ with the colors red, blue, and green, and let $v$ be an arbitrary vertex. Since each vertex has degree 16, it follows that $v$ is incident with at least six edges of the same color. Suppose this color is green and consider the subgraph induced by $v$ and six of its green neighbors. No edge of this induced subgraph may be

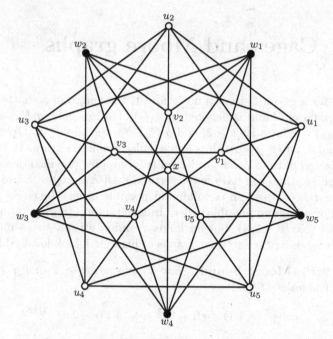

**Figure 9.8.** The Clebsch graph

colored green, because that would create a green triangle with $v$. Thus all of the edges must be colored red or blue. However, we know that $R(3,3) = 6$, so any such coloring must contain a red or blue triangle, and we are done.

To prove that $R(3,3,3) \geq 17$, we must find a coloring of the edges of $K_{16}$ by the colors red, blue, and green with no monochromatic triangle. This can be done by showing that the edges of $K_{16}$ can be partitioned into three disjoint copies of a particular triangle-free graph. This triangle-free graph, which must necessarily be 5-regular of order 16, is known as the **Clebsch graph** (because Clebsch [52] described a related construction in 1868) or as the **Greenwood–Gleason graph** (because they used the graph to show that $R(3,3,3) \geq 17$). To construct this graph, which is shown in Figure 9.8, we start from the Petersen graph on the vertices $u_i$ and $v_i$ for $1 \leq i \leq 5$ and then add six new vertices $x$ and $w_i$ for $1 \leq i \leq 5$ as shown in the figure. This graph has the property that if we remove any vertex and its neighborhood, we are left with the Petersen graph. ■

For over sixty years, $R(3,3,3)$ was the only nontrivial multicolored Ramsey number that was known. Then in 2016, Codish, Frank, Itzhakov, and Miller [53] described a computer search proving that $R(4,3,3) = 30$. The lower bound $R(4,3,3) \geq 30$ had been established fifty years earlier, in Kalbfleisch's thesis [137].

## 9.4    Cages and Moore graphs

Recall that for a graph $G$ that is not a forest, the length of a shortest cycle is called the **girth** of $G$ and is denoted by $g(G)$. For example, $g(K_n) = 3$ for all $n \geq 3$, $g(K_{s,t}) = 4$ for all $s, t \geq 2$, and $g(C_n) = n$ for all $n \geq 3$. Given positive integers $r$ and $g$, an $(\boldsymbol{r}, \boldsymbol{g})$-**graph** is an $r$-regular graph of girth $g$. In this section we are interested in the smallest orders of $(r, g)$-graphs, provided one exists.

No $r$-regular graph can have fewer vertices than $K_{r+1}$ does. Since $K_{r+1}$ has girth 3, it is the $(r, 3)$-graph of minimum possible order. Likewise, $C_g$ is the $(2, g)$-graph of minimum possible order. In general, any $(r, g)$-graph must have order at least $\max\{r + 1, g\}$, but for larger $r$ and $g$, we can give a much better lower bound by using the fact that graphs of high girth look locally like trees.

**Theorem 9.16 (Moore bound).** *For all integers $r \geq 2$ and $g \geq 3$, every $(r, g)$-graph has order at least*

$$1 + r + r(r-1) + r(r-1)^2 + \cdots + r(r-1)^{(g-3)/2}$$

*if $g$ is odd, or*

$$2\left(1 + (r-1) + (r-1)^2 + \cdots + (r-1)^{(g-2)/2}\right)$$

*if $g$ is even.*

*Proof.* Suppose that there is an $(r, g)$-graph $G$ of order $n$. We first consider the case where $g$ is odd. Let $g = 2k + 1$ for some integer $k \geq 1$ and fix an arbitrary vertex $v \in V(G)$. Expanding outward from $v$, we see that for all integers $1 \leq i \leq k$, every vertex of distance $i$ from $v$ is adjacent to one vertex of distance $i - 1$ from $v$ and $r - 1$ vertices of distance $i + 1$ from $v$. From this, we obtain the inequality

$$n \geq 1 + r + r(r-1) + r(r-1)^2 + \cdots + r(r-1)^{k-1}.$$

Since $k - 1 = (g - 3)/2$, this is the bound we sought to prove.

Now suppose that $g$ is even. Let $g = 2k$ for some integer $k \geq 2$ and fix an arbitrary edge $e = uv \in E(G)$. Similarly to the case where $g$ is odd, we see that for all integers $1 \leq i \leq k - 1$, the number of vertices at distance $i$ from $u$ or $v$ and at distance $i + 1$ from the other is $2(r-1)^i$. Thus we obtain the inequality

$$n \geq 2\left(1 + (r-1) + (r-1)^2 + \cdots + (r-1)^{k-1}\right).$$

The result follows by noting that $k - 1 = (g - 2)/2$. ∎

The Moore bound gets its name from the Moore graphs we will encounter shortly.

## ⧝ The existence of cages

An $(r,g)$-graph of minimum possible order is called an **$(r,g)$-cage**, provided that such a graph exists. For which pairs $r,g$ of integers with $r \geq 2$ and $g \geq 3$, does there exist an $(r,g)$-graph? The following result establishes not only that $(r,g)$-cages exist, but also provides an upper bound on their orders. The argument we present is essentially the one credited to Erdős in a paper he wrote jointly with Sachs [83], although we have simplified it a bit at the cost of a slightly worse bound.

**Theorem 9.17.** *For all integers $r \geq 2$ and $g \geq 3$, there is an $(r,g)$-graph of order at most*

$$4 + 2(r-1) + 2(r-1)^2 + \cdots + 2(r-1)^{g-2} + (r-1)^{g-1}.$$

*Proof.* As the cycle $C_g$ satisfies the result when $r = 2$, we may assume that $r \geq 3$. Let $n$ denote the quantity in the statement of the theorem and choose $G$ to be a graph on $n$ vertices with maximum degree $\Delta(G) \leq r$, girth $g(G) = g$, and the maximum number of edges subject to these constraints. We have not assumed that $G$ is a cage, so there is always such a graph; in particular, the union of $C_g$ with $n - g$ isolated vertices has maximum degree 2, which is less than $r$, and girth $g$.

For the purposes of this proof, call a vertex $v \in V(G)$ **deficient** if $\deg v < r$. We may assume that there is at least one deficient vertex, as otherwise we are done. If we could find two deficient vertices $u$ and $v$ such that $d(u,v) \geq g - 1$, then $G + uv$ would have maximum degree at most $r$, girth $g$, and more edges than $G$, a contradiction to our choice of $G$.

Let $u$ and $v$ be deficient vertices, and choose them to be distinct if $G$ has more than one deficient vertex (otherwise, we allow $u = v$). Since $u$ and $v$ are deficient and $\Delta(G) \leq r$, for each integer $i \geq 1$, there are at most $(r-1)^i$ vertices at distance $i$ from each of them. Since

$$n > \left(1 + (r-1) + \cdots + (r-1)^{g-2}\right) + \left(1 + (r-1) + \cdots + (r-1)^{g-1}\right),$$

we can find some vertex $x$ such that $d(u,x) \geq g - 1$ and $d(v,x) \geq g$. If $x$ is itself deficient, then we have a contradiction because $G + ux$ would then have maximum degree at most $r$, girth $g$, and more edges than $G$. Thus we may assume that $\deg x = r \geq 3$. Because of this, there must be some edge $xy \in E(G)$ whose removal does not destroy all of the cycles of length $g$ in $G$, and thus $g(G - xy) = g$.

Because $d(v,x) \geq g$, we know that $d(v,y) \geq g - 1$. This implies that the graph $G - xy + ux + vy$ has girth $g$, but has more edges than $G$. If $\Delta(G - xy + ux + vy) \leq r$, then we have contradicted our choice of $G$. Thus it only remains to consider the case where $\Delta(G - xy + ux + vy) = r + 1$, and this can only occur if $u = v$ and $\deg u = r - 1$. However, a parity argument shows that this case is impossible. If $u = v$, then the other $n - 1$ vertices are not deficient (we chose $v \neq u$ if it was possible). Thus the sum of the degrees of all the vertices of $G$, which equals $2|E(G)|$, is $nr - 1$. However, our choice of $n$ ensures that it is even whenever $r$ is odd, and thus $nr - 1$ is always odd, which guarantees that if $u$ is the unique deficient vertex of $G$, its degree can be at most $r - 2$. ∎

As mentioned earlier, the original argument from Erdős and Sachs [83] gives a slightly better (that is, smaller) bound than Theorem 9.17. This bound was later improved by Sauer [212] to $2(r - 2)^{g-2}$ when $g$ is odd and $4(r - 1)^{g-3}$ when $g$ is even, and this bound is often referred to as the **Sauer bound**. Note that the Sauer (upper) bound is approximately the square of the Moore (lower) bound from Theorem 9.16.

## Specific cages

Now that we know that $(r, g)$-cages exist for all pairs $r, g$ of integers with $r \geq 2$ and $g \geq 3$, we turn to the determination of specific cages. As with Ramsey numbers, there is also a dynamic survey on cages [90] maintained by Exoo and Jajcay at the *Electronic Journal of Combinatorics*.

**Theorem 9.18.** *For each integer $r \geq 2$, the unique $(r, 4)$-cage is the bipartite complete graph $K_{r,r}$ of order $2r$.*

*Proof.* The Moore bound tells us that every $(r, 4)$-graph has order at least $2r$. Obviously, the graph $K_{r,r}$ is $r$-regular with girth 4 and order $2r$. This implies both that $K_{r,r}$ is an $(r, g)$-cage and that every $(r, g)$-cage has order $2r$.

It remains to show that $K_{r,r}$ is the only $(r, 4)$-cage. Let $G$ be any $(r, 4)$-cage, so it has order $2r$. Fix an arbitrary vertex $u_1 \in V(G)$ and denote its neighbors by $v_1, v_2, \ldots, v_r$. Since $G$ has girth 4, it follows that $\{v_1, v_2, \ldots, v_r\}$ must form an independent set. In particular, $v_1$ is not adjacent to $v_2, v_3, \ldots, v_r$, so $G$ must contain at least $r - 1$ additional neighbors of $v_1$. Label these additional neighbors as $u_2, u_3, \ldots, u_r$. By the same girth argument, $\{u_1, u_2, \ldots, u_r\}$ must form an independent set. Moreover, since $G$ has order $2r$, these are all of its vertices, and because it is $r$-regular, $u_i$ must be adjacent to $v_j$ for all $1 \leq i, j \leq r$. ∎

We now turn our attention to $(3, g)$-cages. These graphs, which are by definition cubic and are often simply called **$g$-cages**, have received significant attention. In particular, the number and order of such cages is known for all values of $g$ up to 12. To begin our consideration of these graphs, we note that $K_4$ is the unique $(3, 3)$-cage (Exercise 9.31) and that Theorem 9.18 shows that $K_{3,3}$ is the unique $(3, 4)$-cage.

Since the Petersen graph is a cubic graph of girth 5, it is a $(3, 5)$-graph. Moreover, the Moore bound implies that every $(3, 5)$-graph must have order at least 10, which is the order of the Petersen graph, so the Petersen graph is a $(3, 5)$-cage. Next we show that it is the only $(3, 5)$-cage.

**Theorem 9.19.** *The Petersen graph is the unique $(3, 5)$-cage.*

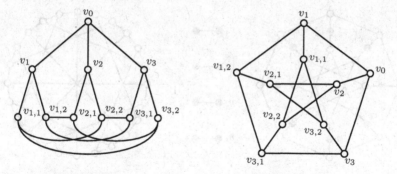

**Figure 9.9.** The Petersen graph, the unique $(3, 5)$-cage

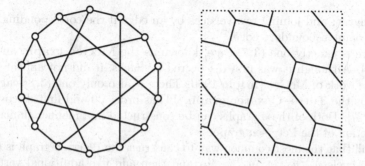

**Figure 9.10.** The Heawood graph (left) can also be seen as the toroidal dual of an embedding of $K_7$ on the torus (right)

*Proof.* Suppose that $G$ is a $(3, 5)$-graph of order 10 and choose an arbitrary vertex $v_0$ of $G$. Since $G$ is cubic, $v_0$ has three neighbors. Label these neighbors $v_1$, $v_2$, and $v_3$. As $G$ has girth 5, no two of these vertices may be adjacent, and no two of them may share a neighbor other than $v_0$. Thus, these vertices together have six neighbors in addition to $v_0$. For $1 \leq i \leq 3$, label the new neighbors of $v_i$ as $v_{i,1}$ and $v_{i,2}$.

Since these are all of the vertices of $G$, and because $G$ is 3-regular and has girth 5, we see that $v_{1,1}$ is adjacent to one of $v_{2,1}$ and $v_{2,2}$ and to one of $v_{3,1}$ and $v_{3,2}$. Without loss of generality, we may assume that $v_{1,1}$ is adjacent to $v_{2,2}$ and $v_{3,1}$. Then, $v_{1,2}$ must be adjacent to $v_{2,1}$ and $v_{3,2}$. Finally, we see that $v_{2,1}$ must be adjacent to $v_{3,2}$, and that $v_{2,2}$ must be adjacent to $v_{3,1}$. Thus, $G$ is as shown on the left of Figure 9.9, and it can be checked that $G$ is isomorphic to the Petersen graph. ∎

There is a unique $(3, 6)$-cage, namely the **Heawood graph** of order 14, and this is shown on the left of Figure 9.10. The Heawood graph is also the toroidal dual of the embedding of $K_7$ on the torus as shown on the right of Figure 9.10. That is, this graph is constructed by inserting a vertex in each of the 14 regions

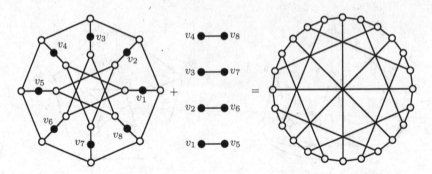

**Figure 9.11.** Two drawings of the McGee graph, the unique $(3,7)$-cage

of this figure, and joining two vertices by an edge if the corresponding regions have a common boundary edge.

There is exactly one $(3,7)$-cage, known as the **McGee graph**, and it has order 24. Although it was first discovered by Sachs, it did not appear in print until the work of McGee [164] in 1960. There is also only one $(3,8)$-cage, often known as the **Tutte–Coxeter graph**. It has order 30 and first discovered by Tutte [237]. Both of these graphs can be constructed in a manner similar to the construction of the Petersen graph.

Recall that the most common way to construct the Petersen graph is to begin with a 5-cycle $u_1, u_2, u_3, u_4, u_5, u_1$ and then add five additional vertices $w_1$, $w_2, w_3, w_4$, and $w_5$, together with the matching $u_i w_i$ for $1 \le i \le 5$. Then each vertex $w_i$ is joined to $w_{i+2}$ (subscripts considered modulo 5), resulting in another 5-cycle $w_1, w_3, w_5, w_2, w_4, w_1$.

To construct the McGee graph, we begin with an 8-cycle $u_1, u_2, \ldots, u_8, u_1$ and then add eight additional vertices $w_1, w_2, \ldots, w_8$, together with the matching $u_i w_i$ for $1 \le i \le 8$. Then each vertex $w_i$ is joined to $w_{i+3}$ (subscripts considered modulo 8), resulting in another 8-cycle $w_1, w_4, w_7, w_2, w_5, w_8, w_3, w_6, w_1$. Each edge $u_i w_i$ for $1 \le i \le 8$ is then subdivided once, introducing new vertices $v_i$. Finally, we add the edges $v_i v_{i+4}$ for $1 \le i \le 4$ as shown on the left of Figure 9.11. Another drawing of this graph is shown on the right of Figure 9.11. The Tutte–Coxeter graph can be constructed in a similar manner, as shown in Figure 9.12.

There are 18 different $(3,9)$-cages, each of order 58. There are three different $(3,10)$-cages, each of order 70. The first $(3,11)$-cage of order 112 was constructed by Balaban [16] in 1973. Twenty-five years later, the **Balaban cage** was shown to be the only $(3,11)$-cage by McKay, Myrvold, and Nadon [167]. There is a unique $(3,12)$-cage of order 126, commonly called the **Benson graph**, named for Benson [22], who first constructed it in 1966. This completes the list of known $(3,g)$-cages.

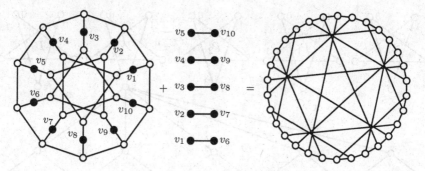

**Figure 9.12.** Two drawings of the Tutte–Coxeter graph, the unique $(3, 8)$-cage

## Moore graphs

Although they are often defined in terms of their diameters and were originally defined only for odd girth (see Exercises 9.33 and 9.34), a **Moore graph** is commonly defined today to be an $(r, g)$-graph that contains only as many vertices as required by the Moore (lower) bound of Theorem 9.16. It follows that every Moore graph is a cage, but the converse is far from true. The term "Moore graph" was coined by Hoffman and Singleton [129] because it was Moore who suggested that they look at such graphs (at an IBM summer workshop, according to Hoffman [128, p. 367]).

For $r = 2$, we see that the cycles $C_g$ are all Moore graphs, and that they are the only such graphs. For larger values of $r$, the situation is more interesting. Exercise 9.32 asks the reader to determine which of the cages we have seen so far are Moore graphs, so the reader may wish to solve this exercise before we spoil the answer here.

Hoffman and Singleton [129] considered the case of $g = 5$. When $r = 2$, we have $C_5$ as the unique Moore graph of degree 2 and girth 5. When $r = 3$, we have already seen that the Petersen graph achieves the Moore bound, and Theorem 9.19 shows that it is the unique $(3, 5)$-cage, so it is the unique Moore graph of degree 3 and girth 5. When $r = 7$, there is a unique 7-regular graph of girth 5 that achieves the Moore bound (and thus has order 50), found by Hoffman and Singleton [129].

To construct this **Hoffman–Singleton graph**, we start with ten pairwise disjoint 5-cycles denoted by $P_1, P_2, \ldots, P_5$, and $Q_1, Q_2, \ldots, Q_5$ whose vertices are labeled as in Figure 9.13, so that the vertices of the cycles $P_j$ are connected in pentagrammic manner as $0 \sim 2 \sim 4 \sim 1 \sim 3 \sim 0$ and the vertices of the cycles $Q_j$ are connected more conventionally as $0 \sim 1 \sim 2 \sim 3 \sim 4 \sim 0$. Then for each $1 \leq i, j, k \leq 5$, we add an edge from vertex $i$ in $P_j$ to vertex $i + jk \pmod 5$ in $Q_k$.

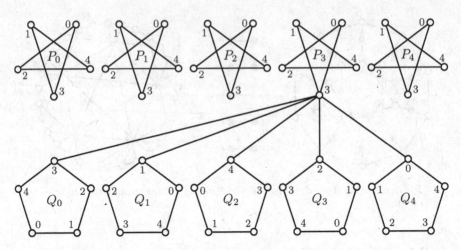

**Figure 9.13.** Constructing the Hoffman–Singleton graph

A 1960 theorem of Hoffman and Singleton [129] shows that there are very few values of $r$ for which there can be an $r$-regular Moore graph of girth 5.

**Theorem 9.20.** *The only values of $r$ for which an $r$-regular Moore graph of girth 5 exists are $r = 2, 3, 7,$ or possibly 57.*

As alluded to in the statement of Theorem 9.20, it is not known if there is a 57-regular Moore graph of girth 5. It is also not known if such a graph would have to be unique if it were to exist, so these are frequently called the **missing Moore graph(s)**. If such a graph were to exist, it would have order 3250 by the Moore bound. Other parameters of such a graph have been estimated as well; for example, the independence number of such a graph would have to be at most 400. For a survey of what is known about these perhaps-nonexistent graphs, we refer to Dalfó [59].

More generally, Damerell [60] and Bannai and Ito [18] have shown that other than the cycles, Moore graphs must have girth 3, 5, 6, 8, or 12, and the following theorem of theirs summarizes all of the possibilities.

**Theorem 9.21.** *For $r \geq 3$, there exists an $r$-regular Moore graph of odd girth $g$ if and only if*

- *$g = 3$ and $r \geq 3$, in which case $K_{r+1}$ is the unique Moore graph;*
- *$g = 5$ and $r = 3$, in which case the Petersen graph is the unique Moore graph;*
- *$g = 5$ and $r = 7$, in which case the Hoffman–Singleton graph is the unique Moore graph; or*
- *$g = 5$ and $r = 57$ (possibly).*

*For $r \geq 3$, there exists an $r$-regular Moore graph of even girth $g$ if and only if*

- *$g = 4$ and $r \geq 4$, in which case $K_{r,r}$ is the unique Moore graph;*
- *$g = 6$ and for all $r$ for which there exists a projective plane of order $r - 1$; or*

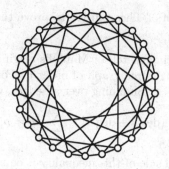

**Figure 9.14.** The 4-regular Moore graph of girth 6

- $g = 8$ or $g = 12$ and for all $r$ for which there exists a certain projective geometry.

The Heawood graph and the Tutte–Coxeter graph are examples of Moore graphs of even girth; an additional example of the unique 4-regular Moore graph of girth 6 and order 26 is shown in Figure 9.14.

In his memoirs, Hoffman wrote about his feelings when Theorem 9.21 was established [128, p. 367], "I felt a twinge of guilt in giving Moore's name to such a small set. But I was wrong: Moore graphs, Moore geometries, etc. continue to be discussed in the profession." Indeed, Moore graphs have proven to be an important and enduring topic in graph theory, with applications ranging from coding theory, where they are used to construct error-correcting codes with optimal parameters, to the study of large graphs and their properties.

# Exercises for Chapter 9

### Section 9.1. Avoiding a complete graph

**9.1.** For each ordered pair

$$(n, r) \in \{(7, 2), (7, 3), (7, 4), (8, 2), (8, 3), (8, 4)\},$$

determine the Turán graph $T(n, r)$ and its number of edges $t(n, r)$.

**9.2.** Assume that $n \geq r \geq 1$ and let $s$ be the remainder when $n$ is divided by $r$. Use the formula given in the text for $t(n, r)$ to show that

$$t(n, r) = \left(1 - \frac{1}{r}\right)\frac{n^2}{2} - \frac{s(r - s)}{2r} \leq \left(1 - \frac{1}{r}\right)\frac{n^2}{2}.$$

**9.3.** From the previous exercise, we see that for all $n \geq r \geq 1$,

$$t(n, r) \leq \left\lfloor \left(1 - \frac{1}{r}\right)\frac{n^2}{2} \right\rfloor.$$

Prove that when $r \geq 8$, there are values of $n$ for which this inequality is strict.

**9.4.** Strengthen our proof of Theorem 9.1 to prove that the only triangle-free graph of order $n$ is $K_{\lceil n/2 \rceil, \lfloor n/2 \rfloor}$.

**9.5.** Complete the following sketch of Mantel's original proof of Theorem 9.1. Let $G = (V, E)$ be a triangle-free graph of order $n$. For every edge $uv \in E$, we must have $\deg u + \deg v \le n$. Summing over all edges, we obtain

$$\sum_{uv \in E} (\deg u + \deg v) \le \sum_{uv \in E} n = n|E|.$$

Explain why the left-hand side of this inequality is equal to

$$\sum_{v \in V} (\deg v)^2,$$

use the Cauchy–Schwarz inequality to show that

$$n \sum_{v \in V} (\deg v)^2 \ge \left( \sum_{v \in V} \deg v \right)^2 = (2|E|)^2,$$

and finally solve to see that $|E| \le n^2/4$.

**9.6.** Prove that every graph of order $n$ with average degree $d = (1/2 + c)n$ contains at least $c\binom{n}{3}$ triangles.

**9.7.** Assume that $n \ge r \ge 1$ and let $s$ be the remainder when $n$ is divided by $r$. Use the definition of $t(n, r)$ as the number of edges in $T(n, r)$ to show that

$$t(n, r) = \binom{n}{2} + (n - r)(r - 1) + t(n - r, r).$$

**9.8.** Turán's original proof of his theorem is sketched in this exercise and the next, and relies on the formula from Exercise 9.7. Let $G$ be a graph of order $n$ and as many edges as possible that does not contain an $(r + 1)$-clique. As the addition of an additional edge to $G$ would result in an $(r + 1)$-clique, $G$ must contain a $r$-clique. Count the edges of $G$ with respect to this $r$-clique to show that $G$ has at most

$$\binom{r}{2} + (n - r)(r - 1) + t(n - r, r)$$

edges. Conclude that $G$ has at most $t(n, r)$ edges.

**9.9.** Strengthen the argument of the previous exercise to get the stronger claim that if $G$ has $t(n, r)$ edges, then $G = T(n, r)$.

**9.10.** Suppose the graph $G$ has order 10 and 26 edges. Prove that $G$ must contain at least 5 triangles.

## Section 9.2. Containing cycles and trees

**9.11.** Show that every graph of order $n = 5$ with $n + 3 = 8$ edges contains two edge-disjoint cycles.

**9.12.** Exhibit a graph of order $n = 5$ and with $n + 2 = 7$ edges that does not contain two edge-disjoint cycles.

**9.13.** For each $n \geq 6$, exhibit a graph of order $n$ and $n + 3$ edges that does not contain two edge-disjoint cycles.

**9.14.** Determine the minimum number of edges $m$ of a graph $G$ of order 6 such that $G$ contains a subdivision of $K_5$ as a subgraph.

**9.15.** By Corollary 9.7, if $G$ is a graph of order $n$ and with at least $(k-1)n - \binom{k}{2} + 1$ edges, where $1 \leq k < n$, then $G$ contains every tree with $k$ edges as a subgraph. Show that this bound on the number of edges of $G$ is not best possible by finding two integers $k$ and $n$ with $1 \leq k < n$ such that every graph of order $n$ and with $m$ edges contains every tree with $k$ edges as a subgraph but $m \leq (k-1)n - \binom{k}{2}$.

**9.16.** Show that every graph of order $n \geq 3$ and with $\lfloor n/2 \rfloor + 1$ edges contains $P_3$ as a subgraph. Describe the extremal graphs.

**9.17.** Show that every graph of order $n \geq 4$ with $n$ edges contains $2K_2$ as a subgraph. Describe the extremal graphs.

**9.18.** Prove that every graph of order $n \geq 4$ with at least $2n - 2$ edges contains a subdivision of $K_4$ as a subgraph.

**9.19.**    (a) Let $G$ be a graph of order $n \geq 4$. Prove that if $\deg v \geq \frac{2n+1}{3}$ for every vertex $v$ of $G$, then every edge of $G$ belongs to a complete subgraph of order 4.

  (b) Show that the result in (a) is best possible in general by showing that $\frac{2n+1}{3}$ cannot be replaced by $\frac{2n}{3}$.

**9.20.** Prove that the minimum number of edges of a graph of order $n \geq 3$ such that every vertex lies on a triangle is $\binom{n-1}{2} + 2$.

**9.21.** For each integer $n \geq 5$, determine the minimum positive integer $m$ such that every graph of order $n$ with $m$ edges contains two edge-disjoint triangles.

**9.22.** Prove that every graph of order at least 3 with $m$ edges contains a 3-partite subgraph with at least $\lceil 2m/3 \rceil$ edges.

## Section 9.3. Ramsey theory

**9.23.** Show that if $G$ is a graph of order $R(k, \ell) - 1$, then
  (a) $K_{k-1} \subseteq G$ or $K_\ell \subseteq \overline{G}$, and
  (b) $K_k \subseteq G$ or $K_{\ell-1} \subseteq \overline{G}$.

**9.24.** Show that $R(1, \ell) = 1$ and $R(2, \ell) = \ell$ for every positive integer $t$.

**9.25.** Prove that if $2 \le i \le k$ and $2 \le j \le \ell$, then $R(i, j) \le R(k, \ell)$. Further prove that equality holds if and only if $i = k$ and $j = \ell$.

**9.26.** Because $R(3, 3) = 6$, we know that every red-blue edge coloring of $K_6$ contains at least one monochromatic triangle. Let $e$ be any edge of $K_6$. Does every red-blue edge coloring of $K_6 - e$ contain a monochromatic triangle?

**9.27.** Prove that every red-blue edge coloring of $K_6$ actually contains at least *two* monochromatic triangles.

**9.28.** We know that $R(3, 4) = 9$. What is $R(2, 3, 4)$?

**9.29.** In one part of the proof of Theorem 9.9 verifying that $R(3, 3) = 6$, it was shown that every red-blue edge coloring of $K_6$ produces a monochromatic $K_3$. This was accomplished by selecting a vertex $v$ of $K_6$ and considering the colors of the five edges incident with $v$. In particular, the situation where three edges of $K_6$ incident with $v$ were colored the same was considered.
  (a) Show that if four edges incident with $v$ are colored the same, then $K_6$ contains at least two monochromatic triangles.
  (b) Show that if the red subgraph $G_R$ or blue subgraph $G_B$ is 3-regular, then $K_6$ contains at least two monochromatic triangles. [Recall that there are only two 3-regular graphs of order 6, namely $K_{3,3}$ or $K_3 \,\square\, K_2$.]
  (c) Show that if neither $G_R$ nor $G_B$ contains a vertex of degree 4 or more or is 3-regular, then either $G_R$ or $G_B$ contains four vertices of degree 3 and two nonadjacent vertices of degree 2. Show in this case that $K_6$ contains two monochromatic triangles.

**9.30.** Prove that every red-blue edge coloring of $K_6$ has a monochromatic $C_4$.

## Section 9.4. Cages and Moore graphs

**9.31.** Prove that $K_4$ is the unique $(3, 3)$-cage.

**9.32.** Compare the orders of the cages presented in Section 9.4 to their corresponding Moore bounds. Which achieve the Moore bound and are therefore Moore graphs?

**9.33.** Prove that every connected $r$-regular graph of diameter $d$ has order at most

$$1 + r + r(r - 1) + r(r - 1)^2 + \cdots + r(r - 1)^{d-1}.$$

**9.34.** Prove that if there is a connected $r$-regular graph of diameter $d$ that achieves the bound of Exercise 9.33, then it is a Moore graph of girth $g = 2d + 1$.

**9.35.**   (a) Prove that $n(3, 6) = 14$.
  (b) Prove that the Heawood graph is the only $(3, 6)$-cage.

**9.36.** Suppose that $G$ is an $(r, g)$-cage for integers $r \geq 2$ and $g \geq 3$. Prove that if $H = G \square K_2$ is an $(s, g)$-graph, then $H$ cannot be an $(s, g)$-cage.

**9.37.** Construct a cubic graph of order 12 with the maximum possible girth.

# Embedddings

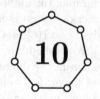

**10**

While only planar graphs can be embedded in the plane, there is a host of common and increasingly complex surfaces on which a nonplanar graph $G$ might possibly be embedded. Determining the simplest of these surfaces on which $G$ can be embedded is the problem of interest in this chapter.

## 10.1 The genus of a graph

We have seen that a graph $G$ is planar if $G$ can be drawn in the plane in such a way that no two edges cross and that such a drawing is called an embedding of $G$ in the plane or a planar embedding. Furthermore, a graph $G$ can be embedded in the plane if and only if $G$ can be embedded on (the surface of) a sphere.

Of course, not all graphs are planar. Indeed, Kuratowski's theorem (Theorem 5.16) and Wagner's theorem (Theorem 5.19) describe conditions (involving the two nonplanar graphs $K_5$ and $K_{3,3}$) under which $G$ can be embedded in the plane. Graphs that are not embeddable in the plane (or on a sphere) can be embedded on other surfaces, however.

A common surface on which a graph may be embedded is the **torus**, a doughnut-shaped surface (see Figure 10.1(a)). Two different embeddings of the (planar) graph $K_4$ on a torus are shown in Figures 10.1(b) and 10.1(c).

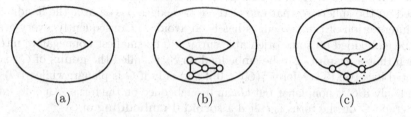

(a)            (b)            (c)

**Figure 10.1.** Embedding $K_4$ on a torus

When a graph $G$ is embedded on a surface, as in the case of plane embedding, the **regions** are the connected pieces of the surface that remain after removing the vertices and edges. Such a region of this embedding is called **2-cell** if every closed curve in that region can be continuously deformed in that region to a single point. (Topologically, a region is a 2-cell if it is homeomorphic to a disk.) For example, the embedding of $K_4$ in Figure 10.1(b) is not a 2-cell embedding as the exterior region is not 2-cell. On the other hand, the embedding of $K_4$ shown in Figure 10.1(c) is a 2-cell embedding.

An embedding of a graph $G$ on some surface is a **2-cell embedding** if every region in the embedding is a 2-cell. Clearly, embeddings of disconnected graphs will never be 2-cell and hence we will restrict our attention to connected graphs. In fact, every embedding of a connected graph on a sphere is necessarily a 2-cell embedding. Of course, such a graph is necessarily planar. If a connected graph is embedded on a surface that is not the sphere, then the embedding may or may not be a 2-cell embedding. For the rest of this chapter, we restrict our attention to 2-cell embeddings, as we will see that they have very nice properties.

One way to construct a torus is to begin with a sphere, insert two holes in its surface (as in Figure 10.2(a)) and attach a handle on the sphere, where the ends of the handle are placed over the two holes (as in Figure 10.2(b)). An embedding of $K_5$ on the torus constructed in this manner is shown in Figure 10.2(c).

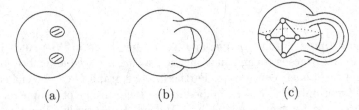

(a)                           (b)                           (c)

**Figure 10.2.** Embedding $K_5$ on a torus

While a torus is a sphere with one handle, a sphere with $k$ handles, $k \geq 0$, is called a **surface of genus $k$** and is denoted by $S_k$. Thus, $S_0$ is a sphere and $S_1$ is a torus. The surfaces $S_k$ are the **orientable surfaces**.

Let $G$ be a nonplanar graph. When drawing $G$ on a sphere, some edges of $G$ will cross. The graph $G$ can always be drawn so that only two edges cross at any point of intersection. At each such point of intersection, a handle can be suitably placed on the sphere so that one of these two edges passes over the handle and the intersection of the two edges has been avoided. Consequently, every graph can be embedded on some orientable surface. The smallest nonnegative integer $k$ such that a graph $G$ can be embedded on $S_k$ is called the **genus** of $G$ and is denoted by $\gamma(G)$. Therefore, $\gamma(G) = 0$ if and only if $G$ is planar; while $\gamma(G) = 1$ if and only if $G$ is nonplanar but $G$ can be embedded on the torus. An embedding of a graph $G$ on the torus is called a **toroidal embedding** of $G$.

We have seen that $K_5$ can be embedded on the torus and thus $\gamma(K_5) = 1$. In order to see that $K_{3,3}$ has a toroidal embedding, another representation of the torus is helpful. Since a torus can be obtained by identifying opposite sides of a rectangle, we often use the representation of the torus given in Figure 10.3(a). The points labeled $A$ in the rectangle in Figure 10.3(a) represent the same point on the torus. This is also true of the points labeled $B$ and the points labeled $C$.

Figure 10.3(b) gives an embedding $K_{3,3}$ on the torus. Thus, since $K_{3,3}$ is not planar, $\gamma(K_{3,3}) = 1$.

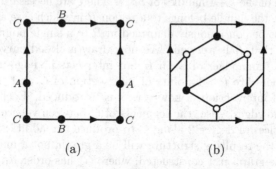

**Figure 10.3.** The rectangular representation of the torus and an embedding of $K_{3,3}$

## The generalized Euler formula

The embeddings of $K_4$, $K_5$ and $K_{3,3}$ on a torus given in Figures 10.1(c), 10.2(c) and 10.3(b) are all 2-cell embeddings with $n$ vertices, $m$ edges and $r$ regions, respectively. Furthermore, in each case, $n - m + r = 0$. As it turns out, if $G$ is a connected graph of order $n$ and size $m$ that is 2-cell embedded on a torus resulting in $r$ regions, then it is always the case that $n - m + r = 0$. This fact together with Euler's formula (Theorem 5.1) are special cases of a more general result given by L'Huilier [157].

**Theorem 10.1 (Generalized Euler formula).** *If $G$ is a connected graph of order $n$ and size $m$ that is 2-cell embedded on a surface of genus $k \geq 0$, resulting in $r$ regions, then*

$$n - m + r = 2 - 2k.$$

*Proof.* We proceed by induction on $k$. If $G$ is a connected graph of order $n$ and size $m$ that is 2-cell embedded on a surface of genus 0, then $G$ is a plane graph. By the Euler Identity, $n - m + r = 2 = 2 - 2 \cdot 0$. Thus the basis step of the induction holds.

Assume, for every connected graph $G'$ of order $n'$ and size $m'$ that is 2-cell embedded on a surface $S_k$ for some nonnegative integer $k$, resulting in $r'$ regions, that

$$n' - m' + r' = 2 - 2k.$$

Let $G$ be a connected graph of order $n$ and size $m$ that is 2-cell embedded on $S_{k+1}$, resulting in $r$ regions. We may assume, without loss of generality, that no vertex of $G$ lies on any handle of $S_{k+1}$ and that the edges of $G$ are drawn on the handles so that a closed curve can be drawn around each handle that intersects no edge of $G$ more than once.

Let $H$ be one of the $k+1$ handles of $S_{k+1}$. There are necessarily edges of $G$ on $H$; for otherwise, the handle belongs to a region $R$ in which case any closed curve around $H$ cannot be continuously deformed in $R$ to a single point, contradicting the assumption that $R$ is a 2-cell. We now draw a closed curve $C$ around $H$, which intersects some edges of $G$ on $H$ but intersects no edge more than once. Suppose that there are $t \geq 1$ points of intersection of $C$ and the edges on $H$. At each point of intersection, a new vertex is introduced. Each of the $t$ edges then becomes two edges. Also, the segments of $C$ between vertices become edges. We add two vertices of degree 2 along $C$ to produce two additional edges. (This guarantees that the resulting structure will be a graph, not a multigraph.)

Let $G_1$ be the graph just constructed, where $G_1$ has order $n_1$, size $m_1$ and $r_1$ regions. Then

$$n_1 = n + t + 2 \text{ and } m_1 = m + 2t + 2.$$

Since each portion of $C$ that became an edge of $G_1$ is in a region of $G$, the addition of such an edge divides that region into two regions, each of which is a 2-cell. Since there are $t$ such edges,

$$r_1 = r + t.$$

We now cut the handle $H$ along $C$ and "patch" the two resulting holes, producing two duplicate copies of the vertices and edges along $C$ (see Figure 10.4). Denote the resulting graph by $G_2$, which is now 2-cell embedded on a surface $S_k$.

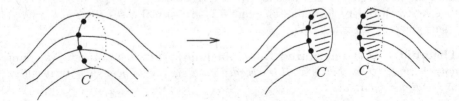

**Figure 10.4.** Converting a 2-cell embedding of $G_1$ on $S_{k+1}$ into a 2-cell embedding of $G_2$ on $S_k$

Let $G_2$ have order $n_2$, size $m_2$ and $r_2$ regions, all of which are 2-cells. Then

$$n_2 = n_1 + t + 2, \, m_2 = m_1 + t + 2 \text{ and } r_2 = r_1 + 2.$$

Furthermore, $n_2 = n + 2t + 4$, $m_2 = m + 3t + 4$ and $r_2 = r + t + 2$. By the induction hypothesis, $n_2 - m_2 + r_2 = 2 - 2k$. Therefore,

$$
\begin{aligned}
n_2 - m_2 + r_2 &= (n + 2t + 4) - (m + 3t + 4) + (r + t + 2) \\
&= n - m + r + 2 = 2 - 2k.
\end{aligned}
$$

Therefore, $n - m + r = 2 - 2(k + 1)$. ∎

We noted that every embedding of a connected planar graph $G$ in the plane is always a 2-cell embedding of $G$. This fact is a special case of a useful result obtained by Youngs [261].

**Theorem 10.2.** *Every embedding of a connected graph $G$ of genus $k$ on $S_k$, where $k$ is a nonnegative integer, is a 2-cell embedding of $G$ on $S_k$.*

## Lower bounds for the genus of a graph

With the aid of Theorems 10.1 and 10.2, we have the following.

**Corollary 10.3.** *If $G$ is a connected graph of order $n$ and size $m$ that is embedded on a surface of genus $\gamma(G)$, resulting in $r$ regions, then*

$$n - m + r = 2 - 2\gamma(G).$$

The following result is a consequence of Corollary 10.3.

**Theorem 10.4.** *If $G$ is a connected graph of order $n \geq 3$ and size $m$, then*

$$\gamma(G) \geq \frac{m}{6} - \frac{n}{2} + 1.$$

*Proof.* Since the result is obviously true for $n = 3$, we may assume that $n \geq 4$. Suppose that $G$ is embedded on a surface of genus $\gamma(G)$, resulting in $r$ regions. By Corollary 10.3, $n - m + r = 2 - 2\gamma(G)$. Let $R_1, R_2, \ldots, R_r$ be the regions of $G$ and let $m_i$ be the number of edges on the boundary of $R_i$ ($1 \leq i \leq r$). Thus, $m_i \geq 3$ for $1 \leq i \leq r$. Since every edge is on the boundary of either one or two regions, it follows that

$$3r \leq \sum_{i=1}^{r} m_i \leq 2m$$

and so $3r \leq 2m$. Therefore,

$$6 - 6\gamma(G) = 3n - 3m + 3r \leq 3n - 3m + 2m = 3n - m.$$

Solving this equation for $\gamma(G)$, we have $\gamma(G) \geq \frac{m}{6} - \frac{n}{2} + 1$. ∎

Theorem 10.4 is a generalization of Theorem 5.3 for the case in which $G$ is planar (and so $\gamma(G) = 0$). The lower bound for $\gamma(G)$ presented in Theorem 10.4 can be improved when more information on cycle lengths in $G$ is available (see Exercise 10.3).

**Theorem 10.5.** *If $G$ is a connected graph of order $n$, size $m$ and girth $k$, then*

$$\gamma(G) \geq \frac{m}{2}\left(1 - \frac{2}{k}\right) - \frac{n}{2} + 1.$$

The following result is a consequence of Theorem 10.5 that includes bipartite graphs as a special case. Recall that a graph is called triangle-free if it contains no triangles.

**Corollary 10.6.** *If $G$ is a connected, triangle-free graph of order $n \geq 3$ and size $m$, then*

$$\gamma(G) \geq \frac{m}{4} - \frac{n}{2} + 1.$$

While there is no general formula for the genus of an arbitrary graph, the following result by Battle, Harary and Kodama [19] implies that, as far as genus formulas are concerned, only 2-connected graphs need be investigated.

**Theorem 10.7.** *If $G$ is a graph having blocks $B_1$, $B_2$, ..., $B_k$, then*

$$\gamma(G) = \sum_{i=1}^{k} \gamma(B_i).$$

The following corollary is a consequence of the preceding result (see Exercise 10.4).

**Corollary 10.8.** *If $G$ is a graph with components $G_1, G_2, \ldots, G_k$, then*

$$\gamma(G) = \sum_{i=1}^{k} \gamma(G_i).$$

Recall that deciding whether a graph is planar can be done in linear time; that is, for a graph $G$, deciding whether $\gamma(G) = 0$ can be done in polynomial time. Thomassen [232] showed that given a graph $G$ and an integer $k \geq 1$, the problem of determining whether $\gamma(G) \leq k$ is NP-complete.

As is often the case when no general formula exists for the value of a parameter for an arbitrary graph, formulas (or partial formulas) are established for certain families of graphs. Ordinarily some of the first classes considered are the complete graphs, the complete bipartite graphs and the $n$-cubes. The genus offers no exception to this statement.

##  The genus of $K_n$ and map-coloring

According to Theorem 10.4,

$$\gamma(K_5) \geq \frac{1}{6}, \quad \gamma(K_6) \geq \frac{1}{2}, \quad \text{and} \quad \gamma(K_7) \geq 1.$$

This says that all three graphs $K_5$, $K_6$ and $K_7$ are nonplanar. Of course, we already knew this by Theorem 5.13. Since $K_5$ is nonplanar, so too are $K_6$ and $K_7$. We have seen that $\gamma(K_5) = 1$. Actually, $\gamma(K_7) = 1$ as well. Figure 10.5 gives an embedding of $K_7$ on the torus where the vertex set of $K_7$ is $\{v_1, v_2, \ldots, v_7\}$.

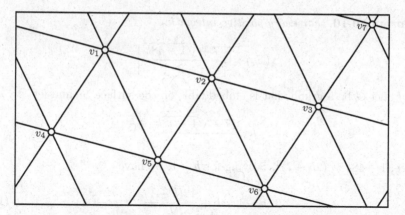

**Figure 10.5.** An embedding of $K_7$ on the torus

Since $K_6$ is nonplanar and $K_6$ is a subgraph of a graph that can be embedded on a torus, $\gamma(K_6) = 1$ as well.

For $n \geq 3$, applying Theorem 10.4 to the complete graph $K_n$ gives

$$\gamma(K_n) \geq \frac{\binom{n}{2}}{6} - \frac{n}{2} + 1 = \frac{(n-3)(n-4)}{12}.$$

Since $\gamma(K_n)$ is an integer,

$$\gamma(K_n) \geq \left\lceil \frac{(n-3)(n-4)}{12} \right\rceil. \tag{10.1}$$

In 1968, Ringel and Youngs [201] completed a proof of a result that has a remarkable history. They solved a problem that became known as the **Heawood map coloring problem**. Their solution involved verifying that the lower bound given in (10.1) is in fact the value of $\gamma(K_n)$.

**Theorem 10.9.** *For every integer $n \geq 3$,*

$$\gamma(K_n) = \left\lceil \frac{(n-3)(n-4)}{12} \right\rceil.$$

In Chapter 6, we saw that every planar graph is 4-colorable. The **chromatic number of a surface** $S$ is defined by

$$\chi(S) = \max\{\chi(G)\},$$

where the maximum is taken over all graphs $G$ that can be embedded on $S$. Since $S_0$ denotes the sphere, the four color theorem implies that $\chi(S_0) = 4$. Heawood [123] obtained an upper bound for the chromatic number of $S_k$, the sphere with $k$ handles, for every positive integer $k$.

**Theorem 10.10.** *For every positive integer* $k$,

$$\chi(S_k) \leq \left\lfloor \frac{7 + \sqrt{1 + 48k}}{2} \right\rfloor.$$

*Proof.* Let $G$ be a graph that is embeddable on the surface $S_k$ and let

$$h = \frac{7 + \sqrt{1 + 48k}}{2}.$$

Hence, $1 + 48k = (2h - 7)^2$. Solving for $h - 1$, we have

$$h - 1 = 6 + \frac{12(k - 1)}{h}. \tag{10.2}$$

Among the subgraphs of $G$, let $H$ be one having the largest minimum degree. We show that $\delta(H) \leq h - 1$. Suppose that $H$ has order $n$ and size $m$. If $n \leq h$, then $\delta(H) \leq h - 1$, as desired. Hence, we may assume that $n > h$. Since $G$ is embeddable on $S_k$, so is $H$. Therefore, $\gamma(H) \leq k$. By Theorem 10.4,

$$k \geq \gamma(H) \geq \frac{m}{6} - \frac{n}{2} + 1.$$

Thus, $m \leq 3n + 6(k - 1)$. We therefore have

$$n\delta(H) \leq \sum_{v \in V(H)} \deg_H v = 2m \leq 6n + 12(k - 1)$$

and so, by (10.2),

$$\delta(H) \leq 6 + \frac{12(k - 1)}{n} < 6 + \frac{12(k - 1)}{h} = h - 1.$$

Hence, $\delta(H) < h - 1$. Thus, $\delta(H) + 1 < h$ and so $\delta(H) + 1 \leq \lfloor h \rfloor$. By Theorem 6.3,

$$\chi(G) \leq 1 + \delta(H) \leq \left\lfloor \frac{7 + \sqrt{1 + 48k}}{2} \right\rfloor,$$

giving the desired result. ∎

In fact, Heawood was under the impression that he had shown that

$$\chi(S_k) = \left\lfloor \frac{7 + \sqrt{1 + 48k}}{2} \right\rfloor, \tag{10.3}$$

for every positive integer $k$, but Heffter [126] pointed out that he had only established the upper bound. Hence, in order to solve the problem, a formula for the genus of the complete graph (Theorem 10.9) was required. The work of many mathematicians over a period of seventy-eight years finally resulted in a solution, primarily through the efforts of Ringel and Youngs [201].

**Theorem 10.11 (Heawood map coloring theorem).** *For every positive integer* $k$,

$$\chi(S_k) = \left\lfloor \frac{7 + \sqrt{1 + 48k}}{2} \right\rfloor.$$

*Proof.* By Theorem 10.10,

$$\chi(S_k) \leq \left\lfloor \frac{7 + \sqrt{1 + 48k}}{2} \right\rfloor.$$

Hence, it remains only to verify the reverse inequality. Define

$$n = \left\lfloor \frac{7 + \sqrt{1 + 48k}}{2} \right\rfloor.$$

Thus, $n \leq (7 + \sqrt{1 + 48k})/2$. Solving this inequality for $k$, we have

$$k \geq (n-3)(n-4)/12$$

and so by Theorem 10.9,

$$k \geq \left\lceil \frac{(n-3)(n-4)}{12} \right\rceil = \gamma(K_n).$$

Therefore, $\gamma(K_n) \leq k$, which implies that

$$\chi(S_{\gamma(K_n)}) \leq \chi(S_k).$$

Since $K_n$ is clearly embeddable on $S_{\gamma(K_n)}$, it follows that $\chi(S_{\gamma(K_n)}) \geq n$ and so

$$\chi(S_k) \geq n = \left\lfloor \frac{7 + \sqrt{1 + 48k}}{2} \right\rfloor,$$

giving the desired result. ∎

## The genus of $K_{s,t}$ and $Q_n$

Ringel [200] was also successful in obtaining a formula for the genus of every complete bipartite graph.

**Theorem 10.12.** *For every two integers* $s, t \geq 2$,

$$\gamma(K_{s,t}) = \left\lceil \frac{(s-2)(t-2)}{4} \right\rceil.$$

In particular, Theorem 10.12 implies that a complete bipartite graph $G$ can be embedded on a torus if and only if $G$ is planar or is a subgraph of $K_{4,4}$ or $K_{3,6}$. In the next section, using rotational embedding schemes, we find a genus embedding of $K_{s,t}$ when $s$ and $t$ are both even.

A formula for the genus of the $n$-cube $Q_n$ was found by Ringel [198] and by Beineke and Harary [20]. We prove this result to illustrate some of the techniques involved.

**Theorem 10.13.** *For $n \geq 2$, the genus of the $n$-cube is given by*

$$\gamma(Q_n) = (n - 4) \cdot 2^{n-3} + 1.$$

*Proof.* Since the $n$-cube is a triangle-free graph of order $2^n$ and size $n \cdot 2^{n-1}$, it follows by Corollary 10.6 that

$$\gamma(Q_n) \geq (n - 4) \cdot 2^{n-3} + 1.$$

To verify the reverse inequality, we employ induction on $n$. In fact, we show for every integer $n \geq 2$, that there is an embedding of $Q_n$ on the surface of genus $(n - 4) \cdot 2^{n-3} + 1$ such that the boundary of every region is a 4-cycle and such that there exist $2^{n-2}$ regions with pairwise disjoint boundaries. Since $Q_2$ and $Q_3$ are planar and $(n - 4) \cdot 2^{n-3} + 1 = 0$ for $n = 2$ and $n = 3$, this is certainly true for $n = 2$ and $n = 3$.

Assume for an integer $k \geq 4$ that there is an embedding of $Q_{k-1}$ on the surface $S$ of genus $(k - 5) \cdot 2^{k-4} + 1$ such that the boundary of every region is a 4-cycle and such that there exist $2^{k-3}$ regions with pairwise disjoint boundaries. Since the order of $Q_{k-1}$ is $2^{k-1}$, each vertex of $Q_{k-1}$ belongs to the boundary of precisely one of the aforementioned $2^{k-3}$ regions. Furthermore, let $Q_{k-1}$ be embedded on another copy $S'$ of the surface of genus $(k - 5) \cdot 2^{k-4} + 1$ such that the embedding of $Q_{k-1}$ on $S'$ is a "mirror image" of the embedding of $Q_{k-1}$ on $S$ (that is, if $v_1, v_2, v_3, v_4$ are the vertices of the boundary of a region of $Q_{k-1}$ on $S$, where the vertices are listed clockwise about the 4-cycle, then there is a region on $S'$ with the vertices $v_1, v_2, v_3, v_4$ on its boundary listed counterclockwise).

We now consider the $2^{k-3}$ distinguished regions of $S$ together with the corresponding regions of $S'$ and join each pair of associated regions by a handle. The addition of the first handle produces the surface of genus $2[(k - 5) \cdot 2^{k-4} + 1]$ while the addition of each of the other $2^{k-3} - 1$ handles results in an increase of 1 to the genus. Thus, the surface just constructed has genus $(k - 4) \cdot 2^{k-3} + 1$. Now each set of four vertices on the boundary of a distinguished region can be joined to the corresponding four vertices on the boundary of the associated region so that the four edges are embedded on the handle joining the regions. It is now immediate that the resulting graph is isomorphic to $Q_k$ and that every region is bounded by a 4-cycle. Furthermore, each added handle gives rise to four regions, "opposite" ones of which have disjoint boundaries, so there exist $2^{k-2}$ regions of $Q_k$ that are pairwise disjoint. ∎

## ⧓ Nonorientable surfaces

Other surfaces of interest are the **nonorientable surfaces** (or the nonorientable 2-dimensional manifolds), the simplest example of which is the projective plane. The **projective plane** can be represented by identifying opposite sides of a rectangle in the manner shown in Figure 10.6(a). Note that $A$ represents the same point in the projective plane, as does $B$. Figure 10.6(b) shows an embedding of $K_5$ on the projective plane.

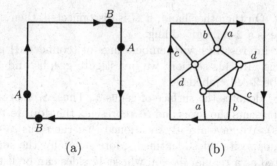

(a)          (b)

**Figure 10.6.** An embedding of $K_5$ on the projective plane

The projective plane can also be represented by a circle in which antipodal pairs of points on the circumference are the same point. Using this representation, an embedding of $K_6$ on the projective plane is shown in Figure 10.7.

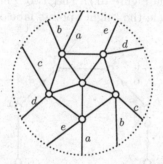

**Figure 10.7.** An embedding of $K_6$ on the projective plane

For the embedding of $K_5$ on the projective plane shown in Figure 10.6(b), $n = 5$, $m = 10$ and $r = 6$; while for the embedding of $K_6$ shown in Figure 10.7, $n = 6$, $m = 15$ and $r = 10$. In both cases, $n - m + r = 1$. In fact, for any connected graph of order $n$ and size $m$ that is 2-cell embedded on the projective plane, resulting in $r$ regions,

$$n - m + r = 1.$$

# 10.2  2-Cell embeddings of graphs

In the preceding section we saw that every graph $G$ has a genus; that is, there exists a surface (a compact orientable 2-manifold) of minimum genus on which $G$ can be embedded. Indeed, by Theorem 10.2, if $G$ is a connected graph that is embedded on the surface of genus $\gamma(G)$, then the embedding is necessarily a

2-cell embedding. On the other hand, if $G$ is disconnected, then no embedding of $G$ on any surface is a 2-cell embedding.

Our primary interest lies with embeddings of (connected) graphs that are 2-cell embeddings. In this section, we investigate graphs and the surfaces on which they can be 2-cell embedded.

Recall that $S_k$ denotes the surface of genus $k$. Thus, $S_0$ represents the sphere (or plane), $S_1$ represents the torus and $S_2$ represents the **double torus** (or sphere with two handles). We have already mentioned that the torus can be represented as a square with opposite sides identified. More generally, the surface $S_k$ $(k > 0)$ can be represented as a regular $4k$-gon whose $4k$ sides can be listed in clockwise order as

$$a_1 b_1 a_1^{-1} b_1^{-1} a_2 b_2 a_2^{-1} b_2^{-1} \ldots a_k b_k a_k^{-1} b_k^{-1}, \tag{10.4}$$

where, for example, $a_1$ is a side directed clockwise and $a_1^{-1}$ is a side also labeled $a_1$ but directed counterclockwise. These two sides are then identified in a manner consistent with their directions. Thus, the double torus can be represented by a regular octagon, as shown in Figure 10.8. The "two" points labeled $X$ are actually the same point on $S_2$, while the "eight" points labeled $Y$ are, in fact, a single point.

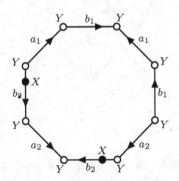

**Figure 10.8.** A representation of the double torus

Although it is probably obvious that there exist numerous graphs that can be embedded on the surface $S_k$ for a given nonnegative integer $k$, it may not be entirely obvious that there always exist graphs for which a 2-cell embedding on $S_k$ exists.

**Theorem 10.14.** *For every nonnegative integer $k$, there exists a connected graph that has a 2-cell embedding on $S_k$.*

*Proof.* For $k = 0$, every connected planar graph has the desired property; thus, we assume that $k > 0$.

We represent $S_k$ as a regular $4k$-gon whose $4k$ sides are described and identified as in (10.4). First, we define a pseudograph $H$ as follows. At each vertex of the $4k$-gon, let there be a vertex of $H$. Actually, the identification process associated with the $4k$-gon implies that there is only one vertex of $H$. Let each side of the

$4k$-gon represent an edge of $H$. The identification produces $2k$ distinct edges, each of which is a loop. This completes the construction of $H$. Hence, the pseudograph $H$ has order 1 and size $2k$. Furthermore, there is only one region, namely the interior of the $4k$-gon; this region is clearly a 2-cell. Therefore, there exists a 2-cell embedding of $H$ on $S_k$.

To convert the pseudograph $H$ into a graph, we subdivide each loop twice, producing a graph $G$ having order $4k + 1$, size $6k$ and again a single 2-cell region. ∎

Figure 10.9 illustrates the construction given in the proof of Theorem 10.14 in the case of the torus $S_1$. The graph $G$ so constructed is shown in Figure 10.9. In Figure 10.9(a)-(d), we see a variety of ways of visualizing the embedding. In Figure 10.9(a), a 3-dimensional embedding is described. In Figures 10.9(b) and (c), the torus is represented as a rectangle with opposite sides identified. (Figure 10.9(b) is the actual drawing described in the proof of the theorem.) In Figure 10.9(d), a portion of $G$ is drawn in the plane. Then two circular holes are made in the plane and a tube (or handle) is placed over the plane joining the two holes. The edge $uv$ is then drawn over the handle, completing the 2-cell embedding.

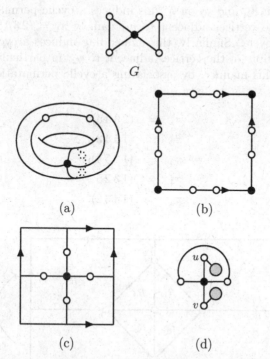

**Figure 10.9.** A graph 2-cell embedded on the torus

The graphs $G$ constructed in the proof of Theorem 10.14 are planar. Hence, for every nonnegative integer $k$, there exist planar graphs that can be 2-cell embedded on $S_k$. It is also true that for every planar graph $G$ and every positive integer $k$, there exists an embedding of $G$ on $S_k$ that is *not* a 2-cell embedding. In general, for a given graph $G$ and *positive* integer $k$ with $k > \gamma(G)$, there always exists an embedding of $G$ on $S_k$ that is not a 2-cell embedding, which can be obtained from an embedding of $G$ on $S_{\gamma(G)}$ by adding $k - \gamma(G)$ handles to the interior of some region of $G$. If $k = \gamma(G)$ and $G$ is connected, then by Theorem 10.2, every embedding of $G$ on $S_k$ is a 2-cell embedding. Of course, if $k < \gamma(G)$, there is no embedding whatsoever of $G$ on $S_k$.

## Rotational embedding schemes

Thus far, whenever we have described a 2-cell embedding (or, in fact, any embedding) of a graph $G$ on a surface $S_k$, we have resorted to a geometric description, such as the ones shown in Figure 10.9. There is a far more useful method that is algebraic in nature which we now discuss. Consider the 2-cell embedding of $K_5$ on $S_1$ shown in Figure 10.10, with the vertices of $K_5$ labeled as indicated. Observe that in this embedding, the edges incident with $v_1$ are arranged cyclically counterclockwise about $v_1$ in the order $v_1v_2, v_1v_3, v_1v_4, v_1v_5$ (or, equivalently, $v_1v_3, v_1v_4, v_1v_5, v_1v_2$, and so on). This induces a cyclic permutation $\pi_1$ of the subscripts of the vertices adjacent to $v_1$, namely $\pi_1 = (2\,3\,4\,5)$, expressed as a permutation cycle. Similarly, this embedding induces a cyclic permutation $\pi_2$ of the subscripts of the vertices adjacent to $v_2$, in particular $\pi_2 = (1\,5\,4\,3)$. Continuing in this manner, by associating a cyclic permutation $\pi_i$ with each vertex $v_i$ for $1 \leq i \leq 5$, we have

$$
\begin{aligned}
\pi_1 &= (2\,3\,4\,5), \\
\pi_2 &= (1\,5\,4\,3), \\
\pi_3 &= (1\,2\,5\,4), \\
\pi_4 &= (1\,3\,2\,5), \\
\pi_5 &= (1\,4\,3\,2).
\end{aligned}
$$

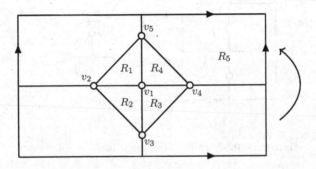

**Figure 10.10.** A 2-cell embedding of $K_5$ on the torus

In the 2-cell embedding of $K_5$ on $S_1$ shown in Figure 10.10, there are five regions, labeled $R_1, R_2, \ldots, R_5$. Each region $R_i$ $(1 \le i \le 5)$ is, of course, a 2-cell. The boundary of the region $R_1$ consists of the vertices $v_1$, $v_2$, and $v_5$ and the edges $v_1v_2$, $v_2v_5$, and $v_5v_1$. If we trace out the edges of the boundary of $R_1$ in a clockwise direction, keeping the boundary to our left and the region to our right (Figure 10.11), beginning with the edge $v_1v_2$, we have $v_1v_2$, followed by $v_2v_5$, and finally $v_5v_1$. This information can also be obtained from the cyclic permutations $\pi_1, \pi_2, \ldots, \pi_5$; indeed, the edge following $v_1v_2 = v_2v_1$ as we trace the boundary edges of $R_1$ in a clockwise direction is precisely the edge incident with $v_2$ that follows $v_2v_1$ if one proceeds counterclockwise about $v_2$; that is, the edge following $v_1v_2$ in the boundary of $R_1$ is $v_2v_{\pi_2(1)} = v_2v_5$. Similarly, the edge following $v_2v_5 = v_5v_2$ as we trace out the edges of the boundary of $R_1$ in a clockwise direction is $v_5v_{\pi_5(2)} = v_5v_1$. Hence with the aid of the cyclic permutations $\pi_1, \pi_2, \ldots, \pi_5$, we can trace out the edges of the boundary of $R_1$. In a similar manner, the boundary of every region of the embedding can be so described.

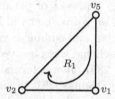

**Figure 10.11.** Tracing out a region

Since the direction (namely, clockwise) in which the edges of the boundary of a region are traced in the above description is of utmost importance, it is convenient to regard each edge of $K_5$ as a symmetric pair of arcs and, thus, to interpret $K_5$ itself as a digraph $D$. With this interpretation, the boundary of the region $R_1$ and thus $R_1$ itself can be described, starting at $v_1$, as

$$(v_1, v_2), (v_2, v_{\pi_2(1)}), (v_5, v_{\pi_5(2)})$$

or

$$(v_1, v_2), (v_2, v_5), (v_5, v_1). \tag{10.5}$$

We now define a mapping $\pi \colon E(D) \to E(D)$ as follows. Let $a \in E(D)$, where $a = (v_i, v_j)$. Then

$$\pi(a) = \pi((v_i, v_j)) = \pi(v_i, v_j) = (v_j, v_{\pi_j(i)}).$$

The mapping $\pi$ is one-to-one and is therefore a permutation of $E(D)$. Thus, $\pi$ can be expressed as a product of disjoint permutation cycles. In this context, each permutation cycle $\pi$ is referred to as an "orbit" of $\pi$. Hence (10.5) corresponds to an orbit of $\pi$ and is often denoted more compactly as $v_1 - v_2 - v_5 - v_1$. (Although this orbit corresponds to a cycle in the graph, this is not always the case for an arbitrary orbit in a graph that is 2-cell embedded.) For the embedding of $K_5$ on $S_1$ shown in Figure 10.10, the five orbits (one for each region) are as follows:

$$
\begin{array}{rcl}
R_1 & : & v_1 - v_2 - v_5 - v_1 \\
R_2 & : & v_1 - v_3 - v_2 - v_1 \\
R_3 & : & v_1 - v_4 - v_3 - v_1 \\
R_4 & : & v_1 - v_5 - v_4 - v_1 \\
R_5 & : & v_2 - v_3 - v_5 - v_2 - v_4 - v_5 - v_3 - v_4 - v_2.
\end{array}
$$

The orbits of $\pi$ form a partition of $E(D)$ and, as such, each arc of $D$ appears in exactly one orbit of $\pi$. Since $D$ is the digraph obtained by replacing each edge of $K_5$ by a symmetric pair of arcs, each edge of $K_5$ appears twice among the orbits of $\pi$, once for each of the two possible directions that are assigned to the edge.

The 2-cell embedding of $K_5$ on $S_1$ shown in Figure 10.10 uniquely determines the collection $\{\pi_1, \pi_2, \ldots, \pi_5\}$ of permutations of the subscripts of the vertices adjacent to the vertices of $K_5$. This set of permutations, in turn, completely describes the embedding of $K_5$ on $S_1$ as shown in Figure 10.10.

This method of describing an embedding is referred to as the **rotational embedding scheme**. Such a scheme was observed and used by von Dyck [246] in 1888 and by Heffter [126] in 1891. It was formalized by Edmonds [69] in 1960 and discussed in more detail by Youngs [261] in 1963.

We now describe the rotational embedding scheme in a more general setting. Let $G$ be a nontrivial connected graph with $V(G) = \{v_1, v_2, \ldots, v_n\}$. Let

$$
V(i) = \{j : v_j \in N(v_i)\}.
$$

For each $i$ $(1 \leq i \leq n)$, let $\pi_i \colon V(i) \to V(i)$ be a cyclic permutation (or rotation) of $V(i)$. Thus, each permutation $\pi_i$ can be represented by a (permutation) cycle of length $|V(i)| = |N(v_i)| = \deg v_i$. The rotational embedding scheme states that there is a one-to-one correspondence between the 2-cell embeddings of $G$ (on all possible surfaces) and the $n$-tuples $(\pi_1, \pi_2, \ldots, \pi_n)$ of cyclic permutations.

**Theorem 10.15 (Rotational embedding scheme).** *Let $G$ be a nontrivial connected graph with $V(G) = \{v_1, v_2, \ldots, v_n\}$. For each 2-cell embedding of $G$ on a surface, there exists a unique $n$-tuple $(\pi_1, \pi_2, \ldots, \pi_n)$, where $\pi_i \colon V(i) \to V(i)$ for $1 \leq i \leq n$ is a cyclic permutation that describes the subscripts of the vertices adjacent to $v_i$ in counterclockwise order about $v_i$.*

*Conversely, for each such $n$-tuple $(\pi_1, \pi_2, \ldots, \pi_n)$, there exists a 2-cell embedding of $G$ on some surface such that, for $1 \leq i \leq n$, the subscripts of the vertices adjacent to $v_i$ in counterclockwise order are given by $\pi_i$.*

*Proof.* Let $G$ be embedded on some surface. For each vertex $v_i$ of $G$, define $\pi_i \colon V(i) \to V(i)$ as follows: If $v_i v_j \in E(G)$ and $v_i v_t$ (possibly $t = j$) is the next edge encountered after $v_i v_j$ as we proceed counterclockwise about $v_i$, then we define $\pi_i(j) = t$. Each function $\pi_i$ so defined is a cyclic permutation.

Conversely, suppose we are given an $n$-tuple $(\pi_1, \pi_2, \ldots, \pi_n)$ such that for each $i$ $(1 \leq i \leq n)$, $\pi_i \colon V(i) \to V(i)$ is a cyclic permutation. We show that this determines a 2-cell embedding of $G$ on some surface. (By necessity, this proof requires the use of properties of compact orientable 2-manifolds.)

Let $D$ denote the digraph obtained from $G$ by replacing each edge of $G$ by a symmetric pair of arcs. We define a mapping $\pi \colon E(D) \to E(D)$ by

$$\pi((v_i, v_j)) = \pi(v_i, v_j) = (v_j, v_{\pi_j(i)}).$$

The mapping $\pi$ is one-to-one and, thus, is a permutation of $E(D)$. Hence, $\pi$ can be expressed as a product of disjoint permutation cycles. These permutation cycles are thus the orbits of $\pi$ and produce a partition of $E(D)$. Assume that

$$R : ((v_i, v_j)(v_j, v_t) \cdots (v_\ell, v_i))$$

is an orbit of $\pi$, which we also write as

$$R : v_i - v_j - v_t - \cdots - v_\ell - v_i.$$

Hence this implies that in the desired embedding, if we begin at $v_i$ and proceed along $(v_i, v_j)$ to $v_j$, then the next arc we must encounter after $(v_i, v_j)$ in a counterclockwise direction about $v_j$ is

$$(v_j, v_{\pi_j(i)}) = (v_j, v_t).$$

Continuing in this manner, we must finally arrive at the arc $(v_\ell, v_i)$ and return to $v_i$, in the process describing the boundary of a (2-cell) region (considered as a subset of the plane) corresponding to the orbit $R$. Therefore, each orbit of $\pi$ gives rise to a 2-cell region in the desired embedding.

To obtain the surface $S$ on which $G$ is 2-cell embedded, pairs of regions, with their boundaries, are "pasted" along certain arcs; in particular, if $(v_i, v_j)$ is an arc on the boundary of $R_1$ and $(v_j, v_i)$ is an arc on the boundary of $R_2$, then $(v_i, v_j)$ is identified with $(v_j, v_i)$ as shown below. The properties of compact orientable 2-manifolds imply that $S$ is indeed an appropriate surface.

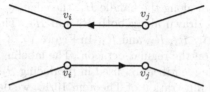

In order to determine the genus of $S$, one only needs to observe that the number $r$ of regions equals the number of orbits. Thus, if $G$ has order $n$ and size $n$, then by Theorem 10.1, $S = S_k$ where $k$ is the nonnegative integer satisfying the equation $n - m + r = 2 - 2k$. ∎

As an illustration of the rotational embedding scheme, we once again consider the complete graph $K_5$, with $V(K_5) = \{v_1, v_2, v_3, v_4, v_5\}$. Consider the 5-tuple $(\pi_1, \pi_2, \pi_3, \pi_4, \pi_5)$ of cyclic permutations given by

$$
\begin{aligned}
\pi_1 &= (2\,3\,4\,5), \\
\pi_2 &= (1\,3\,4\,5), \\
\pi_3 &= (1\,2\,4\,5), \\
\pi_4 &= (1\,2\,3\,5), \\
\pi_5 &= (1\,2\,3\,4).
\end{aligned}
$$

Thus, by Theorem 10.15, this 5-tuple describes a 2-cell embedding of $K_5$ on some surface $S_k$. To evaluate $k$, we consider the digraph $D$ obtained by replacing each edge of $K_5$ by a symmetric pairs of arcs and determine the orbits of the permutation $\pi \colon E(D) \to E(D)$ defined in the proof of Theorem 10.15. The orbits are

$$
\begin{aligned}
R_1 &: \quad v_1 - v_2 - v_3 - v_4 - v_5 - v_1 \\
R_2 &: \quad v_1 - v_3 - v_2 - v_4 - v_3 - v_5 - v_4 - v_1 - v_5 - v_2 - v_1 \\
R_3 &: \quad v_1 - v_4 - v_2 - v_5 - v_3 - v_1
\end{aligned}
$$

and each orbit corresponds to a 2-cell region. Thus, the number of regions in the embedding is $r = 3$. Since $K_5$ has order $n = 5$ and size $m = 10$, and since

$$n - m + r = -2 = 2 - 2k,$$

it follows that $k = 2$, so that the given 5-tuple describes an embedding of $K_5$ on $S_2$.

Given an $n$-tuple of cyclic permutations as we have described, it is not necessarily an easy problem to present a geometric description of the embedding, particularly on surfaces of high genus. For the example just presented, however, we give two geometric descriptions in Figures 10.12 and 10.13. In Figure 10.12, a portion of $K_5$ is drawn in the plane. Two handles are then inserted over the plane, as indicated, and the remainder of $K_5$ is drawn along these handles. The edge $e_1 = v_2 v_5$ is drawn along the handle $H_1$, the edge $e_2 = v_3 v_5$ is drawn along $H_2$ while $e_3 = v_1 v_3$ is drawn along both $H_1$ and $H_2$. The three 2-cell regions produced are denoted by $R_1$, $R_2$, and $R_3$. In Figure 10.13, this 2-cell embedding of $K_5$ on $S_2$ is shown on the regular octagon. The labeling of the eight sides (as in (10.4)) indicates the identification used in producing $S_2$.

As a more general illustration of Theorem 10.15, we determine the genus of the complete bipartite graph $K_{2a,2b}$ where $1 \le a \le b$. By Theorem 10.12,

$$\gamma(K_{2a,2b}) = (a-1)(b-1).$$

That $(a-1)(b-1)$ is a lower bound for $\gamma(K_{2a,2b})$ follows from Corollary 10.6. We will use Theorem 10.15 to show that $K_{2a,2b}$ is 2-cell embeddable on $S_{(a-1)(b-1)}$, thereby proving that $\gamma(K_{2a,2b}) \le (a-1)(b-1)$ and completing the argument.

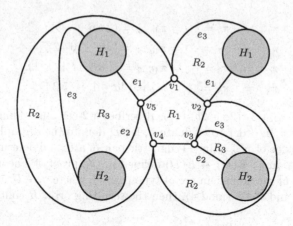

**Figure 10.12.** A geometric representation of a 2-cell embedding of $K_5$ on the double torus

Denote the partite sets of $K_{2a,2b}$ by $U$ and $W$, where $|U| = 2a$ and $|W| = 2b$. Furthermore, label the vertices so that

$$U = \{v_1, v_3, v_5, \ldots, v_{4a-1}\} \quad \text{and} \quad W = \{v_2, v_4, v_6, \ldots, v_{4a}, v_{4a+2}, \ldots, v_{4b}\}.$$

Consider the $(2a + 2b)$-tuple

$$(\pi_1, \pi_2, \ldots, \pi_{4a-1}, \pi_{4a}, \pi_{4a+2}, \pi_{4a+4}, \ldots, \pi_{4b}),$$

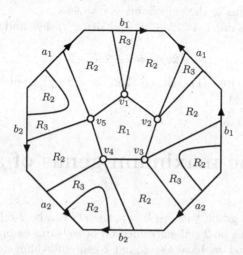

**Figure 10.13.** Another geometric representation of a 2-cell embedding of $K_5$ on the double torus

where

$$\begin{aligned}
\pi_1 &= \pi_5 = \cdots = \pi_{4a-3} = (2\,4\,6 \cdots 4b), \\
\pi_3 &= \pi_7 = \cdots = \pi_{4a-1} = (4b \cdots 6\,4\,2), \\
\pi_2 &= \pi_6 = \cdots = \pi_{4b-2} = (1\,3\,5 \cdots 4a - 1), \\
\pi_4 &= \pi_8 = \cdots = \pi_{4b} = (4a - 1 \cdots 5\,3\,1).
\end{aligned}$$

By Theorem 10.15, this $(2a + 2b)$-tuple describes a 2-cell embedding of $K_{2a,2b}$ on some surface $S_k$. In order to evaluate $k$, let $D$ denote the digraph obtained by replacing each edge of $K_{2a,2b}$ by a symmetric pair of arcs and determine the orbits of the permutation $\pi \colon E(D) \to E(D)$ defined in the proof of Theorem 10.15.

Every orbit of $\pi$ contains an arc of type $(v_s, v_t)$, where $v_s \in U$ and $v_t \in W$. If $s \equiv 1 \pmod 4$ and $t \equiv 2 \pmod 4$, then the resulting orbit $R$ containing $(v_s, v_t)$ is

$$R : v_s - v_t - v_{s+2} - v_{t-2} - v_s,$$

with $s + 2$ expressed modulo $4a$ and $t - 2$ expressed modulo $4b$. Note that $R$ also contains the arc $(v_{s+2}, v_{t-2})$ where necessarily $s + 2 \equiv 3 \pmod 4$ and $t - 2 \equiv 0 \pmod 4$. If $s \equiv 1 \pmod 4$ and $t \equiv 0 \pmod 4$, then the orbit $R'$ containing $(v_s, v_t)$ is

$$R' : v_s - v_t - v_{s-2} - v_{t-2} - v_s,$$

where again $s - 2$ expressed modulo $4a$ and $t - 2$ expressed modulo $4b$. The orbit $R'$ also contains the arc $(v_{s-2}, v_{t-2})$ where $s - 2 \equiv 3 \pmod 4$ and $t - 2 \equiv 2 \pmod 4$. Thus every orbit of $\pi$ is either of type $R$ (where $s \equiv 1 \pmod 4$ and $t \equiv 2 \pmod 4$) or of type $R'$ (where $s \equiv 1 \pmod 4$ and $t \equiv 0 \pmod 4$). Since there are $a$ choices for $s$ and $b$ choices for $t$ in each case, the total number of orbits is $2ab$; therefore, the number of regions in the embedding is $r = 2ab$.

Since $K_{2a,2b}$ has order $n = 2a + 2b$ and size $m = 4ab$, and since $n - m + r = 2 - 2k$, we have

$$(2a + 2b) - 4ab + 2ab = 2 - 2k,$$

so that $k = (a - 1)(b - 1)$. Hence there is a 2-cell embedding of $K_{2a,2b}$ on $S_{(a-1)(b-1)}$ as desired.

 **10.3  The maximum genus of a graph**

If $G$ is a connected graph with $\gamma(G) = p$, then $G$ can be 2-cell embedded on $S_p$. However, $G$ also can be 2-cell embedded on other surfaces $S_k$. For example, we know $\gamma(K_5) = 1$ and we have also found 2-cell embeddings of $K_5$ on $S_2$. This suggests finding the maximum integer $q$ such that $G$ can be 2-cell embedded on $S_q$.

Let $G$ be a connected graph. The **maximum genus** $\gamma_M(G)$ is the maximum among the genera of all surfaces on which $G$ can be 2-cell embedded. At this point in our discussion, it may not even be clear that every graph has a maximum genus since perhaps some graphs may be 2-cell embeddable on infinitely many surfaces. This, however, is impossible as we shall now see. Suppose $V(G) = \{v_1, v_2, \ldots, v_n\}$ with $n \geq 2$. By Theorem 10.15, there exists a one-to-one correspondence between the set of all 2-cell embeddings of $G$ and the $n$-tuples $(\pi_1, \pi_2, \ldots, \pi_n)$ where, for $1 \leq i \leq n$, each $\pi_i \colon V(i) \to V(i)$ is a cyclic permutation. Since the number of such $n$-tuples is finite, and in fact is equal to

$$\prod_{i-1}^{n} (\deg v_i - 1)!,$$

it follows that there are only finitely many 2-cell embeddings of $G$ and thus there exists a surface of maximum genus on which $G$ can be 2-cell embedded. In fact, more is true: Duke [68] proved that $G$ can be 2-cell embedded on $S_k$ for every integer $k$ with $\gamma(G) \leq k \leq \gamma_M(G)$ as we shall now see.

**Theorem 10.16.** *If there exist 2-cell embeddings of a connected graph $G$ on surfaces $S_p$ and $S_q$, where $p \leq q$, and $k$ is any integer with $p \leq k \leq q$, then there exists a 2-cell embedding of $G$ on the surface $S_k$.*

*Proof.* Observe that there exist a 2-cell embedding of $K_1$ only on the sphere; hence, we may assume that $G$ is nontrivial. Let $V(G) = \{v_1, v_2, \ldots, v_n\}$ where $n \geq 2$.

Assume that there exists a 2-cell embedding of $G$ on some surface $S_\ell$. By Theorem 10.15, there exists an $n$-tuple $(\pi_1, \pi_2, \ldots, \pi_n)$ where, for $1 \leq i \leq n$, each $\pi_i \colon V(i) \to V(i)$ is a cyclic permutation of the subscripts of the vertices of $N(v_i)$ in counterclockwise order about $v_i$.

Let $D$ be the symmetric digraph obtained from $G$ by replacing each edge of $G$ by a symmetric par of arcs. Let $\pi \colon E(D) \to E(D)$ be the permutation defined by $\pi(v_i, v_j) = (v_j, v_{\pi_j(i)})$. Denote the number of orbits in $\pi$ by $r$; that is, assume there are $r$ 2-cell regions in the given embedding of $G$ on $S_\ell$.

Suppose there exists some vertex of $G$, say $v_1$, such that $\deg v_1 \geq 3$. Then $\pi_1 = (a\, b\, c \cdots)$ where $a$, $b$, and $c$ are distinct. Let $v_x$ be any vertex adjacent to $v_1$ other than $v_a$, $v_b$, and suppose that $\pi_1(x) = y$. Thus

$$\pi_1 = (a\, b\, c \cdots x\, y \cdots),$$

where, possibly, $x = c$ or $y = a$. Let $E_1$ be the subset of $E(D)$ consisting of the three pairs of arcs

$$(v_a, v_1), (v_1, v_b); \quad (v_b, v_1), (v_1, v_c); \quad (v_x, v_1), (v_1, v_y). \tag{10.6}$$

Note that the six arcs listed in (10.6) are all distinct. By the definition of $\pi$, we have

$$\pi(v_a, v_1) = (v_1, v_b), \quad \pi(v_b, v_1) = (v_1, v_c), \quad \text{and} \quad \pi(v_x, v_1) = (v_1, v_x).$$

This implies that the arc $(v_a, v_1)$ is followed by the arc $(v_1, v_b)$ in some orbit of $\pi$, and that the edge $v_a v_1$ of $G$ is followed by the edge $v_1 v_b$ as we proceed clockwise around the boundary of the corresponding region in the given embedding of $G$ in $S_\ell$. Also, $(v_b, v_1)$ is followed by $(v_1, v_c)$ in some orbit of $\pi$ and $(v_x, v_1)$ is followed by $(v_1, v_y)$ in some orbit.

We now define a new permutation $\pi' : E(D) \to E(D)$ with the aid of the $n$-tuple $(\pi'_1, \pi'_2, \ldots, \pi'_n)$, where for $1 \leq i \leq n$, each $\pi'_i : V(i) \to V(i)$ is a cyclic permutation defined by

$$\pi'_1 = (a\, c \cdots x\, b\, y \cdots)$$

and

$$\pi'_i = \pi_i \quad \text{for} \quad 2 \leq i \leq n.$$

Let $\pi'(v_i, v_j) = (v_j, v_{\pi'_j(i)})$. By Theorem 10.15, the $n$-tuple $(\pi'_1, \pi'_2, \ldots, \pi'_n)$ deter- mines a 2-cell embedding of $G$ on some surface, where for $1 \leq i \leq n$, $\pi'_i$ is a cyclic permutation of the subscripts of the vertices adjacent to $v_i$ in counterclockwise order about $v_i$.

Three cases are now considered, depending on the possible distribution of the pairs (10.6) of arcs in $E_1$ among the orbits of $\pi$.

Case 1. *All arcs of $E_1$ belong to a single orbit $R$ of $\pi$.* Suppose, first, that the orbit of $R$ has the form

$$R : v_1 - v_y - \cdots - v_b - v_1 - v_c - \cdots - v_a - v_1 - v_b - \cdots - v_x - v_1.$$

Here the orbits of $\pi'$ are the orbits of $\pi$ except that the orbit $R$ is replaced by three orbits

$$
\begin{aligned}
R'_1 &: \quad v_1 - v_y - \cdots - v_b - v_1, \\
R'_1 &: \quad v_1 - v_c - \cdots - v_a - v_1, \\
R'_1 &: \quad v_1 - v_b - \cdots - v_x - v_1.
\end{aligned}
$$

Hence, $\pi'$ describes a 2-cell embedding of $G$ with $r + 2$ regions on a surface $S'$. Necessarily, then $S' = S_{\ell-1}$.

The other possible form that the orbit $R$ may take is

$$R : v_1 - v_y - \cdots - v_a - v_1 - v_b - \cdots - v_b - v_1 - v_c - \cdots - v_x - v_1.$$

In this situation, the orbits of $\pi'$ are the orbits of $\pi$ except for $R$, which is replaced by the orbit

$$R' : v_1 - v_y - \cdots - v_a - v_1 - v_c - \cdots - v_x - v_1 - v_b - \cdots - v_b - v_1.$$

Hence $\pi'$ has $r$ orbits and $\pi'$ describes another 2-cell embedding of $G$ on $S_\ell$.

Case 2. *The permutation $\pi$ has two orbits, say $R_1$ and $R_2$, where $R_1$ contains two of the pairs of arcs in $E_1$ and $R_2$ contains the remaining pair of arcs.* In this case, the orbits of $\pi'$ are those of $\pi$, except for $R_1$ and $R_2$, which are replaced by two orbits $R'_1$ and $R'_2$, where one of $R'_1$ and $R'_2$ contains two arcs of $E_1$ and the other contains the remaining four arcs of $E_1$. In this case, $\pi'$ has $r$ orbits.

*Case 3. The permutation $\pi$ has three orbits $R_1$, $R_2$ and $R_3$ such that $(v_a, v_1)$ is followed by $(v_1, v_b)$ in $R_1$, $(v_b, v_1)$ is followed by $(v_1, v_c)$ in $R_2$ and $(v_x, v_1)$ is followed by $(v_1, v_y)$ in $R_3$. In this case, the orbits of $\pi'$ are the orbits of $\pi$ except for $R_1$, $R_2$ and $R_3$, which are replaced by a single orbit $R'$ of the form*

$$R' : v_1 - v_y - \cdots - v_x - v_1 - v_b - \cdots - v_a - v_1 - v_c - \cdots - v_b - v_1.$$

Thus $\pi'$ has $r - 2$ orbits so that $\pi'$ describes a 2-cell embedding of $G$ on $S_{\ell+1}$.

We can now conclude that the shifting of a single term in $\pi_1$ (producing $\pi'_1$) changes the genus of the resulting surface on which $G$ is 2-cell embedded by at most 1.

Let $(\mu_1, \mu_2, \ldots, \mu_n)$ be the $n$-tuple of cyclic permutations associated with a 2-cell embedding of $G$ on $S_p$ and let $(\nu_1, \nu_2, \ldots, \nu_n)$ be the $n$-tuple of cyclic permutations associated with a 2-cell embedding of $G$ on $S_q$. If the degree of $v_i$ is 1 or 2 for each $i$, $1 \leq i \leq n$, then $\mu_i = \nu_i$ so that $p = q$ and the desired result follows. Hence, we may assume that for some $i$, $1 \leq i \leq n$, we have $\deg v_i \geq 3$. For each such $i$, the permutation $\mu_i$ can be transformed into $\nu_i$ by a finite number of single term shifts, as described above. Each such single term shift describes an embedding of $G$ on a surface whose genus differs by at most 1 from the genus of the surface on which $G$ is embedded prior to the shift. Therefore, by performing sequences of single term shifts on those $\mu_i$ for which $\deg v_i \geq 3$, the $n$-tuple $(\mu_1, \mu_2, \ldots, \mu_n)$ can be transformed into $(\nu_1, \nu_2, \ldots, \nu_n)$. Since $p \leq k \leq q$, there must be at least one term $(\pi_1, \pi_2, \ldots, \pi_n)$ in the aforementioned sequence beginning with $(\mu_1, \mu_2, \ldots, \mu_n)$ and ending with $(\nu_1, \nu_2, \ldots, \nu_n)$ that describes a 2-cell embedding of $G$ on $S_k$. ∎

An immediate consequence of Theorem 10.16 is the following.

**Corollary 10.17.** *A connected graph $G$ has a 2-cell embedding on the surface $S_k$ if and only if $\gamma(G) \leq k \leq \gamma_M(G)$.*

We now present an upper bound for the maximum genus of any connected graph. This bound employs a new but very useful concept. The **Betti number** $\mathcal{B}(G)$ of a graph $G$ of order $n$ and size $m$ having $k$ components is defined as

$$\mathcal{B}(G) = m - n + k.$$

Thus, if $G$ is connected, then

$$\mathcal{B}(G) = m - n + 1.$$

The next result is due to Nordhaus, Stewart and White [182].

**Theorem 10.18.** *If $G$ is a connected graph, then*

$$\gamma_M(G) \le \left\lfloor \frac{\mathcal{B}(G)}{2} \right\rfloor.$$

*Furthermore, equality holds if and only if there exists a 2-cell embedding of $G$ on the surface of genus $\gamma_M(G)$ with exactly one or two regions according to whether $\mathcal{B}(G)$ is even or odd, respectively.*

*Proof.* Let $G$ be a connected graph of order $n$ and size $m$ that is 2-cell embedded on the surface of genus $\gamma_M(G)$, producing $r$ (2-cell) regions. By Theorem 10.1,

$$n - m + r = 2 - 2\gamma_M(G).$$

Thus,
$$\mathcal{B}(G) = m - n + 1 = 2\gamma_M(G) + r - 1,$$

and hence
$$\gamma_M(G) = \frac{\mathcal{B}(G) + 1 - r}{2} \le \frac{\mathcal{B}(G)}{2},$$

producing the desired bound.

Moreover, we have

$$\gamma_M(G) = \frac{\mathcal{B}(G) + 1 - r}{2} = \left\lfloor \frac{\mathcal{B})(G)}{2} \right\rfloor,$$

if and only if $r = 1$ (which can only occur when $\mathcal{B}(G)$ is even) or $r = 2$ (which is only possible when $\mathcal{B}(G)$ is odd). ∎

## Upper embeddable graphs

A (connected) graph $G$ is called **upper embeddable** if the maximum genus of $G$ attains the upper bound given in Theorem 10.18; that is, if $\gamma_M(G) = \lfloor \mathcal{B}(G)/2 \rfloor$. The graph $G$ is said to be **upper embeddable on a surface** $S$ if $S = S_{\gamma_M(G)}$. We can now state an immediate consequence of Theorem 10.18.

**Corollary 10.19.** *Let $G$ be a connected graph with even (odd) Betti number. Then $G$ is upper embeddable on a surface $S$ if and only if there exists a 2-cell embedding of $G$ on $S$ with one region (two regions).*

In order to present a characterization of upper embeddable graphs, we introduce a special kind of spanning tree of a connected graph. A spanning tree $T$ of a connected graph $G$ is a **splitting tree** of $G$ if at most one component of $G - E(T)$ has odd size. It then follows that if $G - E(T)$ is connected, then $T$ is a splitting tree. For the graph $G$ of Figure 10.14, the tree $T_1$ is a splitting tree. On the other hand, $T_2$ is not a splitting tree of $G$.

The next elementary observation relates splitting trees and Betti numbers and is highly useful.

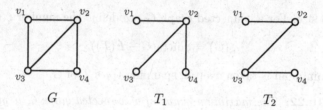

**Figure 10.14.** A splitting tree and a non-splitting tree

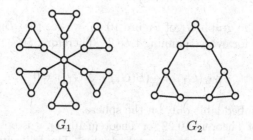

**Figure 10.15.** Graphs that are not upper embeddable

**Theorem 10.20.** *Let $T$ be a splitting tree of a graph $G$ of order $n$ and size $m$. Then every component of $G - E(T)$ has even size if and only if $\mathcal{B}(G)$ is even.*

*Proof.* Suppose that every component of $G - E(T)$ has even size. Then $G - E(T)$ has even size. Since every tree of order $n$ has size $n - 1$, the graph $G - E(T)$ has size $m - (n - 1) = m - n + 1$. Therefore, $\mathcal{B}(G) = m - n + 1$ is even.

Conversely, suppose that $\mathcal{B}(G)$ is even. The graph $G - E(T)$ has size $m - n + 1 = \mathcal{B}(G)$. Since $T$ is a splitting tree of $G$, at most one component of $G - E(T)$ has odd size. Since the sum of the sizes of the components of $G - E(T)$ is even, it is impossible for exactly one such component to have odd size, producing the desired result. ∎

We now state a characterization of upper embeddable graphs, which was discovered independently by Jungerman [135] and Xuong [260].

**Theorem 10.21.** *A graph $G$ is upper embeddable if and only if $G$ has a splitting tree.*

Returning to the graph $G$ of Figure 10.14, we now see that $G$ is upper embeddable since $G$ contains $T_1$ as a splitting tree. On the other hand, neither the graph $G_1$ nor the graph $G_2$ of Figure 10.15 has a single splitting tree. Thus, by Theorem 10.21, neither of these graphs is upper embeddable.

We mentioned earlier that no formula is known for the genus of an arbitrary graph. However, such is not the case with maximum genus. With the aid of Theorem 10.21, Xuong [260] developed a formula for the maximum genus of any connected graph. For a graph $H$, we denote by $\xi_0(H)$ the number of components of $H$ of odd size. For a connected graph $G$, we define the number $\xi(G)$ as

$$\xi(G) = \min \xi_0(G - E(T)),$$

where the minimum is taken over all spanning trees $T$ of $G$.

**Theorem 10.22.** *The maximum genus of a connected graph $G$ is given by*

$$\gamma_M(G) = \frac{1}{2}(\mathcal{B}(G) - \xi(G)).$$

Returning to the graph $G_1$ of Figure 10.15, we see that $\mathcal{B}(G_1) = 6$ and that $\xi_0(G - E(T)) = 6$ for every spanning tree $T$. Therefore, $\xi(G_1) = 6$ so that

$$\gamma_M(G_1) = \frac{1}{2}(\mathcal{B}(G_1) - \xi(G_1)) = 0$$

and $G_1$ is 2-cell embeddable only on the sphere.

With the aid of Theorem 10.22 (or Theorem 10.21), it is possible to show that a wide variety of graphs are upper embeddable. The following result is due to Kronk, Ringeisen and White [154].

**Corollary 10.23.** *For $k \geq 2$, every complete $k$-partite graph is upper embeddable.*

From Corollary 10.23, it follows at once that every complete graph is upper embeddable, a result due to Nordhaus, Stewart and White [182]. We present a proof using Theorem 10.21.

**Corollary 10.24.** *The maximum genus of $K_n$ is given by*

$$\gamma_M(K_n) = \left\lfloor \frac{(n-1)(n-2)}{4} \right\rfloor.$$

*Proof.* If $T$ is a spanning path of $K_n$, then $K_n - E(T)$ contains at most one nontrivial component. Therefore, $T$ is a splitting tree of $K_n$ and, by Theorem 10.21, $K_n$ is upper embeddable. Since $\mathcal{B}(K_n) = (n-1)(n-2)/2$, the result follows. ∎

A formula for the maximum genus of complete bipartite graphs was discovered by Ringeisen [197].

**Corollary 10.25.** *The maximum genus of $K_{s,t}$ is given by*

$$\gamma_M(K_{s,t}) = \left\lfloor \frac{(s-1)(t-1)}{2} \right\rfloor.$$

Zaks [263] discovered a formula for the maximum genus of the $n$-cube.

**Corollary 10.26.** *For $n \geq 2$, the maximum genus of $Q_n$ is given by*

$$\gamma_M(Q_n) = (n-2)2^{n-2}.$$

# 10.4  The graph minor theorem

By Wagner's theorem (Theorem 5.19), a graph $G$ is planar if and only if neither $K_5$ nor $K_{3,3}$ is a minor of $G$. That is, Wagner's theorem is a **forbidden minor** characterization of planar graphs in terms of two forbidden minors: $K_5$ and $K_{3,3}$. A natural question to ask is whether a forbidden minor characterization may exist for graphs that can be embedded on other surfaces. It was shown by Archdeacon and Huneke [12] that there are exactly 35 forbidden minors for graphs that can be embedded on the projective plane. More general results involving minors have been obtained. Wagner conjectured that in every infinite collection of graphs, there are always two graphs where one is isomorphic to a minor of the other. In what may be considered by some as one of the major theorems of graph theory, Robertson and Seymour [204] verified this conjecture. Its lengthy proof is a consequence of a sequence of several papers that required years to complete.

**Theorem 10.27 (Graph minor theorem).** *In every infinite set of graphs, there are two graphs where one is (isomorphic to) a minor of the other.*

One consequence of Theorem 10.27 is another major theorem, also due to Robertson and Seymour [204]. A set $S$ of graphs is said to be **minor-closed** if for every graph $G$ in $S$, every minor of $G$ also belongs to $S$. For example, the set $S$ of planar graphs is minor-closed because every minor of a planar graph is planar; that is, if $G \in S$ and $H$ is a minor of $G$, then $H \in S$ as well.

**Theorem 10.28.** *Let $S$ be a minor-closed set of graphs. Then, there exists a finite set $M$ of graphs such that $G \in S$ if and only if no graph in $M$ is a minor of $G$.*

*Proof.* Let $\overline{S}$ be the set of graphs not belonging to $S$ and let $M$ be the set of all graphs $F$ in $\overline{S}$ such that every proper minor of $F$ belongs to $S$. We claim that this set $M$ has the required properties. First, we show that $G \in S$ if and only if no graph in $M$ is a minor of $G$.

First, suppose that there is some graph $G \in S$ such that there is a graph $F$ that is a minor of $G$ and for which $F \in M$. Since $G \in S$ and $S$ is minor-closed, it follows that $F \in S$. However, since $F \in M$, we have $F \in \overline{S}$, which is a contradiction.

For the converse, assume to the contrary that there is a graph $G \in \overline{S}$ such that no graph in $M$ is a minor of $G$. We consider two cases.

*Case 1. All of the proper minors of $G$ are in $S$.* Then by the defining property of $M$, it follows that $G \in M$. Since $G \in M$ and $G$ is a minor of itself, this contradicts our assumption that no graph in $M$ is a minor of $G$.

*Case 2. Some proper minor of $G$, say $G'$, is not in $S$.* Thus, $G' \in \overline{S}$. Then $G'$ either satisfies the condition of Case 1 or of Case 2. We proceed in this manner as long as we remain in Case 2, producing a sequence $G = G^{(0)}, G^{(1)}, G^{(2)}, \ldots$ of proper minors.

If this process terminates, we have a finite sequence

$$G = G^{(0)}, G' = G^{(1)}, \ldots, G^{(p)},$$

where each graph in the sequence is a proper minor of all those graphs that precede it. Then $G^{(p)} \in \mathcal{M}$, which returns us to Case 1. Otherwise, the sequence $G = G^{(0)}, G' = G^{(1)}, G^{(2)}, \ldots$ is infinite and where each graph $G^{(i)}$, $i \geq 1$, is a proper minor of all those graphs that precede it. This, however, is impossible since for each $i \geq 0$, either the order of $G^{(i+1)}$ is less than that of $G^{(i)}$ or the orders are the same and the size of $G^{(i+1)}$ is less than that of $G^{(i)}$.

It therefore remains only to show that $\mathcal{M}$ is finite. Assume, to the contrary, that $\mathcal{M}$ is infinite. By the graph minor theorem, $\mathcal{M}$ contains two graphs, $H_1$ and $H_2$ say, such that one is a minor of the other. Suppose that $H_2$ is a minor (necessarily a proper minor) of $H_1$. Since every proper minor of $H_1$ belongs to $\mathcal{S}$, it follows that $H_2 \in \mathcal{S}$. However, since $H_2 \in \mathcal{M}$, it follows that $H_2 \in \overline{\mathcal{S}}$, producing a contradiction. ∎

We now return to the question concerning the existence of a forbidden minor characterization for graphs embeddable on a surface $S_k$ of genus $k \geq 0$. Certainly, if $G$ is a sufficiently small graph (in terms of its order and/or size), then $G$ can be embedded on $S_k$. Hence, if we begin with a graph $F$ that cannot be embedded on $S_k$ and perform successive edge contractions, edge deletions and vertex deletions, then eventually we arrive at a graph $F'$ that also cannot be embedded on $S_k$ but such that any additional edge contraction, edge deletion, or vertex deletion of $F'$ produces a graph that *can* be embedded on $S_k$. Such a graph $F'$ is said to be **minimally nonembeddable on $S_k$**. Consequently, a graph $F'$ is minimally nonembeddable on $S_k$ if $F'$ cannot be embedded on $S_k$ but every proper minor of $F'$ can be embedded on $S_k$. Thus, the set of graphs embeddable on $S_k$ is minor-closed. As a consequence of the graph minor theorem, we have the following.

**Theorem 10.29.** *For each integer $k \geq 0$, the set of minimally nonembeddable graphs on $S_k$ is finite.*

Consequently, for each nonnegative integer $k$, there is a finite set $\mathcal{M}_k$ of graphs such that a graph $G$ is embeddable on $S_k$ if and only if no graph in $\mathcal{M}_k$ is a minor of $G$. Of course, the set of minimally nonembeddable graphs on the sphere is $\mathcal{M}_0 = \{K_5, K_{3,3}\}$. Although the number of minimally nonembeddable graphs on the torus is finite, the actual value of this number is not known. However, Myrvold and Woodcock [179] showed that this number is at least 17,523 and so $|\mathcal{M}_1| \geq 17523$.

# Exercises for Chapter 10

## Section 10.1. The genus of a graph

**10.1.** Determine $\gamma = \gamma(K_{4,4})$ without using Theorem 10.12 and label the regions in a 2-cell embedding of $K_{4,4}$ on the surface of genus $\gamma$.

**10.2.** Let $G$ be a graph.
  (a) Show that $\gamma(G) \le \mathrm{cr}(G)$.
  (b) Prove that for every positive integer $k$, there exists a graph $G$ such that $\gamma(G) = 1$ and $\mathrm{cr}(G) = k$.

**10.3.** Prove Theorem 10.5.

**10.4.** Use Theorem 10.7 to prove Corollary 10.8.

**10.5.** Prove or disprove: If $G$ is a graph such that $\chi(G) \le \chi(S_k)$ for some positive integer $k$, then $G$ can be embedded on $S_k$.

**10.6.** Give an example of a graph $G$ with genus 2 and $\chi(G) = \chi(S_2)$. Verify that your example has these properties.

**10.7.** It is known that the Petersen graph $P$ is not planar. Thus, $P$ cannot be embedded on the sphere.
  (a) Show that $P$ can be embedded on the torus.
  (b) How many regions result from a toroidal embedding of $P$?
  (c) What is the minimum number of colors that can be assigned to the regions in (b) so that every two adjacent regions are colored differently?

**10.8.** Show for every two integers $s, t \ge 2$ that

$$\gamma(K_{s,t}) \ge \left\lceil \frac{(s-2)(t-2)}{4} \right\rceil.$$

**10.9.** Consider the graph $K_{3,3} \vee \overline{K}_n$.
  (a) Find a lower bound for $\gamma(K_{3,3} \vee \overline{K}_n)$.
  (b) Determine $\gamma(K_{3,3} \vee \overline{K}_n)$ exactly for $n = 1, 2, 3$.

**10.10.** Determine $\gamma(K_2 \,\square\, C_4 \,\square\, C_6)$.

**10.11.** Prove, for every positive integer $\gamma$, that there exists a connected graph $G$ of genus $\gamma$.

**10.12.** Prove, for each positive integer $k$, that there exists a connected planar graph $G$ such that $\gamma(G \,\square\, K_2) \ge k$.

**10.13.** Show, in a manner similar to the embedding of $K_{3,3}$ shown in Figure 10.3, that $K_5$ can be embedded on the Möbius strip.

**10.14.** By Theorem 10.9, $\gamma(K_7) = 1$. Let there be an embedding of $K_7$ on the torus and let $R_1$ and $R_2$ be two neighboring regions. Let $G$ be the graph obtained by adding a new vertex $v$ in $R_1$ and joining $v$ to the vertices on the boundaries of both $R_1$ and $R_2$. What is $\gamma(G)$?

**10.15.** Does there exist a graph $G$ containing two nonadjacent vertices $u$ and $v$ such that $\gamma(G) = \gamma(G + uv)$ but $cr(G) \neq cr(G + uv)$?

### Section 10.2. 2-Cell embeddings of graphs

**10.16.** For the 2-cell embedding of $K_7$ show in Figure 10.16, determine the 7-tuple of cyclic permutations $\pi_i$ associated with this embedding. Determine the orbits of the resulting permutation $\pi$.

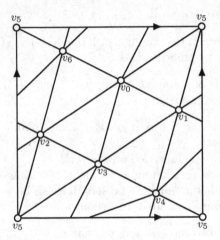

**Figure 10.16.** The embedding of $K_7$ in Exercise 10.16

**10.17.** Let $G = K_4 \square K_2$. Show that $\gamma(G) = 1$ by finding an 8-tuple of cyclic permutations that describes a 2-cell embedding of $G$ on $S_1$. Determine the orbits of the resulting permutations $\pi$.

**10.18.** Let $G$ be a graph with $V(G) = \{v_1, v_2, v_3, v_4, v_5, v_6\}$ and describe a 2-cell embedding of $G$ on the surface $S_k$ by $(\pi_1, \pi_2, \pi_3, \pi_4, \pi_5, \pi_6)$, where

$$\pi_1 = (2\,5\,6\,3), \quad \pi_2 = (3\,6\,1\,4), \quad \pi_3 = (4\,1\,2\,5),$$
$$\pi_4 = (5\,2\,3\,6), \quad \pi_5 = (6\,3\,4\,1), \quad \pi_6 = (1\,4\,5\,2).$$

(a) What familiar graph is $G$?
(b) What is $k$?
(c) Is $k = \gamma(G)$?

**10.19.** A graph $G$ is obtained from $K_{4,4}$ by removing four pairwise nonadjacent edges. Let $V(G) = \{v_1, v_2, \ldots, v_8\}$. The 8-tuple $(\pi_1, \pi_2, \ldots, \pi_8)$ of cyclic permutations defined by

$$\pi_1 = (2\,8\,6), \quad \pi_2 = (1\,5\,3), \quad \pi_3 = (2\,4\,8), \quad \pi_4 = (3\,5\,7),$$
$$\pi_5 = (2\,6\,4), \quad \pi_6 = (7\,5\,1), \quad \pi_7 = (4\,6\,8), \quad \pi_8 = (1\,7\,3)$$

describes a 2-cell embedding of $G$ on some surface $S_k$.
  (a) What is $k$?
  (b) Is $k = \gamma(G)$?

**10.20.** How many of the 2-cell embeddings of $K_4$ are embeddings in the plane? On the torus? On the double torus?

**10.21.** Consider the graph $K_{3,3}$.
  (a) Describe an embedding of $K_{3,3}$ on $S_2$ by means of a 6-tuple of cyclic permutations.
  (b) Show that there exists no 2-cell embedding of $K_{3,3}$ on $S_3$.

## Section 10.3. The maximum genus of a graph

**10.22.** Describe an embedding of $K_5$ on $S_{\gamma_M(K_5)}$ by means of a 5-tuple of cyclic permutations.

**10.23.** Determine the maximum genus of the graph $G_2$ of Figure 10.15.

**10.24.** Determine the maximum genus of the Petersen graph.

**10.25.** Let $G$ be a connected graph with blocks $B_1, B_2, \ldots, B_k$.
  (a) Prove that

$$\gamma_M(G) \geq \sum_{i=1}^{k} \gamma_M(B_i).$$

  (b) Show that the inequality in (a) may be strict.

**10.26.** Prove Theorem 10.21 as a corollary to Theorem 10.22.

**10.27.** Prove Corollary 10.23.

**10.28.** Prove Corollary 10.25.

**10.29.** Prove Corollary 10.26.

**10.30.** Prove or disprove: For every positive integer $k$, there exists a connected graph $G_k$ such that $\lfloor \mathcal{B}(G)/2 \rfloor - \gamma_M(G_k) = k$.

**10.31.** Prove or disprove: If $H$ is a connected spanning subgraph of an upper embeddable graph $G$, then $H$ is upper embeddable.

**10.32.** For $G = C_s \square C_t$ $(s, t \geq 3)$, determine $\gamma(G)$ and $\gamma_M(G)$.

**10.33.** Prove that if each vertex of a connected graph $G$ lies on at most one cycle, then $G$ is only 2-cell embeddable on the sphere.

**10.34.** For positive integers $p$ and $q$ with $p \leq q$, prove that there exists a graph $G$ of genus $p$ that can be 2-cell embedded on $S_q$.

**10.35.** Prove that if $G$ is upper embeddable, then $G \square K_2$ is upper embeddable.

### Section 10.4. The graph minor theorem

**10.36.** Let $\mathcal{F}$ be the set of forests.
  (a) Show that $\mathcal{F}$ is a minor-closed family of graphs.
  (b) What are the forbidden minors of $\mathcal{F}$?

**10.37.** Is the set $B$ of bipartite graphs a minor-closed family of graphs? If so, what are the forbidden minors of $B$?

**10.38.** We have seen that the set of planar graphs is a minor-closed family of graphs. Is the set $O$ of outerplanar graphs a minor-closed family of graphs? If so, what are the forbidden minors of $O$?

**10.39.** Use the Graph Minor Theorem (Theorem 10.27) to show that for any infinite set $S = \{G_1, G_2, G_3, \ldots\}$ of graphs, there exist infinitely many pairwise disjoint 2-element sets $\{i, j\}$ of integers such that one of $G_i$ and $G_j$ is a minor of the other.

# Graphs and algebra  11

In this chapter, we study connections between graph theory and algebra, in particular matrix theory and group theory. We assume the reader has some familiarity with these areas of algebra. We begin with several matrices associated with a graph and some graphical properties that can be discerned from properties of these matrices. We then turn to connections between graphs and groups. We will consider a well-known group associated with a graph as well as how to associate a graph with a given group. Finally, a problem concerning how much of the structure of a graph can be determined from certain subgraphs of a graph is also described.

##  11.1 Graphs and matrices

We have seen that a graph can be defined or described by means of sets (the definition) or diagrams. There are also matrix representations of graphs. Suppose that $G$ is a graph of order $n$, where $V(G) = \{v_1, v_2, \ldots, v_n\}$. The **adjacency matrix** of $G$ is the $n \times n$ $(0,1)$ matrix $A(G) = [a_{ij}]$, or simply $A = [a_{ij}]$, where

$$a_{ij} = \begin{cases} 1 & \text{if } v_i v_j \in E(G) \\ 0 & \text{if } v_i v_j \notin E(G). \end{cases}$$

A graph $G$ and its adjacency matrix are shown in Figure 11.1.

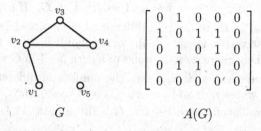

**Figure 11.1.** A graph and its adjacency matrix

There are several observations that can be made about the adjacency matrix $A$ of a graph $G$ of order $n$. First, all entries along the main diagonal of $A$ are 0 since no vertex of $G$ is adjacent to itself. Second, $A$ is a symmetric matrix, that is, $A = A^T$ where $A^T$ is the transpose of $A$.

Whenever $a_{ij} = 1$, this means that $G$ contains the edge $v_i v_j$ and therefore a $v_i$-$v_j$ path of length 1 and, of course, a $v_i$-$v_j$ walk of length 1 as well. Not only can the adjacency matrix of $G$ be used to identify whether $G$ contains a $v_i$-$v_j$ walk of length 1, it can be used to determine whether $G$ contains a $v_i$-$v_j$ walk of length $k$ for an arbitrary positive integer $k$ and, in fact, the number of such walks. Before going further, we need to know when two $u$-$v$ walks in a graph $G$ are the same. Two $u$-$v$ walks $W: u = u_0, u_1, \ldots, u_k = v$ and $W': u = v_0, v_1, \ldots, v_\ell = v$ in a graph are **equal** if $k = \ell$ and $u_i = v_i$ for all $i$ with $0 \le i \le k$.

**Theorem 11.1.** *Let $G$ be a graph with vertex set $V(G) = \{v_1, v_2, \ldots, v_n\}$ and adjacency matrix $A$. For each positive integer $k$, the number of different $v_i$-$v_j$ walks of length $k$ in $G$ is the $(i, j)$-entry in the matrix $A^k$.*

*Proof.* Let $a_{ij}^{(k)}$ denote the $(i, j)$-entry in the matrix $A^k$ for a positive integer $k$. Thus, $A^1 = A$ and $a_{ij}^{(1)} = a_{ij}$. We proceed by induction on $k$. For vertices $v_i$ and $v_j$ of $G$, there can be only one $v_i$-$v_j$ walk of length 1 or no $v_i$-$v_j$ walks of length 1, and this occurs if $a_{ij} = 1$ or $a_{ij} = 0$, respectively. Therefore, the $(i, j)$-entry of the matrix $A$ is the number of $v_i$-$v_j$ walks of length 1 in $G$. Thus, the basis step of the induction is established.

We now verify the inductive step. Assume, for a positive integer $k$, that $a_{ij}^{(k)}$ is the number of different $v_i$-$v_j$ walks of length $k$ in $G$. We show that the $(i, j)$-entry $a_{ij}^{(k+1)}$ in $A^{k+1}$ gives the number of different $v_i$-$v_j$ walks of length $k + 1$ in $G$. First, observe that every $v_i$-$v_j$ walk of length $k + 1$ in $G$ is obtained from a $v_i$-$v_t$ walk of length $k$ for some vertex $v_t$ in $G$ that is adjacent to $v_j$.

Since $A^{k+1} = A^k \cdot A$, it follows that the $(i, j)$-entry $a_{ij}^{(k+1)}$ in $A^{k+1}$ can be obtained by taking the inner product of row $i$ of $A^k$ and column $j$ of $A$. That is,

$$a_{ij}^{(k+1)} = a_{i1}^{(k)} a_{1j} + a_{i2}^{(k)} a_{2j} + \ldots + a_{in}^{(k)} a_{nj} = \sum_{t=1}^{n} a_{it}^{(k)} a_{tj}. \qquad (11.1)$$

By the induction hypothesis, for each integer $t$ with $1 \le t \le n$, the integer $a_{it}^{(k)}$ is the number of different $v_i$-$v_t$ walks of length $k$ in $G$. If $a_{tj} = 1$, then $v_t$ is adjacent to $v_j$ and so there are $a_{it}^{(k)}$ different $v_i$-$v_j$ walks of length $k + 1$ in $G$ whose next-to-last vertex is $v_t$. On the other hand, if $a_{tj} = 0$, then $v_t$ is not adjacent to $v_j$ and there are no $v_i$-$v_j$ walks of length $k + 1$ in $G$ whose next-to-last vertex is $v_t$. In any case, $a_{it}^{(k)} \cdot a_{tj}$ gives the number of different $v_i$-$v_j$ walks of length $k + 1$ in $G$ whose next-to-last vertex is $v_t$. Consequently, the total number of different $v_i$-$v_j$ walks of length $k + 1$ in $G$ is the sum in (11.1), which is $a_{ij}^{(k+1)}$.

Thus, $a_{ij}^{(k)}$ is the number of different $v_i$-$v_j$ walks of length $k$ in $G$ for every positive integer $k$. ■

As an illustration, consider the graph $G$ of Figure 11.2 having the adjacency matrix $A$. We can compute $A^2$ without matrix multiplication by observing that the $(i,i)$-entry of $A^2$, $1 \leq i \leq 4$, is $\deg v_i$, and the $(i,j)$-entry of $A^2$, $i \neq j$, is the number of different $v_i$-$v_j$ paths of length 2. We now turn to $A^3$. Since the different $v_1$-$v_3$ walks of length 3 in $G$ are

$$W_1: v_1, v_3, v_1, v_3, \quad W_2: v_1, v_2, v_1, v_3,$$
$$W_3: v_1, v_3, v_2, v_3, \quad W_4: v_1, v_3, v_4, v_3,$$

the $(1,3)$-entry of $A^3$ is 4. The entire matrix $A^3$ can be computed in this manner.

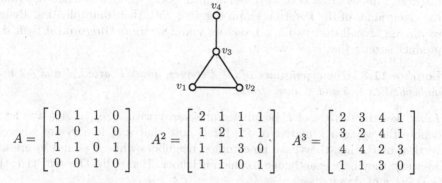

$$A = \begin{bmatrix} 0 & 1 & 1 & 0 \\ 1 & 0 & 1 & 0 \\ 1 & 1 & 0 & 1 \\ 0 & 0 & 1 & 0 \end{bmatrix} \quad A^2 = \begin{bmatrix} 2 & 1 & 1 & 1 \\ 1 & 2 & 1 & 1 \\ 1 & 1 & 3 & 0 \\ 1 & 1 & 0 & 1 \end{bmatrix} \quad A^3 = \begin{bmatrix} 2 & 3 & 4 & 1 \\ 3 & 2 & 4 & 1 \\ 4 & 4 & 2 & 3 \\ 1 & 1 & 3 & 0 \end{bmatrix}$$

**Figure 11.2.** A graph $G$ and powers of its adjacency matrix

Summarizing our observations above together with a few other interesting consequences of Theorem 11.1 yields the following corollary (see Exercise 11.7). The **trace** $\mathrm{tr}(B)$ of a square matrix $B$ is the sum of the entries along its main diagonal.

**Corollary 11.2.** *Let $G$ be a graph with $V(G) = \{v_1, v_2, \ldots, v_n\}$ and adjacency matrix $A = [a_{ij}]$. Let $a_{ij}^{(k)}$ denote the $(i,j)$-entry of the $k$th power of $A$. Then*

1. *$a_{ij}^{(2)}$, $i \neq j$, is the number of $v_i$-$v_j$ paths of length 2,*
2. *$a_{ii}^{(2)} = \deg v_i$,*
3. *$\mathrm{tr}(A^2)/2$ is size of $G$, and*
4. *$\mathrm{tr}(A^3)/6$ is the number of triangles of $G$.*

Note that the rows and columns of an adjacency matrix of a graph correspond to an arbitrary labeling of the vertices, and hence we seek properties of matrices which are invariant under permutations of the rows and columns. Principal among these are the spectral properties of the adjacency matrix. The **eigenvalues of a graph** $G$ of order $n$ are the eigenvalues of the adjacency matrix, that is, if $G$ has adjacency matrix $A$, then the eigenvalues of $G$ are the $n$ eigenvalues of $A$. The eigenvectors of $A$ are also called the **eigenvectors of $G$**. Since every adjacency matrix is a real symmetric matrix, using results from linear algebra and matrix theory, we know that the eigenvalues of every graph are all real numbers, and since the sum of the eigenvalues of a matrix is its trace, we know that the sum of the eigenvalues of given graph is zero. For a given eigenvalue $\lambda$, the set of all

$n \times 1$ eigenvectors corresponding to $\lambda$ together with the zero vector $\mathbf{0}$ is called the **eigenspace** of $\lambda$.

For a graph $G$ of order $n$ with adjacency matrix $A$, the polynomial

$$\det(\lambda I_n - A)$$

in the indeterminant $\lambda$ is called the **characteristic polynomial** of $G$ and is denoted by $\phi(G, \lambda)$. Since different labelings of a graph result in similar adjacency matrices, it follows that $\phi(G, \lambda)$ is well defined. As an example, we now determine the eigenvalues of the Petersen graph together with their multiplicities. Before continuing, recall that two $n \times 1$ vectors $\mathbf{v}$ and $\mathbf{w}$ are **orthogonal** if their dot product is zero, that is, $\mathbf{v} \cdot \mathbf{w} = 0$.

**Lemma 11.3.** *The eigenvalues of the Petersen graph $P$ are 3, 1 and $-2$ with multiplicities 1, 5 and 4, respectively.*

*Proof.* Let the vertices of $P$ be labeled arbitrarily with $v_1, v_2, \ldots, v_{10}$ and let $A$ denote the adjacency matrix of $P$. First, note that if $v_i$ and $v_j$ are adjacent vertices of $P$, then they have no common neighbors while if $v_i$ and $v_j$ are not adjacent, they have exactly one common neighbor. Hence, by Theorem 11.1, the $(i, j)$-entry of $A^2$ satisfies

$$a_{ij}^{(2)} = \begin{cases} 3 & \text{if } i = j, \\ 0 & \text{if } i \neq j \text{ and } v_i v_j \in E(P), \\ 1 & \text{if } i \neq j \text{ and } v_i v_j \notin E(P). \end{cases}$$

Thus,

$$A^2 + A - 2I = J$$

where $J$ is the $10 \times 10$ matrix in which every entry is 1 and $I$ is the $10 \times 10$ identity matrix.

Before continuing, we show that eigenvectors corresponding to distinct eigenvalues are orthogonal. Suppose $\lambda$ and $\mu$ are distinct eigenvalues with corresponding eigenvectors $\mathbf{v}$ and $\mathbf{w}$. Consider

$$(\mathbf{v}^T A \mathbf{w})^T = \mathbf{w}^T A^T \mathbf{v} = \mathbf{w}^T A \mathbf{v}$$

since $A = A^T$. Also, $\mathbf{v}^T A \mathbf{w} = \mathbf{v}^T (\mu \mathbf{w}) = \mu(\mathbf{v} \cdot \mathbf{w})$. Hence, $\mathbf{v}^T A \mathbf{w}$ is a real number so that

$$(\mathbf{v}^T A \mathbf{w})^T = \mathbf{v}^T A \mathbf{w} = \mu(\mathbf{v} \cdot \mathbf{w}).$$

Next

$$\mathbf{w}^T A \mathbf{v} = \mathbf{w}^T (\lambda \mathbf{v}) = \lambda(\mathbf{w} \cdot \mathbf{v}).$$

Hence,

$$\mu(\mathbf{v} \cdot \mathbf{w}) = \lambda(\mathbf{w} \cdot \mathbf{v}) = \lambda(\mathbf{v} \cdot \mathbf{w})$$

since $\mathbf{v} \cdot \mathbf{w} = \mathbf{w} \cdot \mathbf{v}$. So,

$$0 = (\lambda - \mu)(\mathbf{v} \cdot \mathbf{w})$$

and since $\lambda \neq \mu$ it must be that $\mathbf{v} \cdot \mathbf{w} = 0$, and therefore these eigenvectors are orthogonal.

Let $\mathbf{u}$ denote the vector in which every entry is 1. Clearly, every $r$-regular graph has $\mathbf{u}$ as an eigenvector with corresponding eigenvalue $r$. Hence

$$Au = 3u$$

and so $\mathbf{u}$ is an eigenvector for eigenvalue 3. Thus, if $\mathbf{v}$ is an eigenvector of $A$ corresponding to an eigenvalue $\lambda \neq 3$, then $\mathbf{v}$ is orthogonal to $\mathbf{u}$ so that

$$(A^2 + A - 2I)\mathbf{v} = J\mathbf{v} = \mathbf{0}$$

where $\mathbf{0}$ is the zero vector. Hence,

$$\lambda^2 \mathbf{v} + \lambda \mathbf{v} - 2\mathbf{v} = \mathbf{0}$$

and since $\mathbf{v}$ is nonzero, it must be that $\lambda^2 + \lambda - 2 = 0$. Therefore, the other eigenvalues of $A$ are 1 and $-2$.

Suppose the multiplicities of the eigenvalues 3, 1 and $-2$ are $a$, $b$ and $c$ respectively. Thus,

$$a + b + c = 10,$$

$$3a + b - 2c = 0$$

and, since the trace of $A^2$ is the sum of the degrees of the vertices and since the eigenvalues of $A^2$ are precisely the squares of the eigenvalues of $A$ with the same multiplicities,

$$9a + b + 4c = 30.$$

Solving this system of equations yields $a = 1$, $b = 5$, and $c = 4$. ∎

The **spectrum** of a graph $G$ is the set of eigenvalues of $G$, together with their multiplicities as roots of $\phi(G, \lambda)$. For example, by Lemma 11.3, the spectrum of the Petersen graph is 3 with multiplicity 1, 1 with multiplicity 5, and $-2$ with multiplicity 4. Unfortunately, it is possible for two nonisomorphic graphs to have the same spectrum. For example, the graphs $K_{1,4}$ and $C_4 \cup K_1$ have the same spectrum yet are not isomorphic (see Exercise 11.10). Two graphs whose adjacency matrices have the same spectrum are called **cospectral**.

Our example, $K_{1,4}$ and $C_4 \cup K_1$, also shows that the spectrum of the adjacency matrix cannot be used to determine whether a graph is connected. However, the spectrum can be used to determine whether a graph is bipartite. Coulson and Rushbrooke [56] proved that a graph is bipartite if and only if whenever $\lambda$ is an eigenvalue, $-\lambda$ is also an eigenvalue. It also turns out that an eigenvalue of a graph can be no more than the maximum degree as we shall now see.

**Theorem 11.4.** *For a graph $G$ with eigenvalue $\lambda$,*

$$|\lambda| \leq \Delta(G).$$

*Proof.* Let $G$ be a graph of order $n$ with adjacency matrix $A = [a_{ij}]$ and eigenvalue $\lambda$. Let $\mathbf{x} = [x_1\, x_2 \cdots x_n]^T$ be the corresponding $n \times 1$ eigenvector for $\lambda$ so that $A\mathbf{x} = \lambda\mathbf{x}$. Let $x_k$ be the maximum entry in $\mathbf{x}$ in absolute value, that is, $|x_k| = \max\{|x_1|, |x_2|, \ldots, |x_n|\}$. Since $A\mathbf{x} = \lambda\mathbf{x}$, we have $\lambda x_k = \sum_{j=1}^{n} a_{kj}x_j$. Using this fact and the fact that $a_{ij} \geq 0$, we have

$$|\lambda x_k| = \left| \sum_{j=1}^{n} a_{kj}x_j \right| \leq \sum_{j=1}^{n} |a_{kj}x_j| = \sum_{j=1}^{n} a_{kj}|x_j| \leq |x_k| \sum_{j=1}^{n} a_{kj} \leq |x_k|\Delta(G).$$

Hence, $|\lambda| \leq \Delta(G)$. ∎

Given the adjacency matrix of a graph, there are many properties that can be discerned: the order, size, and degree sequence, just to name a few. We now use adjacency matrices in a different way, that is, we will show that $K_{10}$ does not factor into three copies of the Petersen graph, first observed by Schwenk [215] (see also Schwenk and Lossers [216]). Note that the necessary conditions for such a factorization are all satisfied: $K_{10}$ is 9-regular with order 10 and size 45; the Petersen graph is 3-regular with order 10 and size 15. Yet such a factorization does not exist as we shall now see.

**Theorem 11.5.** *The graph $K_{10}$ is not $P$-factorable where $P$ denotes the Petersen graph.*

*Proof.* Suppose, to the contrary, that $\{P_1, P_2, P_3\}$ is a $P$-factorization of $K_{10}$, that is $P_i \cong P$ for $1 \leq i \leq 3$. Then,

$$A(K_{10}) = J - I = A(P_1) + A(P_2) + A(P_3),$$

where again $J$ is the matrix in which every entry is 1 and $I$ is the $10 \times 10$ identity matrix. Recall that, in the proof of Lemma 11.3, the vector $\mathbf{u}$, in which each entry is 1, is an eigenvector of $P$ for eigenvalue 3. Hence $A(P_1)\mathbf{u} = 3\mathbf{u}$ and $A(P_2)\mathbf{u} = 3\mathbf{u}$. Also by Lemma 11.3, $P_1$ has 1 as an eigenvalue with multiplicity 5. It is not difficult to show that the corresponding eigenspace of $A(P_1)$ for eigenvalue 1 has dimension 5, and hence there are five linearly independent eigenvectors for eigenvalue 1, each of which is orthogonal to $\mathbf{u}$. Similarly, the eigenspace of $A(P_2)$ for eigenvalue 1 has five linearly independent eigenvectors, each of which is orthogonal to $\mathbf{u}$. Hence, these two eigenspaces must have nonempty intersection (since they belong to $\mathbb{R}^{10}$), say $\mathbf{z}$ is a vector such that $A(P_1)\mathbf{z} = \mathbf{z}$ and $A(P_2)\mathbf{z} = \mathbf{z}$. Also, since $\mathbf{z}$ is orthogonal to $\mathbf{u}$, it follows that $J\mathbf{z} = \mathbf{0}$. So,

$$(J - I)\mathbf{z} = (A(P_1) + A(P_2) + A(P_3))\mathbf{z}$$

or

$$J\mathbf{z} - I\mathbf{z} = A(P_1)\mathbf{z} + A(P_2)\mathbf{z} + A(P_3)\mathbf{z}$$

or

$$-\mathbf{z} = \mathbf{z} + \mathbf{z} + A(P_3)\mathbf{z}.$$

Hence, $A(P_3)\mathbf{z} = -3\mathbf{z}$ so that $-3$ is an eigenvalue of $A(P_3)$, contradicting the fact that the eigenvalues of the Petersen graph are 3, 1, and $-2$. ∎

Schwenk actually proved something stronger than Theorem 11.5. He proved that if there is a 3-factorization of $K_{10}$ in which two of the 3-factors are isomorphic to the Petersen graph, then the remaining 3-factor must be isomorphic to the graph obtained from $K_{5,5}$ by removing the edges of a 10-cycle.

We now define other matrices associated with a graph. Let $G$ be a graph with $V(G) = \{v_1, v_2, \ldots, v_n\}$. The **degree matrix** $D(G) = [d_{ij}]$ is the $n \times n$ matrix with

$$d_{ij} = \begin{cases} \deg v_i & \text{if } i = j \\ 0 & \text{if } i \neq j. \end{cases}$$

For a graph $G$, the matrix $L(G) = D(G) - A(G)$ is called the **Laplacian matrix** of $G$. Just like the adjacency matrix, the Laplacian matrix is a real symmetric matrix, and it also has the very nice property that the sum of any row or column is zero. Because of this, the vector $\mathbf{u}$, in which every entry is 1, is an eigenvector corresponding to eigenvalue 0. In fact, 0 is the smallest eigenvalue of the Laplacian and its multiplicity is the number of components of $G$. We now turn to another way in which the Laplacian matrix of a graph $G$ can be used, namely to count the number of labeled spanning trees of $G$.

## The matrix-tree theorem

Two labelings of the same graph from the same set of labels are considered **distinct** if they produce different edge sets. Figure 11.3 shows three labelings of a graph of order 9 from the set $\{1, 2, \ldots, 9\}$. Since the first two labelings produce the same edge set, these two labelings are considered the same. The third labeling is different from the first two, however, since $\{2, 6\}$ is an edge of the third labeling while $\{2, 6\}$ is not an edge in either of the first two labelings.

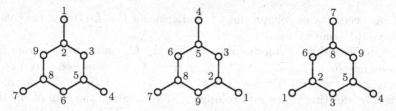

**Figure 11.3.** Labelings of a graph

Determining the number of distinct spanning trees of a labeled graph is an interesting question, with many applications in computer science and chemistry. As an example, determining this parameter for $K_n$ for any given value of $n$ gives the total number of distinct labeled trees of order $n$. An answer to this question can be found from the determinant of a matrix. This result, implicit in the work of Kirchhoff [145], is known as the **matrix-tree theorem**.

The proof we give of the matrix-tree theorem will employ several results from matrix theory, and hence we review those before presenting its proof. Let $M$

be an $r \times s$ matrix and $M'$ an $s \times r$ matrix with $r \leq s$. The product $M \cdot M'$ is therefore an $r \times r$ matrix. Since $M \cdot M'$ is a square matrix, its determinant $\det(M \cdot M')$ exists. An $r \times r$ submatrix $M_0$ of $M$ is said to correspond to the $r \times r$ submatrix $M_0'$ of $M'$ if the column numbers of $M$ determining $M_0$ are the same as the row numbers of $M'$ determining $M_0'$. A result from matrix theory states that

$$\det(M \cdot M') = \sum (\det M_0)(\det M_0'), \qquad (11.2)$$

where the sum is taken over all $r \times r$ submatrices $M_0$ of $M$ and where $M_0'$ is the $r \times r$ submatrix corresponding to $M_0$. The numbers $\det(M_0)$ and $\det(M_0')$ are referred to as **major determinants** of $M$ and $M'$, respectively.

As an illustration, we have

$$\begin{bmatrix} 1 & -2 & 3 \\ 2 & 0 & 4 \end{bmatrix} \begin{bmatrix} 2 & -1 \\ 3 & 1 \\ 0 & 2 \end{bmatrix} = \begin{bmatrix} -4 & 3 \\ 4 & 6 \end{bmatrix},$$

which has a determinant of $-36$. Writing $|A| = \det(A)$ and using 11.2, we also have

$$\begin{vmatrix} 1 & -2 \\ 2 & 0 \end{vmatrix} \begin{vmatrix} 2 & -1 \\ 3 & 1 \end{vmatrix} + \begin{vmatrix} 1 & 3 \\ 2 & 4 \end{vmatrix} \begin{vmatrix} 2 & -1 \\ 0 & 2 \end{vmatrix} + \begin{vmatrix} -2 & 3 \\ 0 & 4 \end{vmatrix} \begin{vmatrix} 3 & 1 \\ 0 & 2 \end{vmatrix} = -36.$$

Suppose that $A$ is an $n \times n$ matrix for some $n \geq 3$. Let $A'$ be the $(n-1) \times (n-1)$ submatrix of $A$ obtained by deleting row $i$ and column $j$ from $A$, where $1 \leq i, j \leq n$. Then $(-1)^{i+j} \det(A')$ is called the $(i, j)$-**cofactor** of $A$. We are now ready to state and prove the matrix-tree theorem.

**Theorem 11.6 (Matrix-tree theorem).** *The number of distinct spanning trees of $G$ is the value of any cofactor of the Laplacian $L(G)$.*

*Proof.* As every row or column sum of the Laplacian $L = L(G)$ is 0, the cofactors of $L$ have the same value.

Assume first that $G$ is a disconnected graph. Of course in this case, $G$ has no spanning trees. Let $G_1$ be a component of $G$ and suppose that $V(G_1) = \{v_1, v_2, \ldots, v_r\}$, where $1 \leq r < n$. Let $M$ be the $(n-1) \times (n-1)$ submatrix of $L$ obtained by deleting row $n$ and column $n$ from $L$. Since the sum of the first $r$ rows of $M$ is the zero vector of length $n - 1$, the rows of $M$ are linearly dependent and so $\det(M) = 0$, as desired.

Henceforth, we assume that $G$ is a connected graph of order $n$ and size $m$ where $E(G) = \{e_1, e_2, \ldots, e_m\}$. Thus $m \geq n - 1$. Let $C = [c_{ij}]$ be an $n \times m$ matrix where $c_{ij} = 1$ or $c_{ij} = -1$ if $v_i$ is incident with $e_j$ and such that each column has one entry that is 1 and one entry that is $-1$, while all other entries in the column are 0. We show that for the transpose $C^t$ of $C$, we have $C \cdot C^t = L$. The $(i, j)$-entry of $C \cdot C^t$ is

$$\sum_{k=1}^{m} c_{ik} c_{jk},$$

which has the value $\deg v_i$ if $i = j$, the value $-1$ if $i \neq j$ and $v_iv_j \in E(G)$ and the value 0 if $i \neq j$ and $v_iv_j \notin E(G)$. Hence, as claimed, $C \cdot C^t = L$.

Consider a spanning subgraph $H$ of $G$ containing $n - 1$ edges of $G$. Let $C'$ be the $(n - 1) \times (n - 1)$ submatrix of $C$ determined by the columns associated with the edges of $H$ and by all rows of $C$ with one exception, say row $k$.

We now determine the absolute value $|\det(C')|$ of the determinant of $C'$. If $H$ is disconnected, then $H$ has a component $H_1$ not containing $v_k$. The sum of the row vectors of $C'$ corresponding to the vertices of $H_1$ is the zero vector of length $n - 1$. Hence the row vectors in $C'$ are linearly dependent and so $|\det(C')| = 0$.

Next, assume that $H$ is connected. Thus, $H$ is a spanning tree of $G$. Let $u_1$ be an end-vertex of $H$ that is distinct from $v_k$ and let $f_1$ be the edge of $H$ that is incident with $u_1$. In the tree $H - u_1$, let $u_2$ be an end-vertex distinct from $v_k$ and let $f_2$ be the edge of $H - u_1$ that is incident with $u_2$. This procedure is continued until only the vertex $v_k$ remains.

A matrix $C'' = \left[ c''_{ij} \right]$ can now be obtained by a permutation of the rows and columns of $C'$ such that $|c''_{ij}| = 1$ if and only if $u_i$ and $f_j$ are incident. From the manner in which $C''$ is defined, any vertex $u_i$ is incident only with edges $f_j$ with $j \leq i$. This, however, implies that $C''$ is a lower triangular matrix and since $|c''_{ii}| = 1$ for all $i$, we conclude that $|\det(C'')| = 1$. Consequently, $|\det(C')| = |\det(C'')| = 1$.

Since every cofactor of $L$ has the same value, it suffices to evaluate the determinant of the matrix obtained by deleting both row $i$ and column $i$ from $L$ for some $i$ $(1 \leq i \leq n)$. Let $C_i$ denote the matrix obtained from $C$ by removing row $i$. Then the cofactor mentioned above equals $\det(C_i \cdot C_i^t)$, which implies by (11.2) that this number is the sum of the products of the corresponding major determinants of $C_i$ and $C_i^t$. However, these corresponding major determinants have the same value, and so their product is 1 if the defining columns correspond to a spanning tree and 0 otherwise. ∎

We now illustrate the matrix-tree theorem for the labeled graph $G$ of Figure 11.4. The Laplacian $L = L(G)$ is

$$
L = \begin{bmatrix}
2 & -1 & -1 & 0 \\
-1 & 3 & -1 & -1 \\
-1 & -1 & 3 & -1 \\
0 & -1 & -1 & 2
\end{bmatrix}.
$$

Next, we may pick any cofactor of $L$ to evaluate, say the $(2, 3)$-cofactor. So, $(2, 3)$-cofactor of $L$ is

$$
(-1)^{2+3} \begin{vmatrix}
2 & -1 & 0 \\
-1 & -1 & -1 \\
0 & -1 & 2
\end{vmatrix} = 8.
$$

Consequently, there are eight distinct spanning trees of the $G$, shown in Figure 11.4.

We now return to the question of determining, for a given integer $n$, the number of distinct trees of order $n$ whose vertices are labeled with the same set of

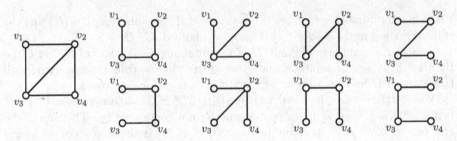

**Figure 11.4.** A labeled graph and its distinct spanning trees

$n$ labels, say $1, 2, \ldots, n$. There are three such labeled trees of order 3, and there are 16 such labeled trees of order 4. These 19 trees are shown in Figure 11.5.

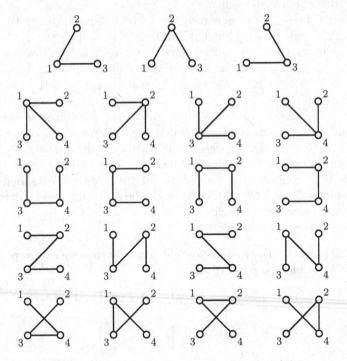

**Figure 11.5.** Labeled trees of orders 3 and 4

In general, the number of distinct trees of order $n$ whose vertices are labeled with the same set of $n$ labels is $n^{n-2}$. This result is due to Cayley [41], and while there are many proof of this fact, it also follows from the matrix-tree theorem (see Exercise 11.14).

**Theorem 11.7 (Cayley's tree formula).** *For each positive integer $n$, there are $n^{n-2}$ distinct labeled trees of order $n$ having the same vertex set.*

# 11.2 The automorphism group

An **automorphism** of a graph $G$ is an isomorphism from $G$ to itself. Thus, an automorphism of $G$ is a permutation of $V(G)$ that preserves adjacency (and nonadjacency). Of course, the identity function $\varepsilon$ on $V(G)$ is an automorphism of $G$. The inverse of an automorphism of $G$ is also an automorphism of $G$, as is the composition of two automorphisms of $G$. These observations lead us to the fact that the set of all automorphisms of a graph $G$ form a group (under the operation of composition), called the **automorphism group** of $G$, denoted by Aut$(G)$.

The automorphism group of the graph $G_1$ of Figure 11.6 is cyclic of order 2, which we write as Aut$(G_1) \cong \mathbb{Z}_2$ (the group of integers modulo 2). In addition to the identity permutation on $V(G_1)$, the group Aut$(G_1)$ contains the *reflection* $\alpha = (u\,y)(v\,x)$, where $\alpha$ is expressed in terms of *permutation cycles*. The graph $G_2$ of Figure 11.6 of order 6 has only the identity automorphism and so Aut$(G_2) \cong \mathbb{Z}_1$. In fact, 6 is the smallest order of a nontrivial graph whose automorphism group consists only of the identity automorphism.

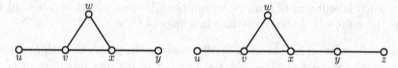

**Figure 11.6.** Graphs with automorphism groups of orders 2 and 1

Every permutation of the vertex set of $K_n$ is an automorphism and so Aut$(K_n)$ is the symmetric group $S_n$ of order $n!$. The automorphism group of $C_n$, $n \geq 3$, is the dihedral group $D_n$ of order $2n$, consisting of $n$ rotations and $n$ reflections. For example, for the 4-cycle $C_4$ as shown in Figure 11.7, the eight automorphisms are $\varepsilon$, $\alpha = (u\,v\,w\,x)$, $\alpha^2 = (u\,w)(v\,x)$, $\alpha^3 = (u\,x\,w\,v)$, $\beta = (u\,w)$, $\beta\alpha = (u\,v)(w\,x)$, $\beta\alpha^2 = (v\,x)$, and $\beta\alpha^3 = (u\,x)(v\,w)$.

**Figure 11.7.** A labeled 4-cycle

Next, we present a few basic facts concerning automorphism groups of graphs. We have already noted that every automorphism of a graph preserves both adjacency and nonadjacency. This leads to the following observation.

**Theorem 11.8.** *For every graph $G$,* Aut$(G) \cong$ Aut$(\overline{G})$.

We mentioned previously that $\mathrm{Aut}(K_n) \cong S_n$ for every positive integer $n$. Certainly, if $G$ is a graph of order $n$ containing adjacent vertices as well as nonadjacent vertices, then $\mathrm{Aut}(G)$ is isomorphic to a proper subgroup of the symmetric group $S_n$. Combining this observation with Theorem 11.8 and Lagrange's Theorem on the order of a subgroup of a finite group (which states that the order of a subgroup of a finite group divides the order of the group), we arrive at the following.

**Theorem 11.9.**  *The order* $|\mathrm{Aut}(G)|$ *of the automorphism group of a graph* $G$ *of order $n$ is a divisor of $n!$ and equals $n!$ if and only if $G = K_n$ or $G = \overline{K}_n$.*

Recall that two labelings of a graph $G$ of order $n$ from the same set of $n$ labels are considered distinct if they do not produce the same edge set. With the aid of the automorphism group of a graph $G$ of order $n$, it is possible to determine the number of distinct labelings of $G$.

**Theorem 11.10.**  *The number of distinct labelings of a graph $G$ of order $n$ from a given set of $n$ labels is $n!/|\mathrm{Aut}(G)|$.*

*Proof.* Let $S$ be a set of $n$ labels. Certainly, there exist $n!$ labelings of $G$ using the elements of $S$ without regard to which labelings are distinct. For a given labeling of $G$, each automorphism of $G$ gives rise to an identical labeling of $G$; that is, each labeling of $G$ from $S$ determines $|\mathrm{Aut}(G)|$ identical labelings of $G$. Hence, there are $n!/|\mathrm{Aut}(G)|$ distinct labelings of $G$. ∎

As an illustration of Theorem 11.10, consider the graph $G = P_3$ and the set $S = \{1, 2, 3\}$. Since $\mathrm{Aut}(G) \cong \mathbb{Z}_2$, the number of distinct labelings of $G$ with the set $S$ is $3!/2 = 3$. The three distinct labelings of $G$ the set $S$ are shown in Figure 11.5. Similarly, when $G = P_4$ and $S = \{1, 2, 3, 4\}$, since $\mathrm{Aut}(G) \cong \mathbb{Z}_2$, there are $4!/2 = 12$ distinct labelings of $G$ with the set $S$. These 12 labelings of $G = P_4$ are shown in Figure 11.5. Finally, when $G = K_{1,3}$ and $S = \{1, 2, 3, 4\}$, since $\mathrm{Aut}(G) \cong \mathbb{S}_3$, there are $4!/6 = 4$ distinct labelings of $G$ with the set $S$. These four labelings of $G = K_{1,3}$ are shown in Figure 11.5.

Let $R$ be a relation defined on the vertex set of a graph $G$ by $u\, R\, v$ if $\phi(u) = v$ for some automorphism $\phi$. Then $R$ is reflexive, symmetric and transitive and therefore is an equivalence relation. Hence, the relation $R$ then produces a partition of $V(G)$ into equivalence classes, referred to as the **orbits** of $G$. Two vertices belonging to the same orbit are called **similar vertices**. Therefore, two similar vertices have the same degree. The automorphism group of the graph $G$ of Figure 11.8 is cyclic of order 3 and $G$ has four orbits: $\{v_1\}$, $\{v_2, v_3, v_4\}$, $\{v_5, v_7, v_9\}$, and $\{v_6, v_8, v_{10}\}$.

## Vertex-transitive graphs

A graph that contains a single orbit is called **vertex-transitive**. Thus, a graph $G$ is vertex-transitive if and only if for every two vertices $u$ and $v$ of $G$, there exists an automorphism $\phi$ of $G$ such that $\phi(u) = v$. Necessarily then, every

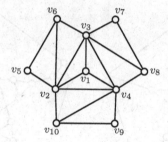

**Figure 11.8.** A graph with four orbits of similar vertices

vertex-transitive graph is regular. The graphs $K_n$ $(n \geq 1)$, $C_n$ $(n \geq 3)$ and $K_{r,r}$ $(r \geq 1)$ are all vertex-transitive. Also, the regular graphs $G_1 = C_5 \,\square\, K_2$ and $G_2 = K_{2,2,2}$, shown in Figure 11.9, are vertex-transitive. The regular graphs $G_3$ and $G_4$ in Figure 11.9 are not vertex-transitive, however (see Exercise 11.26).

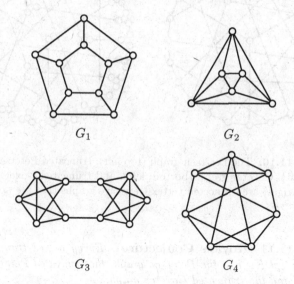

**Figure 11.9.** Vertex-transitive graphs and regular graphs that are not vertex-transitive

The two vertex-transitive graphs $G_1$ and $G_2$ of Figure 11.9 are hamiltonian. In fact, there are many examples of vertex-transitive hamiltonian graphs. Indeed, other than $K_1$ and $K_2$, there are only four known connected vertex-transitive graphs that are not hamiltonian, namely the Petersen graph and the **Coxeter graph** (shown in Figure 11.10) and the two graphs obtained from these by replacing each vertex by a triangle. These are called the **truncated Petersen graph** and the **truncated Coxeter graph**, also shown in Figure 11.10. In fact, Royle made the following conjecture that these are the only connected vertex-transitive graphs that are not hamiltonian.

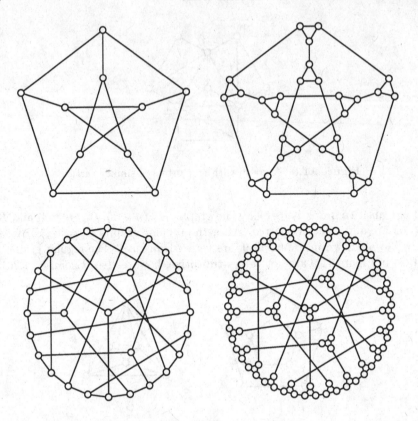

**Figure 11.10.** The Petersen graph (top left), truncated Petersen graph (top right), Coxeter graph (bottom left), and truncated Coxeter graph (bottom right) are connected vertex-transitive graphs that are not hamiltonian

**Conjecture 11.11 (Royle's Conjecture).** *Every vertex-transitive graph is hamiltonian except $K_1, K_2$, the Petersen graph, the truncated Petersen graph, the Coxeter graph and the truncated Coxeter graph.*

All four graphs in Figure 11.10 fail to contain a hamiltonian cycle; yet all four contain a hamiltonian path. Indeed, Lovász made the following conjecture.

**Conjecture 11.12 (Lovász's conjecture).** *Every connected vertex-transitive graph contains a hamiltonian path.*

Every digraph also has an automorphism group. An **automorphism** of a digraph $D$ is an isomorphism from $D$ to itself; that is, an automorphism of $D$ is a permutation $\alpha$ on $V(D)$ such that $(u, v)$ is an arc of $D$ if and only if $(\alpha(u), \alpha(v))$ is an arc of $D$. The set of all automorphisms under composition forms a group, called the **automorphism group** of $D$, which, as expected, is denoted by $\mathrm{Aut}(D)$. While $\mathrm{Aut}(G) \cong \mathbb{Z}_3$ for the graph $G$ of Figure 11.8, there is an even simpler digraph $D$ having $\text{Aut}(D) \cong \mathbb{Z}_3$. In particular, the digraphs $D_1$ and $D_2$ of Figure 11.11 have cyclic automorphism groups, namely $\text{Aut}(D_1) \cong \mathbb{Z}_3$ and $\text{Aut}(D_2) \cong \mathbb{Z}_5$.

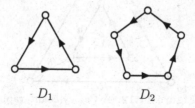

$D_1$　　　　$D_2$

**Figure 11.11.** Digraphs with cyclic automorphism groups

# 11.3　Cayley color graphs

We have seen that with every graph and every digraph, there is a finite group that can be associated with it. We now consider the reverse question of associating a digraph and a graph with a given finite group.

A nontrivial group $\Gamma$ is said to be **generated** by nonidentity elements $h_1$, $h_2$, ..., $h_k$ (and these elements are called **generators**) of $\Gamma$ if every element of $\Gamma$ can be expressed as a (finite) product of generators. Every nontrivial finite group has a finite **generating set** (often several such sets) since the set of all nonidentity elements of the group is always a generating set for $\Gamma$.

Let $\Gamma$ be a given nontrivial finite group having the generating set $\Delta = \{h_1, h_2, \ldots, h_k\}$. We associate a digraph with $\Gamma$ and $\Delta$, commonly called the **Cayley color graph of $\Gamma$ with respect to $\Delta$** and denoted by $D_\Delta(\Gamma)$. The vertex set of $D_\Delta(\Gamma)$ is the set of group elements of $\Gamma$ and so the order of $D_\Delta(\Gamma)$ is $|\Gamma|$. Each generator $h_i$ is now regarded as a color. For $g_1, g_2 \in \Gamma$, there exists an arc $(g_1, g_2)$ colored $h_i$ in $D_\Delta(\Gamma)$ if $g_2 = g_1 h_i$. If $h_i$ is a group element of order 2 (and is therefore self-inverse) and $g_2 = g_1 h_i$, then necessarily $g_1 = g_2 h_i$. When a Cayley color graph $D_\Delta(\Gamma)$ contains each of the arcs $(g_1, g_2)$ and $(g_2, g_1)$, both colored $h_i$, then, for simplicity, it is customary to represent this symmetric pair of arcs by the single edge $g_1 g_2$ colored $h_i$. As we have now seen, a Cayley color graph is actually a digraph, each arc of which is assigned a color, where a color is a generator in $\Delta$.

For example, let $\Gamma$ denote the symmetric group $S_3$ of all permutations on the set $\{1, 2, 3\}$, and let $\Delta = \{a, b\}$, where $a = (1\,2\,3)$ and $b = (1\,2)$. The Cayley color graph $D_\Delta(\Gamma)$ in this case is shown in Figure 11.12.

If the generating set $\Delta$ of a given nontrivial finite group $\Gamma$ with $n$ elements is chosen to be the set of all nonidentity group elements, then for every two vertices $g_1, g_2$ of $D_\Delta(\Gamma)$, both $(g_1, g_2)$ and $(g_2, g_1)$ are arcs (although not necessarily of

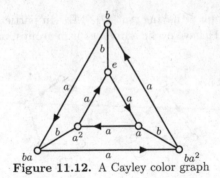

**Figure 11.12.** A Cayley color graph

the same color) and $D_\Delta(\Gamma)$ is the complete symmetric digraph $K_n^*$ of order $n$ in this case.

Let $\Gamma$ be a nontrivial finite group with generating set $\Delta$. Every element $\alpha$ in the automorphism group $\text{Aut}(D_\Delta(\Gamma))$ of the Cayley color graph $D_\Delta(\Gamma)$ has the property that if $(g_1, g_2)$ is an arc of $D_\Delta(\Gamma)$, then $(\alpha(g_1), \alpha(g_2))$ is also an arc of $D_\Delta(\Gamma)$. If for every arc $(g_1, g_2)$ of $D_\Delta(\Gamma)$, the arcs $(g_1, g_2)$ and $(\alpha(g_1), \alpha(g_2))$ have the same color, then $\alpha$ is said to be **color-preserving**. For a given nontrivial finite group $\Gamma$ with generating set $\Delta$, the set of all color-preserving automorphisms of $D_\Delta(\Gamma)$ forms a subgroup of $\text{Aut}(D_\Delta(\Gamma))$. A useful characterization of color-preserving automorphisms is given in the next result (see Exercise 11.29).

**Theorem 11.13.** *Let $\Gamma$ be a nontrivial finite group with generating set $\Delta$ and let $\alpha$ be a permutation of $V(D_\Delta(\Gamma))$. Then $\alpha$ is a color-preserving automorphism of $D_\Delta(\Gamma)$ if and only if $\alpha(gh) = \alpha(g)h$ for every $g \in \Gamma$ and $h \in \Delta$.*

The major significance of the group of color-preserving automorphisms of a Cayley color graph is next, which we prove with the aid of Theorem 11.13.

**Theorem 11.14.** *Let $\Gamma$ be a nontrivial finite group with generating set $\Delta$. Then the group of color-preserving automorphisms of $D_\Delta(\Gamma)$ is isomorphic to $\Gamma$.*

*Proof.* Let $\Gamma = \{g_1, g_2, \ldots, g_n\}$. For $i = 1, 2, \ldots, n$, define $\alpha_i : V(D_\Delta(\Gamma)) \to V(D_\Delta(\Gamma))$ by $\alpha_i(g_s) = g_i g_s$ for $1 \le s \le n$. Since $\Gamma$ is a group, the mapping $\alpha_i$ is one-to-one and onto. Let $h \in \Delta$. Then for each $i$ ($1 \le i \le n$) and for each $s$ ($1 \le s \le n$),

$$\alpha_i(g_s h) = g_i(g_s h) = (g_i g_s)h = (\alpha_i(g_s))h.$$

Hence, by Theorem 11.13, $\alpha_i$ is a color-preserving automorphism of $D_\Delta(\Gamma)$.

Let $\alpha$ be an arbitrary color-preserving automorphism of $D_\Delta(\Gamma)$ and let $g_1$ be the identity of $\Gamma$. Suppose that $\alpha(g_1) = g_r$. Let $g_s \in \Gamma$. The element $g_s$ of $\Gamma$ can be expressed as a product of generators, say $g_s = h_1 h_2 \cdots h_t$, where $h_j \in \Delta$ and $1 \le j \le t$. Therefore,

$$\begin{aligned} \alpha(g_s) &= \alpha(g_1 h_1 h_2 \cdots h_t) = \alpha(g_1 h_1 h_2 \cdots h_{t-1})h_t \\ &= \alpha(g_1 h_1 \cdots h_{t-2})h_{t-1}h_t = \cdots = \alpha(g_1)h_1 h_2 \cdots h_t = g_r g_s. \end{aligned}$$

Thus, $\alpha = \alpha_r$.

We now show that the mapping $\phi$ defined by $\phi(g_i) = \alpha_i$ is an isomorphism from $\Gamma$ to the group of color-preserving automorphisms of $D_\Delta(\Gamma)$. The mapping $\phi$ is already one-to-one and onto. It remains to show that $\phi$ is operation-preserving, namely that $\phi(g_i g_j) = \phi(g_i)\phi(g_j)$ for $g_i, g_j \in \Gamma$. Let $g_i g_j = g_k$. Then $\phi(g_i g_j) = \phi(g_k) = \alpha_k$ and $\phi(g_i)\phi(g_j) = \alpha_i \alpha_j$. Now,

$$\alpha_k(g_s) = g_k g_s = (g_i g_j)g_s = g_i(g_j g_s) = \alpha_i(g_j g_s) = \alpha_i(\alpha_j(g_s)) = (\alpha_i \alpha_j)g_s$$

and so $\alpha_k = \alpha_i \alpha_j$. ∎

## Frucht's theorem

In 1936 the first book on graph theory was published in which the author Kőnig [151, p. 5] posed the problem of determining all the finite groups $\Gamma$ for which there exists a graph $G$ such that $\mathrm{Aut}(G) = \Gamma$. In 1938, Frucht [98] proved that *every* finite group has this property as we shall now see.

If $\Gamma$ is the trivial group, then for $G = K_1$, we have $\mathrm{Aut}(G) \cong \Gamma$. Therefore, we may assume that $\Gamma$ is nontrivial and so $\Gamma = \{g_1, g_2, \ldots, g_n\}$ where $n \geq 2$. Let $\Delta = \{h_1, h_2, \ldots, h_t\}$, $1 \leq t \leq n$, be a generating set for $\Gamma$. We first construct the Cayley color graph $D_\Delta(\Gamma)$ of $\Gamma$ with respect to $\Delta$, which, recall, is actually a digraph. By Theorem 11.14, the group of color-preserving automorphisms of $D_\Delta(\Gamma)$ is isomorphic to $\Gamma$. We now transform the digraph $D_\Delta(\Gamma)$ into a graph $G$ by the following technique. Let $(g_i, g_j)$ be an arc of $D_\Delta(\Gamma)$ colored $h_k$. Delete this arc and replace it by the *graphical* path $g_i, u_{ij}, u'_{ij}, g_j$. At the vertex $u_{ij}$ we construct a new path $P_{ij}$ of length $2k - 1$ and at the vertex $u'_{ij}$ we construct a path $P'_{ij}$ of length $2k$. This construction is now performed for every arc of $D_\Delta(\Gamma)$. This is illustrated in Figure 11.13 for $k = 1, 2, 3$.

The addition of the paths $P_{ij}$ and $P'_{ij}$ in the formation of $G$ is equivalent, in a sense, to assigning a direction and a color to each arc in the construction of $D_\Delta(\Gamma)$. Observing that every color-preserving automorphism of $D_\Delta(\Gamma)$ induces an automorphism of $G$, and conversely, results in a proof of Frucht's theorem.

**Theorem 11.15 (Frucht's Theorem).** *For every finite group $\Gamma$, there exists a graph $G$ such that $\mathrm{Aut}(G) \cong \Gamma$.*

The condition of having a given group prescribed as the automorphism group of a graph is not a particularly stringent one. For example, Izbicki [133] showed that for every finite group $\Gamma$ and integer $r \geq 3$, there exists an $r$-regular graph $G$ with $\mathrm{Aut}(G) \cong \Gamma$.

## Cayley graphs

We have now seen that for every finite group $\Gamma$ and generating set $\Delta$, there is an associated digraph, namely the Cayley color graph $D_\Delta(\Gamma)$. The underlying graph of a Cayley color graph $D_\Delta(\Gamma)$ is called a **Cayley graph** and is denoted

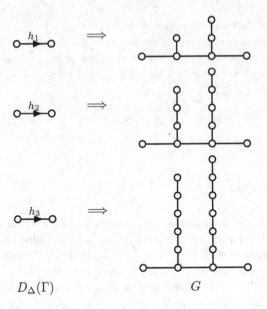

**Figure 11.13.** Constructing a graph $G$ from a given group $\Gamma$

by $G_\Delta(\Gamma)$. Thus, a graph $G$ is a Cayley graph if and only if there exists a finite group $\Gamma$ and a generating set $\Delta$ for $\Gamma$ such that $G \cong G_\Delta(\Gamma)$; that is, the vertices of $G$ are the elements of $\Gamma$ and two vertices $g_1$ and $g_2$ of $G$ are adjacent if and only if either $g_1 = g_2 h$ or $g_2 = g_1 h$ for some $h \in \Delta$.

As observed earlier, the complete symmetric digraph $K_n^*$ is a Cayley color graph; consequently, every complete graph is a Cayley graph. In fact, other families of graphs we have encountered are also Cayley graphs including cycles and $n$-cubes. Every Cayley graph is necessarily regular. Indeed, every Cayley graph is vertex-transitive (see Exercise 11.35). The converse is not true, however. The Petersen graph (Figure 11.10), for example, is vertex-transitive but it is not a Cayley graph. In fact, none of the vertex-transitive graphs in Figure 11.10 is a Cayley graph. This suggests the following conjecture.

**Conjecture 11.16.** *Every connected Cayley graph with at least three vertices is hamiltonian.*

This conjecture has been around for over fifty years and has been attributed to several mathematicians with no agreement on who originally posed the question. A large class of groups for which the conjecture is known to be true is the class of abelian groups, due to a stronger result of Chen and Quimpo [46] (see Exercise 11.36).

**Theorem 11.17.** *Every connected Cayley graph on an abelian group with at least three vertices is hamiltonian.*

**The reconstruction problem**

If $\phi$ is an automorphism of a nontrivial graph $G$ and $u$ is a vertex of $G$, then $G - u \cong G - \phi(u)$, that is, if $u$ and $v$ are similar vertices of a graph $G$, then $G - u \cong G - v$. The converse of this statement is not true, however. Indeed, the vertices $u$ and $v$ of the graph $G$ of Figure 11.14 are not similar; yet $G - u \cong G - v$.

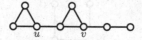

**Figure 11.14.** A graph with nonsimilar vertices whose vertex-deleted subgraphs are isomorphic

This brings up a question. Suppose that $G$ and $H$ are two graphs of the same order with $V(G) = \{v_1, v_2, \ldots, v_n\}$ and $V(H) = \{u_1, u_2, \ldots, u_n\}$, say. If it should occur that $G - v_1 \cong H - u_1$, then this does not imply that $G \cong H$. But what if, in addition to having $G - v_1 \cong H - u_1$, we also know that $G - v_2 \cong H - u_2$, $G - v_3 \cong H - u_3$ and so on, up to $G - v_n \cong H - u_n$. Can we then conclude that $G \cong H$? This question is related to the problem of determining how much structure of a graph $G$ can be recovered from its vertex-deleted subgraphs. This, in fact, brings us to a famous problem in graph theory.

 **Reconstructible graphs**

A graph $G$ with $V(G) = \{v_1, v_2, \ldots, v_n\}$, $n \geq 2$, is said to be **reconstructible** if, for every graph $H$ having $V(H) = \{u_1, u_2, \ldots, u_n\}$, $G - v_i \cong H - u_i$ for $i = 1, 2, \ldots, n$ implies $G \cong H$. Hence, if $G$ is a reconstructible graph, then the subgraphs $G - v$, $v \in V(G)$, uniquely determine $G$. It is believed by many but has never been verified that every graph of order at least 3 is reconstructible.

**Conjecture 11.18 (Reconstruction conjecture).** *Every graph of order at least 3 is reconstructible.*

This conjecture is believed to have been made in 1941 and is often attributed jointly to Kelly [140] and Ulam [242]. The **reconstruction problem** is the problem of determining the truth or falsity of the Reconstruction Conjecture. The condition on the order in the Reconstruction Conjecture is necessary for if $G_1 = K_2$, then $G_1$ is not reconstructible. This is because if $G_2 = 2K_1$, then the subgraphs $G_1 - v$, $v \in V(G_1)$, and the subgraphs $G_2 - v$, $v \in V(G_2)$, are precisely the same. Thus, $G_1$ is not uniquely determined by its subgraphs $G_1 - v$, $v \in V(G_1)$. By the same reasoning, $G_2 = 2K_1$ is also not reconstructible. The Reconstruction Conjecture claims that $K_2$ and $2K_1$ are the only non-reconstructible graphs.

If there is a counterexample to the Reconstruction Conjecture, then it must have order at least 14, for, with the aid of computers, McKay [165] has shown that all graphs of orders ranging from 2 to 13 are reconstructible. The graph $G$ of Figure 11.15 is therefore reconstructible since its order is less than 12. Hence the graphs $G - v_i$ $(1 \leq i \leq 6)$ uniquely determine $G$. However, there exists a graph $H$ with $V(H) = \{v_1, v_2, \ldots, v_6\}$ such that $G - v_i \cong H - v_i$ for $1 \leq i \leq 5$, but $G - v_6 \not\cong H - v_6$. Therefore, the graphs $G - v_i$ $(1 \leq i \leq 5)$ do not uniquely determine $G$. On the other hand, the graphs $G - v_i$ $(4 \leq i \leq 6)$ do uniquely determine $G$.

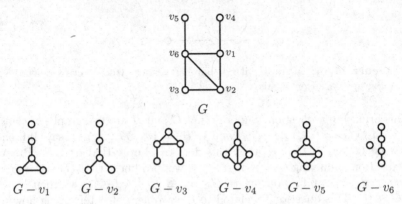

**Figure 11.15.** A reconstructible graph

Digraphs are not reconstructible, however. The vertex-deleted subdigraphs of the tournaments $D_1$ and $D_2$ of Figure 11.16 are the same; yet $D_1 \not\cong D_2$. Indeed, Stockmeyer [223] showed that there are infinitely many pairs of counterexamples for digraphs (see [148, 224] as well).

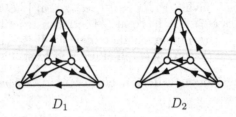

**Figure 11.16.** Two non-reconstructible digraphs

## ⬡ Recognizable properties

There are several properties of a graph $G$ that can be identified with the aid of the subgraphs $G - v$, $v \in V(G)$. We begin with the most elementary of these.

**Theorem 11.19.** *If $G$ is a graph of order $n \geq 3$ and size $m$, then $n$ and $m$ as well as the degrees of the vertices of $G$ are determined from the $n$ subgraphs $G - v$, $v \in V(G)$.*

*Proof.* It is trivial to determine the number $n$, which is necessarily one greater than the order of any subgraph $G - v$. Also, $n$ is equal to the number of subgraphs $G - v$. To determine $m$, label these subgraphs by $G_i$, $i = 1, 2, \ldots, n$. Let $V(G) = \{v_1, v_2, \ldots, v_n\}$ and suppose that $G_i = G - v_i$, where $v_i \in V(G)$. Let $m_i$ denote the size of $G_i$. Consider an arbitrary edge $e$ of $G$, say $e = v_j v_k$. Then $e$ belongs to $n - 2$ of the subgraphs $G_i$, namely all except $G_j$ and $G_k$. Since $\sum_{i=1}^{n} m_i$ counts each edge $n - 2$ times, it follow that $\sum_{i=1}^{n} m_i = (n-2)m$ and so

$$m = \frac{\sum_{i=1}^{n} m_i}{n - 2}. \tag{11.3}$$

The degrees of the vertices of $G$ can be determined by simply noting that $\deg v_i = m - m_i$, $i = 1, 2, \ldots, n$. ∎

We illustrate Theorem 11.19 with the six subgraphs $G - v$ shown in Figure 11.17 of some unspecified graph $G$. From these subgraphs we determine $n$, $m$ and $\deg v_i$ for $i = 1, 2, \ldots, 6$. Clearly, $n = 6$. By calculating the integers $m_i$ ($1 \leq i \leq 6$), we find that $m = 9$. Thus, $\deg v_1 = \deg v_2 = 2$, $\deg v_3 = \deg v_4 = 3$ and $\deg v_5 = \deg v_6 = 4$.

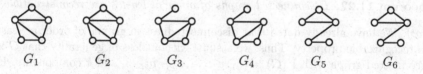

$$G_1 \qquad G_2 \qquad G_3 \qquad G_4 \qquad G_5 \qquad G_6$$

**Figure 11.17.** The subgraphs $G - v$ of a graph $G$

We say that a graphical parameter or graphical property is **recognizable** if, for each graph $G$ of order at least 3, it is possible to determine the value of the parameter for $G$ or whether $G$ has the property from the subgraphs $G - v$, $v \in V(G)$. Theorem 11.19 thus states that for a graph of order at least 3, the order, the size and the degrees of its vertices are recognizable parameters. From Theorem 11.19, it also follows that the property of a graph being regular is recognizable; indeed, the degree of regularity is a recognizable parameter. For regular graphs, much more can be said.

**Theorem 11.20.** *Every regular graph of order at least 3 is reconstructible.*

*Proof.* As we have already mentioned, regularity and the degree of regularity are recognizable. Thus, without loss of generality, we may assume that $G$ is an $r$-regular graph with $V(G) = \{v_1, v_2, \ldots, v_n\}$, for some $n \geq 3$. It remains to show that $G$ is uniquely determined by its subgraphs $G - v_i$, $i = 1, 2, \ldots, n$. Consider $G - v_1$, say. The graph $G - v_1$ then has order $n - 1$, where $r$ vertices have degree $r - 1$ and the remaining $n - r - 1$ vertices have degree $r$. Adding the vertex $v_1$ to $G - v_1$ together with all those edges $v_1v$ where $\deg_{G-v_1} v = r - 1$ produces the graph $G$. ∎

If $G$ has order $n \geq 3$, then it can be determined whether $G$ is connected from the $n$ subgraphs $G - v$, $v \in V(G)$.

**Theorem 11.21.** *For graphs of order at least 3, connectedness is a recognizable property. In particular, if $G$ is a graph with $V(G) = \{v_1, v_2, \ldots, v_n\}$, $n \geq 3$, then $G$ is connected if and only if at least two of the subgraphs $G - v_i$ are connected.*

*Proof.* Let $G$ be a connected graph. By Theorem 3.1, $G$ contains at least two vertices that are not cut-vertices, implying the results.

Conversely, assume that there exist vertices $v_1, v_2 \in V(G)$ such that both $G - v_1$ and $G - v_2$ are connected. Thus, in $G - v_1$ and also in $G$, the vertex $v_2$ is connected to each vertex $v_i$ for $i \geq 3$. Moreover, in $G - v_2$ (and thus in $G$), $v_1$ is connected to each vertex $v_i$ for $i \geq 3$. Hence, every pair of vertices of $G$ are connected and so $G$ is connected. ∎

Since connectedness is a recognizable property, it is possible to determine from the subgraphs $G - v$, $v \in V(G)$, whether a graph $G$ of order at least 3 is disconnected. We now show that disconnected graphs are reconstructible. There have been several proofs of this fact. The proof given here is due to Manvel [162].

**Theorem 11.22.** *Disconnected graphs of order at least 3 are reconstructible.*

*Proof.* We have already noted that disconnectedness in graphs of order at least 3 is a recognizable property. Thus, we assume without loss of generality that $G$ is a disconnected graph with $V(G) = \{v_1, v_2, \ldots, v_n\}$, $n \geq 3$, and $k$ components. Further, let $G_i = G - v_i$ with $k_i$ components for $i = 1, 2, \ldots, n$. From Theorem 11.19, the degrees of the vertices $v_i$, $i = 1, 2, \ldots, n$, can be determined from the graphs $G - v_i$. Hence, if $G$ contains an isolated vertex, then $G$ is reconstructible. Assume then that $G$ has no isolated vertices.

Since every component of $G$ is nontrivial, it follows that $k_i \geq k$ for $i = 1, 2, \ldots, n$ and that $k_j = k$ for some integer $j$ satisfying $1 \leq j \leq n$. Hence, the number of components of $G$ is $\min\{k_i : i = 1, 2, \ldots, n\}$. Suppose that $F$ is a component of $G$ of maximum order. Necessarily, $F$ is a component of maximum order among the components of the graphs $G_i$, that is, $F$ is recognizable. Delete a vertex that is not a cut-vertex from $F$, obtaining $F'$.

Assume that there are $r$ ($\geq 1$) components of $G$ isomorphic to $F$. The number $r$ is recognizable, as we shall see. Let

$$S = \{G_i : k(G_i) = k(G)\}$$

and let $S'$ be the subset of $S$ consisting of all those graphs $G_i$ having a minimum number $\ell$ of components isomorphic to $F$. (Observe that if $r = 1$, then there exist graphs $G_i$ in $S$ containing no components isomorphic to $F$, that is, $\ell = 0$.) In general, then, $r = \ell + 1$. Next, let $S''$ denote the set of those graphs $G_i$ in $S'$ having a maximum number of components isomorphic to $F'$.

Assume that $G_1, G_2, \ldots, G_t$ $(t \geq 1)$ are the elements of $S''$. Each graph $G_i$ in $S''$ has $k(G)$ components. Since each graph $G_i$ $(1 \leq i \leq t)$ has a minimum number of components isomorphic to $F$, each vertex $v_i$ $(1 \leq i \leq t)$ belongs to a component $F_i$ of $G$ isomorphic to $F$, where the components $F_i$ of $G$ $(1 \leq i \leq t)$ are not necessarily distinct. Further, since each graph $G_i$ $(1 \leq i \leq t)$ has a maximum number of components isomorphic to $F'$, it follows that $F_i - v_i = F'$ for each $i = 1, 2, \ldots, t$. Hence, every two of the graphs $G_1, G_2, \ldots, G_t$ are isomorphic and $G$ can be produced from $G_1$, say, by replacing a component of $G_1$ isomorphic to $F'$ by a component isomorphic to $F$. ∎

# Exercises for Chapter 11

### Section 11.1. Graphs and matrices

**11.1.** Determine the adjacency matrix of the graph $G_1$ of Figure 11.18. Then determine $A^2$ and $A^3$ without multiplying matrices.

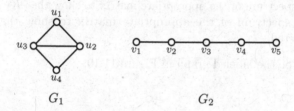

$G_1$ $\qquad\qquad\qquad\qquad$ $G_2$

**Figure 11.18.** Graphs $G_1$ and $G_2$ in Exercises 11.1 and 11.2

**11.2.** Determine the adjacency matrix of the graph $G_2$ of Figure 11.18. Then determine $A^2$, $A^3$ and $A^4$ without multiplying matrices.

**11.3.** Determine the graph $G$ with adjacency matrix $A$ for which

$$A^2 = \begin{bmatrix} 2 & 1 & 1 & 1 & 0 \\ 1 & 2 & 1 & 1 & 0 \\ 1 & 1 & 3 & 0 & 1 \\ 1 & 1 & 0 & 2 & 0 \\ 0 & 0 & 1 & 0 & 1 \end{bmatrix} \text{ and } A^3 = \begin{bmatrix} 2 & 2 & 3 & 1 & 1 \\ 2 & 2 & 3 & 1 & 1 \\ 3 & 3 & 2 & 4 & 0 \\ 1 & 1 & 4 & 0 & 2 \\ 1 & 1 & 0 & 2 & 0 \end{bmatrix}.$$

**11.4.** Show that a graph $G$ is bipartite if and only if $G$ can be labeled in such a way that its adjacency matrix can be represented in the form

$$A = \left[ \begin{array}{c|c} \mathbf{0} & A_{12} \\ \hline A_{21} & \mathbf{0} \end{array} \right]$$

where $A_{12}$ and $A_{21}$ are submatrices of $A$ with $A_{12}^T = A_{21}$.

**11.5.** Prove that a graph $G$ is disconnected if and only if $G$ can be labeled in such a way that its adjacency matrix can be represented in the form

$$A = \left[ \begin{array}{c|c} A_{11} & 0 \\ \hline 0 & A_{22} \end{array} \right].$$

**11.6.** Prove that a graph $G$ does not contain a 4-cycle if and only if the dot product of any row of $A(G)$ with any column of $A(G)$ is at most 1.

**11.7.** Let $G$ be a graph with $V(G) = \{v_1, v_2, \ldots, v_n\}$ and let $A$ be its adjacency matrix.
   (a) Show that the $(i, i)$-entry of $A^3$ is twice the number of triangles containing vertex $v_i$.
   (b) Show that the trace of $A^3$ divided by 6 is the number of triangles in $G$.

**11.8.** For a graph $G$ of order $n$ with adjacency matrix A, prove $G$ is connected if and only if $(A + I)^{n-1}$ has no zero entries where $I$ is the $n \times n$ identity matrix.

**11.9.** Determine the eigenvalues of $K_3$, $K_{1,2}$ and $K_{1,3}$.

**11.10.** Consider the graphs $K_{1,4}$ and $C_4 \cup K_1$.
   (a) Show that $K_{1,4}$ and $C_4 \cup K_1$ are cospectral.
   (b) Use the spectrum of the appropriate matrix to show that $K_{1,4}$ is connected.
   (c) Use the spectrum of the appropriate matrix to show that $C_4 \cup K_1$ is disconnected.

**11.11.** Let $G$ be the labeled graph in Figure 11.19.

**Figure 11.19.** The graph $G$ in Exercise 11.11

   (a) Use the Matrix Tree Theorem to compute the number of distinct labeled spanning trees of $G$.
   (b) Draw all the distinct labeled spanning trees of $G$.

**11.12.** Let $G = K_4$ with $V(G) = \{v_1, v_2, v_3, v_4\}$. Draw all labeled spanning trees of $G$ in which $v_4$ is an end-vertex.

**11.13.** Let $v$ be a fixed vertex of $G = K_n$. Determine the number of labeled spanning trees of $G$ in which $v$ is an end-vertex.

**11.14.** Prove Cayley's Tree Formula (Theorem 11.7) as a corollary of the matrix-tree theorem (Theorem 11.6).

**11.15.** For the graph $G$ in Figure 11.20, determine

(a) the Laplacian matrix $L$ of $G$,
(b) a cofactor of the matrix $L$,
(c) the matrix $C$ described in the proof of Theorem 11.6,
(d) the matrix $C_3$ described in the proof of Theorem 11.6,
(e) the matrix $C_3 \cdot C_3^t$,
(f) the major determinants of $C_3$ and $C_3^t$.

Illustrate (11.2) in the case where $M = C_3$ and $M' = C_3^t$, and where $\det(M_0)$ and $\det(M_0')$ are the corresponding major determinants of $C_3$ and $C_3^t$. Then show that the value of $\det(C_3 \cdot C_3^t)$ obtained is the expected value.

**Figure 11.20.** The graph $G$ in Exercise 11.15

## Section 11.2. The automorphism group

**11.16.** For the graphs $G_1$ and $G_2$ in Figure 11.21, describe the automorphisms of $G_1$ and of $G_2$ in terms of permutation cycles.

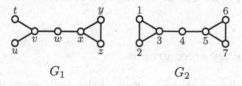

**Figure 11.21.** The graphs $G_1$ and of $G_2$ in Exercise 11.16

**11.17.** Figure 11.7 shows a 4-cycle $C_4$ and the elements of $\text{Aut}(C_4)$.
(a) Construct the group table for $\text{Aut}(C_4)$.
(b) Draw the graph $G$ where $V(G) = \text{Aut}(C_4)$ such that two vertices $\gamma_1$ and $\gamma_2$ of $G$ are adjacent if and only if $\gamma_1$ and $\gamma_2$ commute in $\text{Aut}(C_4)$.
(c) Draw the graph $\overline{G}$ for the graph $G$ in (b).

**11.18.** Does there exist a graph $H$ of order 4 such that the graph $G$ with $V(G) = \text{Aut}(H)$ where $\gamma_1\gamma_2 \in E(G)$ if and only if $\gamma_1$ and $\gamma_2$ commute in $\text{Aut}(H)$ has order 4?

**11.19.** Describe the elements of $\text{Aut}(C_5)$.

**11.20.** Find a 2-connected graph $G$ whose automorphism group is isomorphic to the cyclic group of order 4.

**11.21.** Determine the number of distinct labelings of $K_{r,r}$.

**11.22.** For which pairs $k, n$ of positive integers with $k \leq n$ does there exist a graph $G$ of order $n$ having $k$ orbits?

**11.23.** For which pairs $k, n$ of positive integers does there exist a graph $G$ of order $n$ and a vertex $v$ of $G$ such that there are exactly $k$ vertices similar to $v$?

**11.24.** Show for every even integer $n \geq 2$ that there exists a graph $G$ of order $n$ such that $G$ has $n/2$ pairs of similar vertices.

**11.25.** Let $G$ be a graph that is not vertex-transitive and let $H$ be the graph where $V(H) = V(G)$ and $xy \in E(H)$ if and only if $x$ and $y$ are similar vertices of $G$. Describe the complement $\overline{H}$ of $H$.

**11.26.** Show that the graphs $G_3$ and $G_4$ in Figure 11.9 are not vertex-transitive.

**11.27.** Describe the automorphism groups of the digraphs in Figure 11.22.

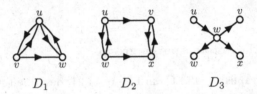

**Figure 11.22.** The digraphs in Exercise 11.27

### Section 11.3. Cayley color graphs

**11.28.** Construct the Cayley color graph of the cyclic group of order 4 when the generating set $\Delta$ has (a) one element and (b) three elements.

**11.29.** Prove Theorem 11.13: Let $\Gamma$ be a nontrivial finite group with generating set $\Delta$ and let $\alpha$ be a permutation of $V(D_\Delta(\Gamma))$. Then $\alpha$ is a color-preserving automorphism of $D_\Delta(\Gamma)$ if and only if $\alpha(gh) = \alpha(g)h$ for every $g \in \Gamma$ and $h \in \Delta$.

**11.30.** Determine the group of color-preserving automorphisms for the Cayley color graph $D_\Delta(\Gamma)$ of Figure 11.23.

**11.31.** For a given finite group $\Gamma$, determine an infinite number of mutually nonisomorphic graphs whose automorphism groups are isomorphic to $\Gamma$.

**11.32.** Show that every $n$-cycle is a Cayley graph.

**11.33.** Show that the cube $Q_3$ is a Cayley graph.

**11.34.** For $\Gamma = \mathbb{Z}_{12}$ and $\Delta = \{3, 4\}$, show that $D_\Delta(\Gamma)$ is not hamiltonian.

**11.35.** Let $\Gamma$ be a group.

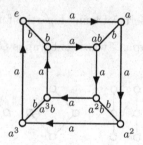

**Figure 11.23.** The Cayley color graph $D_\Delta(\Gamma)$ in Exercise 11.30

1. Let $h \in \Gamma$. Prove that the function $\phi_h \colon \Gamma \to \Gamma$ defined by $\phi_h(x) = hx$ is an automorphism of $G_\Delta(\Gamma)$ for any generating set $\Delta$.
2. If $\Delta$ is a generating set for $\Gamma$, prove that $G_\Delta(\Gamma)$ is vertex-transitive.

**11.36.** Let $\Gamma = \mathbb{Z}_m \oplus \mathbb{Z}_n$, $m, n \geq 2$, where

$$(a, b) \oplus (c, d) = (a + c \bmod m, b + d \bmod n).$$

If $\Delta = \{(1, 0), (0, 1)\}$, show that $G_\Delta(\Gamma)$ is hamiltonian.

## Section 11.4. The reconstruction problem

**11.37.** Reconstruct the graph $G$ whose subgraphs $G - v, v \in V(G)$ are given in Figure 11.17.

**11.38.** Reconstruct the graph $G$ whose subgraphs $G - v, v \in V(G)$ are given in Figure 11.24.

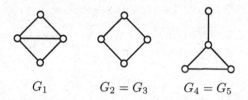

$$G_1 \qquad\qquad G_2 = G_3 \qquad G_4 = G_5$$

**Figure 11.24.** The subgraphs $G - v$ of the graph $G$ in Exercise 11.38

**11.39.** Let $G$ be a graph with $V(G) = \{v_1, v_2, \ldots, v_7\}$ such that $G - v_i = K_{2,4}$ for $i = 1, 2, 3$ and $G - v_i = K_{3,3}$ for $i = 4, 5, 6, 7$. Show that $G$ is reconstructible.

**11.40.** Show that the tournaments of Figure 11.16 are not isomorphic.

**11.41.** Let $G$ be a graph.
(a) Prove that if $G$ is reconstructible, then $\overline{G}$ is reconstructible.
(b) Prove that every graph of order $n$ $(\geq 3)$ whose complement is disconnected is reconstructible.

**11.42.** Prove that the property of being bipartite for a graph is recognizable.

**11.43.** Reconstruct the graph $G$ whose subgraphs $G - v, v \in V(G)$ are given in Figure 11.25.

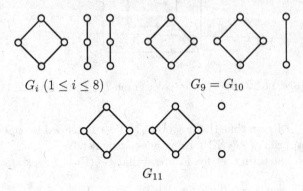

$$G_i \ (1 \le i \le 8) \qquad\qquad G_9 = G_{10}$$

$$G_{11}$$

**Figure 11.25.** The subgraphs $G - v$ of the graph $G$ in Exercise 11.43

**11.44.** Show that no graph of order at least 3 can be reconstructed from exactly two of the subgraphs $G - v, v \in V(G)$.

# Hints to selected exercises

## Hints for Chapter 1

**1.2.** The degree of each remaining vertex is 3.

**1.5.** (a) $G_1 \cong G_2$. (b) $H_1 \not\cong H_2$.

**1.10.** Consider the average degree $2m/n$ of $G$.

**1.12.** Let $V(G) = \{u, v, w, x\}$ and $E(G) = \{uv, vw, wx, xu, vx\}$. Let $e = vx$. Then $G - e = C_4$ and $G - u = C_3$.

**1.16.** If $G$ is not $r$-regular, then let $G'$ be another copy of $G$ and join corresponding vertices whose degrees are less than $r$. Continue, replacing $G$ with $G'$, until an $r$-regular graph is obtained.

**1.18.** $n = 22$.

**1.21.** Let the partite sets of a 3-partite graph $G$ of order $n = 3k$ and size $m$ be $A, B$ and $C$, where $|A| = a$, $|B| = b$ and $|C| = c$ and $a + b + c = 3k$. We may assume that $a \geq b \geq c$ and let $a = k + x, c = k - y$ and $b = 3k - a - c = k - x + y$. Then show that $m \leq ab + ac + bc \leq 3k^2 - x^2 + xy - y^2 \leq 3k^2$.

**1.23.** Observe that if $G$ has a partite set with three or more vertices, then $\overline{G}$ is not bipartite.

**1.25.** Observe that $\overline{G}$ in (a) is $C_7$ or $C_3 \cup C_4$, while $\overline{G}$ in (b) is one of $C_9, C_6 \cup C_3$, $C_5 \cup C_4$, or $3C_3$.

**1.27.** Let $C: v_1, v_2, v_3, v_4, v_5, v_1$, replace $v_i$ by a copy $H_i$ of $C_5$ and join every vertex of $H_i$ to every vertex of $H_j$ if $v_i v_j$ is an edge of $C$.

**1.29.** Use induction.

**1.32.** Since there is a self-complementary graph of order $n$ for every integer $n$ with $n \equiv 0 \pmod 4$ or $n \equiv 1 \pmod 4$ by Exercise 1.29, we may assume that $n \equiv 2 \pmod 4$ or $n \equiv 3 \pmod 4$. Use induction on $n$.

**1.33.**
  (a) The degree of $u_1$ in $G_1 \cup G_2$ is $\deg_{G_1} u_1$.
  (b) The degree of $u_1$ in $G_1 \vee G_2$ is $\deg_{G_1} u_1 + n_2$.
  (c) The degree of $(u_1, u_2)$ in $G_1 \square G_2$ is $\deg_{G_1} u_1 + \deg_{G_2} u_2$.

**1.35.**
  (a) $rK_1 \cup (s/2)K_2$.
  (b) The minimum size is $s/2$.
  (c) The maximum size is $\binom{n}{2} - s/2$.

**1.36.** Assume that for each graph $H$ with $V(H) = \{v_1, v_2, \ldots, v_n\}$ where $\deg v_i = d_i$ for $1 \leq i \leq n$, there is a vertex $v_k$ $(1 \leq k \leq n)$ such that both (1) and (2) fail.

**1.37.** (b), (c), and (e) are graphical.

**1.39.** If a graph $G$ of order 6 contains two vertices of degree 5, it has no vertex of degree 1.

**1.41.** Observe that $d_1 + d_2 + d_3 = 17$ and

$$3(3-1) + \sum_{i=1}^{7} \min\{3, d_i\} = 16.$$

**1.42.** Let $S = \{a_1, a_2, \cdots, a_n\}$ and $k = \mathrm{lcm}(a_1 + 1, a_2 + 1, \cdots, a_n + 1)$. For $S = \{2, 6, 7\}$, $k = 4$.

**1.43.** In one direction, assume that $s_1$ and $s_2$ are bigraphical sequences. Let $V_1 = \{u_1, u_2, \ldots, u_r\}$ and $V_2 = \{w_1, w_2, \ldots, w_t\}$. Suppose that there is no bipartite graph $H$ with partite sets $V_1$ and $V_2$ and a vertex $u$ of $H$ of degree $a_1$ in $V_1$ adjacent to vertices of degrees $b_1, b_2, \ldots, b_{a_1}$.

**1.45.** Let $G = K_{2,3}$ and let $u$ and $v$ be the two vertices of $G$ with $\deg u = \deg v = 3$

**1.48.** Consider a longest path in $G$.

**1.49.** There are three properties to prove, for all vertices $u, v, w \in V(G)$: reflexivity ($u$ is connected to $u$); symmetry (if $u$ is connected to $v$, then $v$ is connected to $u$); and transitivity (if $u$ is connected to $v$ and $v$ is connected to $w$, then $u$ is connected to $w$).

**1.50.** Let $V(G) = \{u = v_1, v_2, \cdots, v_n = v\}$ and let $P_i$ be a $v_i$-$v_{i+1}$ path for $i = 1, 2, \cdots, n-1$.

**1.57.** The statement is false. Consider the graph $2K_2 \vee K_1$.

**1.58.** The statement is false. Consider cycles.

**1.59.** For part (a), prove that every pair of vertices must have a common neighbor.

**1.61.** Let $P_{k+1}: u_1, u_2, \ldots, u_{k+1}$ be a path of order $k+1$ and let $G$ be the graph obtained from $P_{k+1}$ by adding $k$ new vertices $v_1, v_2, \ldots, v_k$ and joining $v_i$ to $u_i$ and $u_{i+1}$ for $1 \leq i \leq k$.

**1.62.** Suppose that the statement is false and consider a longest path in $G$.

**1.64.**

(a) Consider a path connecting two vertices of distinct degrees.

(b) No.

**1.66.** For every nonempty proper subset $S$ of $V(G)$, there is a vertex in $S$ adjacent to a vertex in $G - S$.

**1.69.** Observe that

$$d(u, v) + d(u, w) + d(v, w) = \Big(d(u, v) + d(v, w)\Big) + d(u, w) \geq 2d(u, w).$$

**1.78.** There are 20 forests of order 6.

**1.81.** These are precisely the forests.

**1.86.** Apply Theorem 1.20.

**1.88.** The 4-cycle $C_4$ is the only graph with this property.

**1.93.** Only the sequence in (c) is a degree sequence of an irregular multigraph.

**1.94.** Each of the sequences in (a), (c), (e), (g), and (h) is the degree sequence of a multigraph.

**1.96.**
  (a) $m = 20$.
  (b) $9 \le m \le 11$.
  (c) Let $U = \{u_1, u_2, \ldots, u_s\}$ and $W = \{w_1, w_2, \ldots, w_t\}$ where

$$\deg u_1 \ge \deg u_2 \ge \cdots \ge \deg u_s$$

and

$$\deg w_1 \ge \deg w_2 \ge \cdots \ge \deg w_t.$$

Then

$$m = \min \left\{ \sum_{i=1}^{s} i \deg u_i, \sum_{i=1}^{t} i \deg w_i \right\}.$$

# Hints for Chapter 2

**2.1.** Let $D$ be the digraph with $V(D) = \{v_1, v_2, v_3, v_4, v_5\}$ where $(v_i, v_j) \in E(D)$ if and only if $1 \le i < j \le 5$.

**2.2.** Yes.

**2.4.** Let $D$ be the digraph of order $2k$ such that $V(D)$ is partitioned into two sets $U$ and $W$ where $|U| = |W| = k$, where every vertex $u \in U$ is adjacent to every vertex of $W$.

**2.6.** The statement is false.

**2.8.** The outdegree and indegree of a vertex in a regular tournament of order $n$ is $(n-1)/2$.

**2.12.** For one direction, all orientations of an odd cycle contain directed paths of length 2.

**2.14.** Follow the proof of Theorem 1.6.

**2.16.**
  (a) By Robbins' theorem (Theorem 2.5), $G$ has a strong orientation $D$. Consider $D$ and $\vec{D}$.
  (b) Consider $G = C_n$ where $n \ge 3$.

**2.17.**

(a) Consider two cases: (1) $G$ contains exactly one bridge and (2) $G$ contains exactly two bridges.

(b) Consider $G = K_{1,3}$.

**2.19.** See the tournaments below.

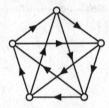 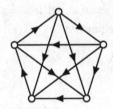

**2.21.** If $D$ is an $r$-regular tournament of order $n$, then $n = 2r + 1$.

**2.23.** Follow the proof of Theorem 2.7.

**2.24.** Use Theorem 2.10.

**2.25.** If $\widetilde{T}$ is not transitive, then $\widetilde{T}$ contains a triangle $(S_1, S_2, S_3, S_1)$. Consider $v_i \in V(S_i)$ for $i = 1, 2, 3$.

**2.27.**

(a) One.

(b) Suppose that there is a tournament that contains three vertices not having both positive outdegree and positive indegree.

(c) None.

**2.29.** Consider a shortest $u$-$v$ path.

**2.31.** Consider the set of vertices to which $v$ is adjacent.

**2.34.** Assume that $(u, v) \in E(T)$ and then consider the set of vertices to which $v$ is adjacent.

**2.36.** Note that every vertex of maximum outdegree in a tournament is a king.

**2.38.** Let $T$ be the tournament with $V(T) = \{u, v, w, x\}$ and

$$E(T) = \{(v, u), (u, w), (u, x), (w, v), (x, v), (w, x)\}.$$

**2.40.** Assume that $n \geq 6$ is even. One possible construction of $T$: Let $T_{n-3}$ be a regular tournament of order $n - 3$. Then the tournament $T$ of order $n$ has

$$V(T) = \{v_1, v_2, v_3\} \cup V(T_{n-3})$$

and

$$E(T) = \{(v_1, v_2), (v_2, v_3), (v_3, v_1)\} \cup E(T_{n-3}) \cup E_1 \cup E_2 \cup E_3,$$

where $E_1 = \{(v, v_1) : v \in E(T_{n-3})\}$, $E_2 = \{(v, v_2) : v \in E(T_{n-3})\}$, and $E_3 = \{(v_3, v) : v \in E(T_{n-3})\}$.

**2.42.** (b) and (c) are score sequences of tournaments.

**2.44.** The tournament $T$ is the transitive tournament of order $n$.

**2.46.** Consider the sets of vertices to which $u$ is adjacent and from which $u$ is adjacent.

# Hints for Chapter 3

**3.1.** Yes.

**3.2.** This is possible and one such route is $A$, $B$, $A$, $B$, $A$, $C$, $A$, $C$, $A$, $D$, $B$, $D$, $C$, $D$, $A$.

**3.3.** Show that each vertex of $G$ has even degree.

**3.5.** The graph $G \square H$ is eulerian if and only if both $G$ and $H$ are eulerian or every vertex of $G$ and $H$ has odd degree.

**3.7.** No.

**3.10.** The statement is false.

**3.11.** (a) $m - n/2$. (b) The Petersen graph.

**3.13.** Prove that every graph with the property is of the form $H \vee K_2$ where $H$ is a graph that also has this property. Then argue by induction or minimal counterexample.

**3.14.** Follow the proof of Theorem 3.1.

**3.16.** Add a new vertex $w$ to $D$ together with the arcs $(v, w)$ and $(w, u)$. Then apply Theorem 3.3.

**3.21.** (a) Yes. (b) No.

**3.23.** Assume that $\text{od}\, v_i \geq \text{id}\, v_i$ for $1 \leq i \leq k$ and that $\text{od}\, v_i < \text{id}\, v_i$ for $k + 1 \leq i \leq n$. Construct a digraph $D'$ by adding $t$ new vertices $x_i$ $(1 \leq i \leq t)$ so that $D'$ is eulerian.

**3.33.** Assume, to the contrary, that the result is false. Among all counterexamples of order $2k$, let $G$ be one of maximum size and consider a hamiltonian path in $G$. The bound is sharp.

**3.36.** Apply Dirac's theorem (Theorem 3.10) and Theorem 3.8.

**3.37.** The graph $G$ is hamiltonian if and only if $n_1 \leq n_2 + n_3 + \cdots + n_k$.

**3.38.** Since $\delta(G) > n/2$, it follows by Dirac's theorem (Theorem 3.10) that $G$ has a hamiltonian cycle $C : v_1, v_2, \ldots, v_{101}, v_1$. Consider the vertex $v_1$ and the set $S_i = \{v_i, v_{i+25}, v_{i+50}, v_{i+75}\}$ for $2 \leq i \leq 26$. The generalization of this result to graphs of order $4k + 1$ is as follows.

**Theorem.** *Let $k$ be a positive integer. If $G$ is a graph of order $4k + 1$ such that $\delta(G) \geq 2k + 1$, then every vertex of $G$ lies on a cycle of length $k + 2$.*

**3.40.** If $uv \notin E(D)$, then $\text{od}\, u + \text{id}\, v \geq n$, which implies that there is a vertex $w$ with $uw, wv \in E(D)$.

**3.41.** Suppose that the hypotheses of Woodall's theorem hold for a digraph $D$. One must show that the hypotheses of Meyniel's theorem hold. Part of this is Exercise 3.40.

**3.43.** Let $D_1$ and $D_2$ be copies of the digraph $K_k^*$ obtained by replacing each edge $uv$ of $K_k$ by the symmetric pair $(u, v)$ and $(v, u)$ of arcs. Let $D$ be obtained by identifying a vertex of $D_1$ and a vertex of $D_2$.

**3.45.** If $T$ is strongly connected, then it is hamiltonian. If $T$ is not strongly connected, then it has at least two strong components, at least one of which must be nontrivial.

**3.47.** The strong component $S_2$ has $k/15$ hamiltonian paths.

**3.49.** The statement is false.

**3.50.** The statement is false.

**3.52.** Apply Theorem 3.26.

**3.53.** Suppose that the statement was false. This would mean that there are two vertices, $u$ and $v$, such that there is no hamiltonian path from $u$ to $v$. Consider the hamiltonian graph $G - v$.

**3.56.** Let $H = K_3 \square K_2$, where $\{u, v, w\}$ are the vertices in one copy of $K_3$ and $\{u', v', w'\}$ are the vertices in the other copy of $K_3$, where $uu', vv', ww' \in E(G)$. Let $G$ be the graph obtained by adding a new vertex $x$ to $H$ where $x$ is joined to $w$ and $w'$.

**3.59.** $c = 1$

**3.60.** Let $u, v \in V(G)$ such that $d_G(u, v) = \text{diam}(G) = \ell$. Consider a $u$-$v$ geodesic $P = (u = v_0, v_1, \ldots, v_\ell = v)$.

**3.61.** Apply Exercise 3.26.

**3.63.** For the second part of the exercise, if $\text{diam}(G) = 2$, then $G^2$ is complete. Hence, it may be assumed that $\text{diam}(G) = 3$. If $G$ does not have a cut-vertex, we can apply Theorem 3.32, so we can assume that $G$ does have a cut-vertex. Remove this cut-vertex and apply induction.

# Hints for Chapter 4

**4.1.** Apply Theorem 4.1 or argue from first principles.

**4.2.** Apply Theorem 4.1.

**4.3.** Either $u$ or $v$ must have degree at least 2. Suppose that $\deg v \geq 2$ and consider a neighbor of $v$.

**4.4.** One way is to suppose otherwise, delete the bridge, and count vertices of odd degree in the connected components of the resulting graph.

**4.6.** Apply Proposition 4.5.

**4.7.** In one direction, suppose that at least one of $e_1$ or $e_2$ is not a bridge and show that $G - e_1 - e_2$ has at most two components.

**4.8.**

**Theorem.** *A vertex $v$ of a graph $G$ is a cut-vertex if and only if there are vertices $u$ and $w$ in the same component of $G$ and distinct from $v$ such that $v$ lies on every $u$-$w$ path.*

**4.9.** Consider the end-vertices of the edge.

**4.10.** We must have $k \geq \ell + 1$. This can be proved by induction on $\ell$.

**4.11.** Let $u, w \in V(\overline{G} - v)$. Consider two cases, depending on whether $u$ and $w$ lie in the same component of $G - v$. In fact, this approach will show that $\operatorname{diam}(\overline{G} - v) \leq 3$.

**4.13.** The statement is false; consider cycles.

**4.14.** The statement is false.

**4.15.** The statement is false.

**4.17.** The statement is false.

**4.20.** Apply Theorem 4.1.

**4.21.** Apply Theorems 4.1 and 4.8.

**4.22.** No.

**4.24.** Let $H$ be the graph obtained by adding two vertices $u$ and $w$ to $G$ and joining $u$ to the two vertices of $U$ and joining $w$ to the two vertices of $W$.

**4.25.** No. One can find a counterexample in the graph $K_4 - e$, where $e$ is any edge.

**4.26.** The graph $H$ is complete if and only if $G$ has order 1 or 2 or $G$ is 2-connected.

**4.27.** There are three statements to prove, depending on what the elements are. If the elements are both vertices, this is Theorem 4.8. Now suppose we are given that $G$ is 2-connected and we have two edges $e = uv$ and $f = xy$. Subdivide these edges by adding new vertices $s$ and $t$, and replacing the edges $e$ and $f$ by the edges $us$, $sv$, $xt$, and $ty$. One can show that the resulting graph is 2-connected, so the new vertices $s$ and $t$ lie on a common cycle by Theorem 4.8. For the converse direction in this case, to show that $G$ is 2-connected, given vertices $u$ and $v$, choose edges $e = ux$ and $f = vy$.

**4.29.** For one inequality, show that if $X$ is a set of vertices of $G \vee K_1$ with $|X| \leq k$, then $(G \vee K_1) - X$ is connected. For the other inequality, show how to obtain a vertex-cut of $G \vee K_1$ from a vertex-cut of $G$.

**4.31.** Observe that $k \leq \kappa(G) \leq \delta(G)$ and apply Theorem 1.20.

**4.36.** Every $k$-connected graph of order $n$ has minimum degree at least $k$ and so has size at least $kn/2$. Show that there exists a $k$-connected graph $G$ of (even) order $n$ and size $kn/2$.

**4.41.** If $G$ is a complete $k$-partite graph of order $n$ whose largest partite set contains $n_k$ vertices, then $\kappa(G) = \kappa'(G) = \delta(G) = n - n_k$.

**4.43.** Show that if $X$ is a set of edges of $G \vee K_1$ with $|X| \leq k$, then $(G \vee K_1) - X$ is connected. Equality does not necessarily hold.

**4.44.** First, determine the diameter of $G \vee K_1$.

**4.45.** There are many possibilities, but one might start with two copies of $K_{d+1}$.

**4.46.** Let $G_k$ be the graph obtained by adding an edge to the disjoint union of two copies of $K_k$.

**4.48.** Appeal to Theorem 4.15.

**4.51.** Both directions follow from the edge form of Menger's theorem (Theorem 4.20). First suppose that $G$ is $k$-edge-connected, and consider a $u$-$v$ separating subset of $E(G)$. For the converse, assume that $G$ contains $k$ pairwise edge-disjoint $u$-$v$ paths for every pair of distinct vertices $u$ and $v$.

**4.53.** The statement is true.

**4.57.** Let $u$ and $v$ be two adjacent vertices of a 3-connected graph $G$. The global form of Menger's theorem (Theorem 4.19) states that there are three pairwise internally disjoint $u$-$v$ paths. At least two of these paths have length at least 2.

**4.58.** By the global form of Menger's theorem (Theorem 4.19), $\kappa(G) = k$. Let $X$ be a vertex-cut of $G$ with $|X| = k$ and let $u$ and $v$ be vertices belonging to different components of $G - X$. Take $w \in V(G) \setminus (X \cup \{u, v\})$ and consider $X \cup \{w\}$.

**4.60.** Let $u$ and $v$ be two vertices of $G$ with $d(u, v) = \operatorname{diam} G = k$ and consider $k$ internally disjoint $v$-$u$ paths.

# Hints for Chapter 5

**5.1.** To prove that a graph $G$ is planar if each block of $G$ is planar, employ induction on the number of blocks of $G$. Consider an end-block.

**5.3.** Let $G$ be the graph consisting of two paths $P$: $u, v, w, x$ and $Q$: $y_1, y_2, z_2, z_1$ where the end-vertices $y_1$ and $z_1$ are joined to every vertex of $P$.

**5.5.** Apply Euler's formula (Theorem 5.1).

**5.6.** Use Exercise 5.5.

**5.11.** Consider $K_{2,2,2}$ and $P_4 \vee K_2$.

**5.13.** Suppose that there is a planar graph of order $n \geq 3$ and size $m = 3n - 6$ that is not maximal planar.

**5.15.** The statement is false.

**5.18.** Suppose that the result is false. Assign to each vertex $v$ of $G$ a charge of $6 - \deg v$. For each vertex $v$ of $G$ having degree 5, distribute its charge of $+1$ equally to four neighbors of $v$ not having degree 5 or 6.

**5.20.**
  (a) The statement is true.
  (b) $G[U] = \overline{K}_k$.

**5.22.** If $G$ is nonplanar, then $G$ contains a subgraph $H$ that is a subdivision of $K_5$. Compute the size of $H$.

**5.24.** The only such graph is $G = K_5$.

**5.26.**
  (a) Give a proof by contradiction.
  (b) Let $G$ be the graph obtained from $K_{3,3}$ by subdividing one edge of $K_{3,3}$.

**5.28.** The graph $C_n^2$ is nonplanar if and only if $n \geq 5$ and $n$ is odd.

**5.31.** The statement is false.

**5.33.** No conclusion can be made.

**5.36.** Give a proof by contradiction.

**5.41.** The graph $G = P_n$.

**5.43.** Consider a planar embedding of $K_{2,3}$ and suppose that $G$ contains a Hamiltonian cycle $C$.

**5.45.** Suppose that the Grinberg graph $G$ is Hamiltonian and observe that $3(r_5 - r_5') + 6(r_8 - r_8') + 7(r_9 - r_9') = 0$.

**5.47.** Suppose that this graph $G$ has a Hamiltonian cycle $C$ that contains both $e$ and $f$. Then one of the regions having $e$ on its boundary is in the interior of $C$ and the other is in the exterior of $C$.

**5.48.** See the figure below.

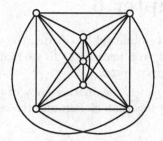

**5.50.** See the figure below.

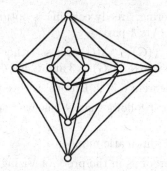

**5.52.** $\operatorname{cr}(K_{1,2,3}) = 1$.

**5.54.** See the figure below.

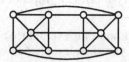

**5.55.** The statement is false.

**5.57.** See the figure below.

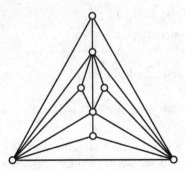

# Hints for Chapter 6

**6.1.** Proper colorings can be obtained by coloring the vertices in clockwise order. Clearly $\chi(C_{2k})$ cannot be 1, so it must be 2. To see that $\chi(C_{2k-1}) \geq 3$, one might progress outward from a given vertex.

**6.5.** If $G$ is not $r$-regular, let $G'$ be another copy of $G$ and join corresponding vertices whose degrees are less than $r$. Permute the colors of the vertices of $G'$. Continue in this manner.

**6.6.** There is a $\chi(G)$ coloring of $G$; order the vertices by their color in this coloring.

**6.8.** With any vertex ordering, greedy coloring is optimal on complete $k$-partite graphs; if $\chi(G) = k$, then $G$ is $k$-partite.

**6.9.** For every graph $G$, $\chi(G) \leq \Delta(G)$ unless either $\Delta(G) = 2$ and $G$ has a component that is an odd cycle or $\Delta(G) \geq 3$ and $G$ has $K_{\Delta(G)+1}$ as a component.

**6.10.** Theorem 6.2: 2; Theorem 6.3: 2; Brooks' theorem: 5.

**6.16.** Since $G$ is 3-regular, it follows that $G$ has even order. Consider a Hamiltonian cycle $C$ of $G$.

**6.19.** The graph has edge chromatic number 4.

**6.20.** Give an inductive proof as in the proof of Vizing's theorem (Theorem 6.5) from Lemma 6.6.

**6.22.** A self-complementary regular graph is overfull.

**6.23.** First observe that every 3-critical graph of order $n \geq 3$ contains an odd cycle $C$.

**6.28.** If $H$ is a connected and noncomplete induced subgraph of the graph of the octahedron, then $H$ is one of the three graphs. In each case, $\chi(H) = \omega(H)$.

**6.30.** For $n = 7$, let $G_7 = \overline{C}_7$.

**6.31.** The bound simplifies to $\chi(G) \leq 5/2$, which implies that $\chi(G) \leq 2$.

**6.37.** This graph is clearly cubic and bridgeless, and Exercise 6.19 asks the reader to show that it has edge chromatic number 4, so it satisfies the definition of snark we have given. That said, it contains a triangle, so it would not satisfy stricter definitions of snarks.

# Hints for Chapter 7

**7.2.**
   (a) $a = 1$, $b = 2$ and $c = 1$.
   (b) $\mathrm{val}(f) = 4$.

**7.4.** Show that for the flow $f'$, the net flow out of each vertex in $N'$ equals the net flow out of this vertex for the flow $f$.

**7.6.**
   (a) Let $\mathcal{U}$ be the set of all maximal paths $P$ in $D$ whose initial vertex is $u$ and such that no arc of $P$ belongs to $A$. Consider the set $X = \bigcup_{P \in \mathcal{U}} V(P)$.
   (b) Let $D$ be a $u$-$v$ path of length 2.

**7.8.** Show that $\mathrm{val}(f) = \mathrm{cap}(K)$ and apply Corollary 7.3.

**7.9.** Use the proof of Theorem 7.6 to show that $f(\overline{X}, X) = 0$ and $c(X, \overline{X}) = f(X, \overline{X})$.

**7.11.** A maximum flow $f$ has $\mathrm{val}(f'') = 6$ with minimum cut $[X, \overline{X}]$ where $X = \{u, s\}$.

**7.13.**
   (a) The statement is false.
   (b) The statement is true.

**7.15.** The maximum number of pairwise arc-disjoint $u$-$v$ paths in $D$ is 1.

**7.17.** Suppose that $S = \{s_1, s_2, \ldots, s_r\}$ and $T = \{t_1, t_2, \ldots, t_q\}$. Add a new vertex $s$ and the arcs $(s, s_i)$, $1 \le i \le r$, and add a new vertex $t$ and the arcs $(t_i, t)$, $1 \le i \le q$. Let $C = \sum_{e \in E(D)} c(e)$.

**7.18.** For vertices $u$ and $v$ in $G$, observe that there is a one-to-one correspondence between the $u$-$v$ paths in $G$ and the (directed) $u$-$v$ paths in $D$.

# Hints for Chapter 8

**8.1.** Suppose that there exists a tree with two distinct perfect matchings. Apply Theorem 8.1.

**8.3.** Consider the graph $H$ obtained from $G$ by adding two new vertices $u$ and $w$ where $u$ is joined to every vertex of $U$ and $w$ is joined to every vertex of $W$.

**8.6.**
   (a) Proceed by contradiction and apply Hall's theorem 8.3.
   (b) Consider $G = K_{k,k-1} + K_1$.

**8.7.** After observing that $\alpha'(G) \le |U|$, consider the cases where $\operatorname{def}(U) \le 0$ and $\operatorname{def}(U) > 0$.

**8.9.**

(a) The statement is true.

(b) The statement is false.

**8.13.** Let $S = \{u, v\}$.

**8.14.** For (a), proceed by contradiction and apply Tutte's theorem (Theorem 8.6).

**8.16.** Apply the first theorem of graph theory (Theorem 1.1).

**8.17.** Proceed by contradiction and apply Tutte's theorem (Theorem 8.6).

**8.20.** In one direction, let $G$ be a graph that is not bipartite and let $C$ be a smallest odd cycle of $G$.

**8.22.** By Exercise 8.21, $\frac{n}{1+\Delta(G)} \le \alpha'(G) \le \lfloor \frac{n}{2} \rfloor$. Apply Gallai's theorem (Theorem 8.8).

**8.26.** $\gamma(Q_3) = 2$ and $\gamma(Q_4) = 4$.

**8.29.** Let $G$ be a graph of order $n$. Then $\gamma(G) = 1$ if and only if $\Delta(G) = n - 1$.

**8.31.** Let $d = \operatorname{diam}(G) \ge 3$ and let $v$ be a vertex of $G$ such that $e(v) = d$. For a vertex $w$ of $G$ with $d(v, w) \ge 3$, consider $S = \{v, w\}$.

**8.32.** Use the fact that $\alpha(G) + \beta(G) = \alpha'(G) + \beta'(G) = n$.

**8.34.** The statement is true.

**8.35.** Suppose that there is a 1-factorization $\{F_1, F_2, F_3\}$ of the Petersen graph and consider the spanning subgraph with edge set $E(F_1) \cup E(F_2)$.

**8.37.**

(a) Apply Theorem 8.17.

(b) Observe that if $Q_n$ is $k$-factorable, then $k \mid n$. For the converse, apply (a).

**8.40.** Let $n = 4k$ for some integer $k \ge 2$. Let $H_1 = K_{2k-1}$ and $H_2' = K_{2k+1}$, let $C$ be a Hamiltonian cycle in $H_2'$ and let $H_2 = H_2' - E(C)$. Consider $G = H_1 \cup H_2$.

**8.42.** See the figures below.

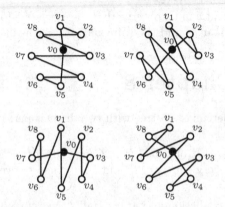

**8.45.** Proceed by contradiction and consider the size of each factor.

**8.47.** Use Theorem 8.24.

**8.49.** Let $C' = (v_1, v_2, v_4, v_6, v_8, v_7, v_5, v_3, v_1)$ and $C'' = (v_1, v_7, v_5, v_6, v_4, v_3, v_2, v_8, v_1)$. Then $C'$ and $C''$ form a Hamiltonian-factorization of $C_8^2$.

**8.51.** The graph of the octahedron in Figure 8.13 can be decomposed into three copies of $P_5$, namely $P = (u, x, z, w, v)$, $P' = (v, z, y, u, w)$ and $P''' = (w, y, x, v, u)$.

**8.53.** If there exists a Steiner triple system of order $n \geq 3$, then the complete graph $K_n$ is $K_3$-decomposable and so $3 \mid \binom{n}{2}$.

**8.54.** The possible values of $k$ are $k = 3, 4, 9$.

**8.57.** Show that it is not possible for three edges to be labeled $1, 3, 5$.

**8.61.** Label the seven vertices of $K_7$ by $0, 1, 2, \ldots, 6$ and observe that the vertex labels $0, 1$ and $3$ produce a graceful labeling of $K_3$.

**8.64.** Let the vertices of $K_{10}$ be labeled with $v_\infty, v_0, v_1, \ldots, v_8$. Let $P_0 : v_\infty, v_0, v_1, v_8, v_2, v_7$. Let $P_i : v_\infty, v_{0+i}, v_{1_i}, v_{8+i}, v_{2+i}, v_{7+i}$ where all arithmetic is modulo 9. Then, $\{P_0, p_1, \ldots, P_8\}$ is a $P_6$-decomposition of $K_{10}$.

**8.66.** Let $m = 2k$. Then $K_{3m+1} = K_{6k+1}$. Arrange the vertices $v_1, v_2, \cdots, v_{6k+1}$ cyclically about a regular $(6k + 1)$-gon and join every two vertices by a straight line segment, producing the complete graph $K_{6k+1}$. Let $C = (v_1, v_2, \cdots, v_{6k+1}, v_1)$. Assign each edge $xy$ of $K_{6k+1}$ the value $d_C(x, y)$. Use the fact that $G$ is graceful to place $G$ in $K_{6k+1}$ so that the edge labels coincide.

**8.68.** Use the drawing of the Petersen graph shown in Exercise 8.67.

**8.71.** A $\rho$-labeling of $C_6$ is given by the subscripts in the figure below.

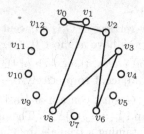

**8.73.** See the figures below.

**8.75.** If $F$ is a forest of order $n$ with $k \geq 2$ components, then the size of $F$ is $m = n - k \leq n - 2$.

**8.76.** The graph $P_3 \cup C_3$ is not graceful.

**8.78.** Apply the definition of a graceful labeling to $f$ and $\overline{f}$.

# Hints for Chapter 9

**9.1.** $T(7,2) = K_{4,3}$, $t(7,2) = 12$; $T(7,3) = K_{3,2,2}$, $t(7,3) = 16$; $T(7,4) = K_{2,2,2,1}$, $t(7,4) = 18$; $T(8,2) = K_{4,4}$, $t(8,2) = 16$; $T(8,3) = K_{3,3,2}$, $t(8,3) = 21$; $T(8,4) = K_{2,2,2,2}$, $t(8,4) = 24$.

**9.3.** The inequality is strict whenever $s(r-s)/2k \geq 1$. For fixed $r$, $s(r-s)/2r$ is maximized when $s = r/2$.

**9.4.** If $|E| = \lfloor n^2/4 \rfloor$, then the final inequality of the proof shows that we must have $|E| = |A||B|$. To see that the cardinalities of the parts must be as close to $n/2$ as possible, maximize the function $x(n-x)$.

**9.6.** Let $G = (V,E)$ be a graph of order $n$ and average degree $d = 2|E|/n = (1/2 + c)n$. The number of triangles containing a given edge $uv \in E$ is at least $\deg u + \deg v - n$. Thus the number of triangles in $G$ is at least

$$\frac{1}{3} \sum_{uv \in E} (\deg u + \deg v - n) = \frac{1}{3} \left( \sum_{v \in V} (\deg v)^2 - \frac{n^2 d}{2} \right).$$

Now apply the Cauchy–Schwarz inequality to see that this is at least $cdn^2/3$, and then use the fact that $d \geq n/2$.

**9.8.** Each of the other $n - r$ vertices of $G$ is adjacent to at most $r - 1$ vertices of this $r$-clique, and those vertices have at most $t(n-r, r)$ edges amongst them.

**9.9.** If $G$ has precisely $t(n,r)$ edges, then each of the $n - r$ vertices of $G$ that is not in the $r$-clique must be adjacent to precisely $r - 1$ vertices of this clique.

**9.10.** By Mantel's theorem, $G$ must contain at least one triangle. Let $X$ denote the vertices of this triangle and $Y = V(G) \setminus X$. The induced subgraph $G[Y]$ could contain 12 edges without containing a triangle, but beyond that, every additional edge contributes at least one additional triangle (why?). Similarly bound the number of edges and number of triangles between $X$ and $Y$.

**9.11.** By Theorem 9.1, such a graph must contain a triangle. Removing this triangle leaves a graph of order 5 with 5 edges.

**9.13.** For $n = 6$, consider the bipartite graph $K_{3,3}$.

**9.14.** $m = 13$.

**9.16.** Show that $G$ contains a vertex of degree 2 or greater.

**9.21.** $m = \lfloor n^2/4 \rfloor + 2$.

**9.23.** For (a), consider $G \vee K_1$. For (b), consider $G + K_1$.

**9.25.** To prove that $R(i,j) \leq R(k,\ell)$, let $G$ be a graph of order $R(k,\ell)$ and appeal to the definition of a Ramsey number.

**9.26.** No.

**9.28.** $R(2,3,4) = 9$.

**9.30.** Let $V(K_6) = \{v_1, v_2, v_3, v_4, v_5, v_6\}$. Because $R(3,3) = 6$, there is a monochromatic $K_3$; suppose that its vertex set is $\{v_1, v_2, v_3\}$ and that it is red. Precisely one of the edges from each of the vertices $v_4$, $v_5$, and $v_6$ to the set $\{v_1, v_2, v_3\}$ is red (why?).

**9.32.** $K_{r+1}$, $K_{r,r}$, the Petersen graph, the Heawood graph, and the Tutte–Coxeter graph achieve their corresponding Moore bounds. The McGee graph, all of the $(3,9)$- and $(3,10)$-cages, the Balaban cage, and the Benson graph do not achieve their corresponding Moore bounds.

**9.33.** Examine the proof of the Moore bound, Theorem 9.16.

**9.36.** Suppose that $H$ is an $(s,g)$-graph and assume, to the contrary, that $H$ is an $(s,g)$-cage. Since $H$ contains 4-cycles, either $g = 3$ or $g = 4$.

# Hints for Chapter 10

**10.1.** Since $K_{3,3}$ is a subgraph of $K_{4,4}$, it follows that $\gamma(K_{4,4}) \geq 1$. Exhibit an embedding of $K_{4,4}$ on the torus to show that $\gamma(K_{4,4}) = 1$.

**10.3.** Since the boundary of every region contains at least $k$ edges and every edge is on the boundary of at most two regions, $kr \leq 2m$. The result follows by applying Corollary 10.3.

**10.8.** Apply Corollary 10.6.

**10.10.** $\gamma(C_4 \,\square\, C_6 \,\square\, K_2) = 7$.

**10.12.** For $n \geq 6k + 6$, let $G$ be a maximal planar graph of order $n$. Apply Theorem 10.4.

**10.14.** $\gamma(G) = 2$.

**10.15.** Yes.

**10.16.** $\pi_0 = (1\,5\,4\,6\,2\,3)$, $\pi_1 = (0\,3\,4\,2\,6\,5)$, $\pi_2 = (0\,6\,1\,4\,5\,3)$, $\pi_3 = (0\,2\,5\,6\,4\,1)$, $\pi_4 = (0\,5\,2\,1\,3\,6)$, $\pi_5 = (0\,1\,6\,3\,2\,4)$, $\pi_6 = (0\,4\,3\,5\,1\,2)$. There are 14 orbits (regions).

**10.18.**
  (a) $G = K_{2,2,2}$.
  (b) $k = 1$.
  (c) No.

**10.20.** Use the generalized Euler identity (Theorem 10.1) to conclude that for a 2-cell embedding of $K_4$ on $S_0$ wee have $r = 4$, for a 2-cell embedding of $K_4$ on $S_1$ we have $r = 2$ and for a 2-cell embedding of $K_4$ on $S_2$ we have $r = 0$. There are two 2-cell embeddings of $K_4$ on $S_0$, 14 2-cell embeddings of $K_4$ on $S_1$, and no 2-cell embeddings on $K_4$ on $S_2$.

**10.22.** $\pi_1 = (2\,5\,3\,4)$, $\pi_2 = (1\,3\,4\,5)$, $\pi_3 = (1\,2\,4\,5)$, $\pi_4 = (1\,2\,3\,5)$, $\pi_5 = (4\,3\,2\,1)$.

**10.24.** $\gamma_M(P) = 3$.

**10.26.** If $G$ is upper embeddable, then $\xi(G) = 0$ or $\xi(G) = 1$. This implies that $G$ has a splitting tree.

**10.28.** Apply Corollary 10.23.

**10.30.** The statement is true.

**10.32.** $\gamma(C_s \Box C_t) = 1$ and $\gamma_M(C_s \Box C_t) = \lfloor \frac{st+1}{2} \rfloor$.

**10.34.** If $p \leq q \leq 3p$, let $G$ be constructed from $p$ copies $F_1, F_2, \ldots, K_p$ of $K_5$ by joining a vertex of $F_i$ to a vertex of $F_{i+1}$ for $1 \leq i \leq p-1$. If $q > 3p$, similarly construct $G$ from $p$ copies of $K_5$ and $q - 3p$ copies of $K_{2,3}$.

**10.36.**
  (a) Apply the definition of a minor-closed family of graphs.
  (b) $\{K_3\}$.

**10.38.** Yes; $\{K_4, K_{2,3}\}$.

# Hints for Chapter 11

**11.1.** For $i \neq j$, the $(i, j)$-entry of $A^2$ is the number of different $v_i$-$v_j$ paths of length 2 and the $(i, j)$-entry of $A^3$ is the number of different $v_i$-$v_j$ walks of length 3. The vertices $v_1$ and $v_4$ belong to one triangle while the vertices $v_2$ and $v_3$ belong to two triangles.

**11.3.** $V(G) = \{v_1, v_2, v_3, v_4, v_5\}$ and $E(G) = \{v_1v_2, v_1v_3, v_2v_3, v_3v_4, v_4v_5\}$.

**11.12.** There are six paths and three stars.

**11.13.** $(n-1)^{n-2}$.

**11.16.** $\text{Aut}(G_1) = \{\varepsilon, (t\,u), (y\,z), (t\,u)(y\,z)\}$.
  $\text{Aut}(G_2) = \{\varepsilon, (1\,2), (6\,7), (1\,2)(6\,7), (3\,5)(1\,6)(2\,7),$
$(3\,5)(2\,6)(1\,7), (3\,5)(1\,7\,2\,6), (3\,5)(1\,6\,2\,7)\}$.

**11.18.** Yes; $K_4 - e$.

**11.20.** See the graph below.

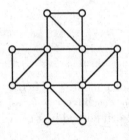

**11.22.** Such a pair $(k, n)$ is realizable except when $k = n$ and $2 \le n \le 5$.

**11.24.** Let $n = 2k$ for some positive integer $k$ and let $G$ be the bipartite graph with partite sets $U = \{u_1, u_2, \ldots, u_k\}$ and $W = \{w_1, w_2, \ldots, w_k\}$ where $u_i$ $(1 \le i \le k)$ is adjacent to $w_j$ if $i + j > k$.

**11.25.** The graph $\overline{H}$ is a complete $k$-partite graph where $k$ is the number of orbits of $G$. Since the orbits of $G$ produce a partition of $V(G)$ into $k$ subsets, the graph $H$ is a union of $k$ complete graphs and so $\overline{H}$ is a complete $k$-partite graph.

**11.26.** The graph $G_3$ has two cut-vertices. In $G_4$, consider the subgraphs induced by the neighborhoods of its vertices.

**11.28.**

(a) Let $\Delta_1 = \{a\}$. The Cayley color graph $D_{\Delta_1}(\Gamma)$ is shown below.

(b) Let $\Delta_2 = \{a, a^2, a^3\}$.

**11.30.** The group of color-preserving automorphisms consists of

$$\varepsilon,$$
$$(e\, a^2)(a\, a^3)(b\, a^2 b)(ab\, a^3 b),$$
$$(e\, b)(a\, ab)(a^2\, a^2 b)(a^3\, a^3 b),$$
$$(e\, a^2 b)(a\, a^3 b)(a^2\, b)(a^3\, ab), \quad \text{and}$$

$$(e\, a\, a^2\, a^3)(b\, ab\, a^2 b\, a^3 b),$$
$$(e\, a^3\, a^2\, a)(b\, a^3 b\, a^2 b\, ab),$$
$$(e\, ab\, a^2\, a^3 b)(a\, a^2 b\, a^3\, b),$$
$$(e\, a^3 b\, a^2\, ab)(a\, b\, a^3\, a^2 b).$$

**11.32.** Let $\Gamma \cong \mathbb{Z}_n = \{e, a, a^2, \cdots, a^{n-1}\}$ and $\Delta = \{a\}$.

**11.37.** See the graph below.

**11.39.** $G = K_{3,4}$.

**11.41.**

(a) Observe that $\overline{G} - v = \overline{G - v}$.

(b) Apply Theorem 11.22 and (a).

**11.25.** Apply the proof of Theorem 11.22 to obtain $G = 2C_4 + P_3$.

# Bibliography

*The translations of titles provided are original, except in a few cases such as [79] where an official English translation was provided by the journal at the time of publication, in which case that translation is given.*

## A

[1] H. L. Abbott and B. Zhou. On small faces in 4-critical planar graphs. *Ars Combin.*, 32:203–207, 1991. Cited on p. 163.

[2] O. Aichholzer. On the rectilinear crossing number. `http://www.ist.tugraz.at/staff/aichholzer/research/rp/triangulations/crossing/`. Cited on p. 131.

[3] M. Aigner and G. M. Ziegler. *Proofs from THE BOOK.* Springer, Berlin, Germany, 2018. Cited on p. 222.

[4] V. A. Aksenov. The extension of a 3-coloring on planar graphs. *Diskret. Analiz*, 26:3–19, 1974. Cited on p. 161.

[5] B. Alspach. The wonderful Walecki construction. *Bull. Inst. Combin. Appl.*, 52:7–20, 2008. Cited on p. 207.

[6] I. Anderson. Perfect matchings of a graph. *J. Combin. Theory Ser. B*, 10:183–186, 1971. Cited on p. 194.

[7] V. Angeltveit and B. D. McKay. $R(5,5) \leq 48$. *J. Graph Theory*, 89(1):5–13, 2018. Cited on p. 236.

[8] M. Aouchiche and P. Hansen. A survey of Nordhaus–Gaddum type relations. *Discrete Appl. Math.*, 161(4-5):466–546, 2013. Cited on p. 152.

[9] K. I. Appel and W. Haken. Every planar map is four colorable. *Bull. Amer. Math. Soc.*, 82(5):711–712, 1976. Cited on p. 159.

[10] K. I. Appel and W. Haken. Every planar map is four colorable. Part I: Discharging. *Illinois J. Math.*, 21(3):429–490, 1977. Cited on p. 159.

[11] K. I. Appel, W. Haken, and J. A. Koch. Every planar map is four colorable. Part II: Reducibility. *Illinois J. Math.*, 21(3):491–567, 1977. Cited on p. 159.

[12] D. Archdeacon and P. Huneke. A Kuratowski theorem for nonorientable surfaces. *J. Combin. Theory Ser. B*, 46(2):173–231, 1989. Cited on p. 277.

[13] V. I. Arnautov. Abschätzung der äußeren Stabilitätszahl eines Graphen mit Hilfe des Minimalgrades der Ecken (Estimation of the exterior stability number of a graph by means of the minimal degree of the vertices). *Prikl. Mat. i Programmirovanie*, 11:3–8, 1974. Cited on p. 202.

# B

[14] L. Babai. Graph isomorphism in quasipolynomial time [extended abstract]. In *STOC'16: Proceedings of the 48th Annual ACM SIGACT Symposium on Theory of Computing*, pages 684–697. Association for Computing Machinery, New York, New York, 2016. Cited on p. 5.

[15] L. Babai. Group, graphs, algorithms: the graph isomorphism problem. In *Proceedings of the International Congress of Mathematicians—Rio de Janeiro 2018. Vol. IV. Invited lectures*, pages 3319–3336. World Sci. Publ., Hackensack, New Jersey, 2018. Cited on p. 5.

[16] A. T. Balaban. Trivalent graphs of girth nine and eleven, and relationships among cages. *Rev. Roumaine Math. Pures Appl.*, 18:1033–1043, 1973. Cited on p. 242.

[17] J. Balogh, B. Lidický, and G. Salazar. Closing in on Hill's conjecture. *SIAM J. Discrete Math.*, 33(3):1261–1276, 2019. Cited on p. 125.

[18] E. Bannai and T. Ito. On finite Moore graphs. *J. Fac. Sci. Univ. Tokyo Sect. IA Math.*, 20:191–208, 1973. Cited on p. 244.

[19] J. Battle, F. Harary, and Y. Kodama. Every planar graph with nine points has a nonplanar complement. *Bull. Amer. Math. Soc.*, 68:569–571, 1962. Cited on p. 256.

[20] L. W. Beineke and F. Harary. The genus of the $n$-cube. *Canadian J. Math.*, 17:494–496, 1965. Cited on p. 259.

[21] L. W. Beineke and R. J. Wilson. On the edge-chromatic number of a graph. *Discrete Math.*, 5:15–20, 1973. Cited on p. 148.

[22] C. T. Benson. Minimal regular graphs of girths eight and twelve. *Canadian J. Math.*, 18:1091–1094, 1966. Cited on p. 242.

[23] C. J. Berge. Two theorems in graph theory. *Proc. Nat. Acad. Sci. U.S.A.*, 43:842–844, 1957. Cited on p. 190.

[24] C. J. Berge. Perfect graphs. In *Six Papers on Graph Theory*, pages 1–21. Indian Statistical Institute, Calcutta, India, 1963. Cited on p. 154.

[25] D. Bienstock. Some provably hard crossing number problems. *Discrete Comput. Geom.*, 6(5):443–459, 1991. Cited on p. 131.

[26] D. Blanuša. Problem cetiriju boja (The problem of four colors). *Hrvatsko Prirod. Društvo. Glasnik Mat.-Fiz. Astr. Ser. II*, 1:31–42, 1946. Cited on p. 165.

[27] J. Blažek and M. Koman. A minimal problem concerning complete plane graphs. In *Theory of Graphs and its Applications (Proc. Sympos. Smolenice, 1963)*, pages 113–117. Publ. House Czechoslovak Acad. Sci., Prague, Czechoslovakia, 1964. Cited on p. 124.

[28] J. A. Bondy. Bounds for the chromatic number of a graph. *J. Combin. Theory*, 7:96–98, 1969. Cited on p. 153.

[29] J. A. Bondy. Pancyclic graphs. I. *J. Combin. Theory Ser. B*, 11:80–84, 1971. Cited on p. 71.

[30] J. A. Bondy. Small cycle double covers of graphs. In G. Hahn, G. Sabidussi, and R. E. Woodrow, editors, *Cycles and Rays*, volume 301 of *NATO Adv. Sci. Inst. Ser. C Math. Phys. Sci.*, pages 21–40. Kluwer Acad. Publ., Dordrecht, The Netherlands, 1990. Cited on p. 57.

[31] J. A. Bondy and V. Chvátal. A method in graph theory. *Discrete Math.*, 15(2):111–135, 1976. Cited on p. 62.

[32] J. A. Bondy and C. Thomassen. A short proof of Meyniel's theorem. *Discrete Math.*, 19(2):195–197, 1977. Cited on p. 65.

[33] O. V. Borodin, A. N. Glebov, A. Raspaud, and M. R. Salavatipour. Planar graphs without cycles of length from 4 to 7 are 3-colorable. *J. Combin. Theory Ser. B*, 93(2):303–311, 2005. Cited on p. 162.

[34] O. V. Borodin, A. V. Kostochka, B. Lidický, and M. P. Yancey. Short proofs of coloring theorems on planar graphs. *European J. Combin.*, 36:314–321, 2014. Cited on p. 161.

[35] R. C. Brigham and R. D. Dutton. A compilation of relations between graph invariants. *Networks*, 15(1):73–107, 1985. Cited on p. 153.

[36] A. Brodsky, S. Durocher, and E. Gethner. Toward the rectilinear crossing number of $K_n$: new drawings, upper bounds, and asymptotics. *Discrete Math.*, 262(1-3):59–77, 2003. Cited on p. 130.

[37] R. L. Brooks. On colouring the nodes of a network. *Math. Proc. Cambridge Philos. Soc.*, 37(2):194–197, 1941. Cited on p. 143.

[38] L. E. Bush. The William Lowell Putnam mathematical competition. *Amer. Math. Monthly*, 60(8):539–542, 1953. Cited on p. 234.

# C

[39] P. F. Camion. Chemins et circuits hamiltoniens des graphes complets (Hamiltonian paths and circuits of complete graphs). *C. R. Acad. Sci. Paris*, 249(2):2151–2152, 1959. Cited on p. 67.

[40] A. Cayley. On the colouring of maps. *Proc. Roy. Geographic Soc. and Monthly Rec. of Geography*, 1(4):259–261, 1879. Cited on p. 157.

[41] A. Cayley. A theorem on trees. *Quart. J. Pure and Appl. Math.*, 23:376–378, 1889. Cited on p. 292.

[42] M. Cetina, C. Hernández-Vélez, J. Leaños, and C. Villalobos. Point sets that minimize ($\leq k$)-edges, 3-decomposable drawings, and the rectilinear crossing number of $K_{30}$. *Discrete Math.*, 311(16):1646–1657, 2011. Cited on p. 131.

[43] G. T. Chartrand and F. Harary. Graphs with prescribed connectivities. In P. Erdős and G. Katona, editors, *Theory of Graphs, Proc. Colloq. Tihany 1966*, pages 61–63. Akadémiai Kiadó, Budapest, Hungary, 1968. Cited on pp. 85 and 89.

[44] G. T. Chartrand, A. M. Hobbs, H. A. Jung, S. F. Kapoor, and C. S. J. A. Nash-Williams. The square of a block is Hamiltonian connected. *J. Combin. Theory Ser. B*, 16:290–292, 1974. Cited on p. 73.

[45] G. T. Chartrand and J. A. Mitchem. Graphical theorems of the Nordhaus–Gaddum class. In M. F. Capobianco, J. B. Frechen, and M. Krolik, editors, *Recent Trends in Graph Theory (Jamaica, NY, USA 1970)*, volume 186 of *Lecture Notes in Math.*, pages 55–61. Springer, Berlin, Germany, 1971. Cited on p. 153.

[46] C. C. Chen and N. F. Quimpo. On strongly hamiltonian abelian group graphs. In K. L. McAvaney, editor, *Combinatorial Mathematics, VIII (Geelong, 1980)*, volume 884 of *Lecture Notes in Math.*, pages 23–34. Springer-Verlag, Berlin, Germany, 1981. Cited on p. 300.

[47] A. G. Chetwynd and A. J. W. Hilton. Star multigraphs with three vertices of maximum degree. *Math. Proc. Cambridge Philos. Soc.*, 100(2):303–317, 1986. Cited on pp. 149 and 206.

[48] M. Chudnovsky, N. Robertson, P. D. Seymour, and R. Thomas. The strong perfect graph theorem. *Ann. of Math. (2)*, 164(1):51–229, 2006. Cited on p. 155.

[49] V. Chvátal. On Hamilton's ideals. *J. Combin. Theory Ser. B*, 12:163–168, 1972. Cited on p. 64.

[50] V. Chvátal. Tough graphs and Hamiltonian circuits. *Discrete Math.*, 5:215–228, 1973. Cited on p. 59.

[51] V. Chvátal. A combinatorial theorem in plane geometry. *J. Combin. Theory Ser. B*, 18:39–41, 1975. Cited on p. 130.

[52] R. F. A. Clebsch. Ueber die Flächen vierter Ordnung, welche eine Doppelcurve zweiten Grades besitzen (On fourth-order surfaces, which have a double curve of the second degree). *J. Reine Angew. Math.*, 69:142–184, 1868. Cited on p. 237.

[53] M. Codish, M. Frank, A. Itzhakov, and A. Miller. Computing the Ramsey number $R(4,3,3)$ using abstraction and symmetry breaking. *Constraints*, 21(3):375–393, 2016. Cited on p. 237.

[54] V. Cohen-Addad, M. Hebdige, D. Král', Z. Li, and E. Salgado. Steinberg's conjecture is false. *J. Combin. Theory Ser. B*, 122:452–456, 2017. Cited on p. 162.

[55] S. A. Cook. The complexity of theorem-proving procedures. In *STOC '71: Proceedings of the Third Annual ACM Symposium on the Theory of Computing*, pages 151–158. Association for Computing Machinery, New York, New York, 1971. Cited on p. 5.

[56] C. A. Coulson and G. S. Rushbrooke. Note on the method of molecular orbitals. *Math. Proc. Cambridge Philos. Soc.*, 36(2):193–200, 1940. Cited on p. 287.

[57] D. W. Cranston and L. Rabern. Brooks' theorem and beyond. *J. Graph Theory*, 80(3):199–225, 2015. Cited on p. 143.

[58] B. Csaba, D. Kühn, A. Lo, D. Osthus, and A. Treglown. Proof of the 1-factorization and Hamilton decomposition conjectures. *Mem. Amer. Math. Soc.*, 244(1154), 2016. Cited on pp. 72, 206, and 209.

# D

[59] C. Dalfó. A survey on the missing Moore graph. *Linear Algebra Appl.*, 569:1–14, 2019. Cited on p. 244.

[60] R. M. Damerell. On Moore graphs. *Proc. Cambridge Philos. Soc.*, 74:227–236, 1973. Cited on p. 244.

[61] C. F. A. de Jaenisch. *Traité des Applications de L'analyse Mathématique au Jeu des Échecs (Treatise on the Applications of Mathematical Analysis to the Game of Chess)*, volume 1. Imperial St. Petersburg Academy of Sciences, St. Petersburg, Russia, 1862. Cited on p. 199.

[62] E. de Klerk, D. V. Pasechnik, and A. Schrijver. Reduction of symmetric semidefinite programs using the regular ∗-representation. *Math. Program.*, 109(2-3, Ser. B):613–624, 2007. Cited on p. 127.

[63] A. De Morgan. Review of *The Philosophy of Discovery, Chapters Historical and Critical*. *Athenaeum*, (1694):501–503, 1860. Cited on p. 156.

[64] B. Descartes. Network-colourings. *Math. Gazette*, 32:67–69, 1948. Cited on p. 165.

[65] G. A. Dirac. Some theorems on abstract graphs. *Proc. London Math. Soc. (3)*, 2:69–81, 1952. Cited on p. 61.

[66] G. A. Dirac. The structure of $k$-chromatic graphs. *Fund. Math.*, 40:42–55, 1953. Cited on p. 150.

[67] G. A. Dirac. In abstrakten Graphen vorhandene vollständige 4-Graphen und ihre Unterteilungen (Complete 4-graphs and their subdivisions present in abstract graphs). *Math. Nachr.*, 22:61–85, 1960. Cited on p. 97.

[68] R. A. Duke. The genus, regional number, and Betti number of a graph. *Canadian J. Math.*, 18:817–822, 1966. Cited on p. 271.

# E

[69] J. R. Edmonds. A combinatorial representation for polyhedral surfaces. *Notices Amer. Math. Soc.*, 7(5):646, Oct. 1960. Abstract 572-1. Cited on p. 266.

[70] J. R. Edmonds and R. M. Karp. Theoretical improvements in algorithmic efficiency for network flow problems. *J. Assoc. Comput. Mach.*, 19:248–264, 1972. Cited on p. 178.

[71] K. Edwards, D. P. Sanders, P. D. Seymour, and R. Thomas. Three-edge-colouring doublecross cubic graphs. *J. Combin. Theory Ser. B*, 119:66–95, 2016. Cited on p. 166.

[72] J. Egerváry. Matrixok kombinatorius tulajdonságairól (On combinatorial properties of matrices). *Math. Fiz. Lapok*, 38:16–28, 1931. Cited on p. 199.

[73] A. Ehrenfeucht, V. Faber, and H. A. Kierstead. A new method of proving theorems on chromatic index. *Discrete Math.*, 52(2-3):159–164, 1984. Cited on p. 146.

[74] P. Elias, A. Feinstein, and C. E. Shannon. A note on the maximum flow through a network. *Inst. Radio Engrs. Trans. Inform. Theory*, 2(4):117–119, 1956. Cited on p. 176.

[75] P. Erdős. Some remarks on the theory of graphs. *Bull. Amer. Math. Soc.*, 53(4):292–294, 1947. Cited on p. 233.

[76] P. Erdős. Extremal problems in graph theory. In *Theory of Graphs and its Applications (Proc. Sympos. Smolenice, 1963)*, pages 29–36. Publ. House Czechoslovak Acad. Sci., Prague, Czechoslovakia, 1964. Cited on p. 228.

[77] P. Erdős. Extremal problems in graph theory. In F. Harary and L. W. Beineke, editors, *A Seminar on Graph Theory*, pages 54–59. Holt, Rinehart and Winston, New York, New York, 1967. Cited on pp. 225 and 227.

[78] P. Erdős. Problems. In P. Erdős and G. Katona, editors, *Theory of Graphs, Proc. Colloq. Tihany 1966*, pages 361–362. Akadémiai Kiadó, Budapest, Hungary, 1968. Cited on p. 151.

[79] P. Erdős. Turán Pál gráf tételéről (On the graph-theorem of Turán). *Math. Lapok*, 21(3-4):249–251, 1970. Cited on pp. 222 and 329.

[80] P. Erdős. Problems and results in combinatorial analysis and graph theory. *Discrete Math.*, 72(1-3):81–92, 1988. Cited on p. 234.

[81] P. Erdős and T. Gallai. Graphen mit Punkten vorgeschriebenen Grades (Graphs with prescribed degrees of vertices). *Math. Lapok*, 11:264–274, 1960. Cited on p. 15.

[82] P. Erdős and L. Pósa. On the maximal number of disjoint circuits of a graph. *Publ. Math. Debrecen*, 9:3–12, 1962. Cited on pp. 224, 225, and 227.

[83] P. Erdős and H. Sachs. Reguläre Graphen gegebener Taillenweite mit minimaler Knotenzahl (Regular graphs with given girth and minimal number of knots). *Wiss. Z. Martin-Luther-Univ. Halle-Wittenberg Math.-Natur. Reihe*, 12:251–257, 1963. Cited on pp. 239 and 240.

[84] P. Erdős and G. Szekeres. A combinatorial problem in geometry. *Compositio Math.*, 2:463–470, 1935. Cited on p. 230.

[85] P. Erdős and R. J. Wilson. On the chromatic index of almost all graphs. *J. Combin. Theory Ser. B*, 23(2-3):255–257, 1977. Cited on p. 148.

[86] L. Euler. Solutio problematis ad geometriam situs pertinentis (The solution of a problem relating to the geometry of position). *Comment. Acad. Sci. Petrop.*, 8:128–140, 1741. Cited on p. 54.

[87] L. Euler. Demonstratio nonnullarum insignium proprieatatum, quibus solida hedris planis inclusa sunt praedita (Proof of some of the properties of solid bodies enclosed by planes). *Novi Comment. Acad. Sci. Petrop.*, 4:140–160, 1758. Cited on p. 103.

[88] L. Euler. Elementa doctrinae solidorum (Elements of the doctrine of solids). *Novi Comment. Acad. Sci. Petrop.*, 4:109–140, 1758. Cited on p. 103.

[89] G. Exoo. A lower bound for $R(5,5)$. *J. Graph Theory*, 13(1):97–98, 1989. Cited on p. 236.

[90] G. Exoo and R. Jajcay. Dynamic cage survey. *Electron. J. Combin.*, Dynamic Surveys:DS16, 55pp., 2013. Cited on p. 240.

# F

[91] I. Fáry. On straight line representation of planar graphs. *Acta Univ. Szeged. Sect. Sci. Math.*, 11:229–233, 1948. Cited on p. 128.

[92] S. T. Fisk. A short proof of Chvátal's watchman theorem. *J. Combin. Theory Ser. B*, 24(3):374, 1978. Cited on p. 130.

[93] H. Fleischner. The square of every two-connected graph is Hamiltonian. *J. Combin. Theory Ser. B*, 16:29–34, 1974. Cited on p. 73.

[94] L. R. Ford, Jr. and D. R. Fulkerson. Maximal flow through a network. *Canadian J. Math.*, 8:399–404, 1956. Cited on pp. 175 and 176.

[95] L. R. Ford, Jr. and D. R. Fulkerson. A simple algorithm for finding maximal network flows and an application to the Hitchcock problem. *Canadian J. Math.*, 9:210–218, 1957. Cited on p. 177.

[96] J.-C. Fournier. Colorations des arêtes d'un graphe (Colorings of the edges of a graph). *Cahiers Centre Études Rech. Opér.*, 15:311–314, 1973. Cited on p. 148.

[97] P. Franklin. The four color problem. *Amer. J. Math.*, 44(3):225–236, 1922. Cited on p. 110.

[98] R. W. Frucht. Herstellung von Graphen mit vorgegebener abstrakter Gruppe (Creation of graphs with given abstract group). *Compositio Math.*, 6:239–250, 1939. Cited on p. 299.

# G

[99] T. Gallai. Über extreme Punkt- und Kantenmengen (On extreme sets of points and edges). *Ann. Univ. Sci. Budapest. Eötvös Sect. Math.*, 2:133–138, 1959. Cited on p. 198.

[100] M. Gardner. Mathematical games. *Sci. Amer.*, 234(4):126–131, April 1976. Cited on p. 165.

[101] M. R. Garey and D. S. Johnson. Crossing number is NP-complete. *SIAM J. Algebraic Discrete Methods*, 4(3):312–316, 1983. Cited on p. 131.

[102] A. Georgakopoulos. A short proof of Fleischner's theorem. *Discrete Math.*, 309(23-24):6632–6634, 2009. Cited on p. 73.

[103] A. Ghouila-Houri. Une condition suffisante d'existence d'un circuit hamiltonien (A sufficient condition for the existence of a hamiltonian circuit). *C. R. Acad. Sci. Paris*, 251(1):495–497, 1960. Cited on p. 65.

[104] S. W. Golomb. How to number a graph. In *Graph Theory and Computing*, pages 23–37. Academic Press, New York, New York, 1972. Cited on p. 211.

[105] F. Göring. Short proof of Menger's Theorem. *Discrete Math.*, 219(1-3):295–296, 2000. Cited on p. 93.

[106] J. E. Graver and J. W. Yackel. Some graph theoretic results associated with Ramsey's theorem. *J. Combin. Theory*, 4(2):125–175, 1968. Cited on p. 235.

[107] R. E. Greenwood, Jr. and A. M. Gleason. Combinatorial relations and chromatic graphs. *Canadian J. Math.*, 7:1–7, 1955. Cited on pp. 230, 234, and 236.

[108] E. Grinberg. Plane homogeneous graphs of degree three without Hamiltonian circuits. In *Latvian Math. Yearbook*, volume 4, pages 51–58. Izdat. "Zinatne", Riga, Latvia, 1968. Cited on p. 120.

[109] C. M. Grinstead and S. M. Roberts. On the Ramsey numbers $R(3, 8)$ and $R(3, 9)$. *J. Combin. Theory Ser. B*, 33(1):27–51, 1982. Cited on p. 235.

[110] H. Grötzsch. Zur Theorie der diskreten Gebilde. VII. Ein Dreifarbensatz für dreikreisfreie Netze auf der Kugel (On the theory of discrete entities. VII. A three-color theorem for three-circle-free nets on the sphere). *Wiss. Z. Martin-Luther-Univ. Halle-Wittenberg Math.-Natur. Reihe*, 8:109–120, 1958/59. Cited on p. 160.

[111] B. Grünbaum. Grötzsch's theorem on 3-colorings. *Michigan Math. J.*, 10:303–310, 1963. Cited on p. 161.

[112] R. P. Gupta. The chromatic index and the degree of a graph. *Notices Amer. Math. Soc.*, 13(6):719, Oct. 1966. Abstract 66T-429. Cited on p. 145.

[113] F. Guthrie. Note on the colouring of maps. *Proc. Roy. Soc. Edinburgh*, 10:727–728, 1880. Cited on p. 155.

[114] R. K. Guy. A combinatorial problem. *NABLA (Bull. Malayan Math. Soc.)*, 7:68–72, 1960. Cited on p. 124.

[115] R. K. Guy. Crossing numbers of graphs. In Y. Alavi, D. R. Lick, and A. T. White, editors, *Graph Theory and Applications (Kalamazoo, MI, USA 1972)*, volume 303 of *Lecture Notes in Math.*, pages 111–124. Springer, Berlin, Germany, 1972. Cited on pp. 123, 125, and 130.

# H

[116] H. Hadwiger. Über eine Klassifikation der Streckenkomplexe (On a classification of line complexes). *Vierteljschr. Naturforsch. Ges. Zürich*, 88(2):133–142, 1943. Cited on p. 159.

[117] S. L. Hakimi. On realizability of a set of integers as degrees of the vertices of a linear graph. I. *SIAM J. Appl. Math.*, 10:496–506, 1962. Cited on p. 13.

[118] P. Hall. On representatives of subsets. *J. London Math. Soc. (2)*, 10(1):26–30, 1935. Cited on p. 192.

[119] Z. Hamaker and V. Vatter. Three coloring via triangle counting. *Australas. J. Combin.*, 87(2):352–356, 2023. Cited on p. 163.

[120] F. Harary. The maximum connectivity of a graph. *Proc. Nat. Acad. Sci. U.S.A.*, 48:1142–1146, 1962. Cited on p. 84.

[121] F. Harary and L. Moser. The theory of round robin tournaments. *Amer. Math. Monthly*, 73:231–246, 1966. Cited on pp. 48 and 68.

[122] V. Havel. Eine Bemerkung über die Existenz der endlichen Graphen (A remark on the existence of finite graphs). *Časopis Pěst. Mat.*, 80:477–480, 1955. Cited on p. 13.

[123] P. J. Heawood. Map-colour theorem. *Quart. J. Pure Appl. Math.*, 24:332–338, 1890. Cited on pp. 157, 160, and 257.

[124] P. J. Heawood. On the four-colour map theorem. *Quart. J. Pure Appl. Math.*, 29:270–285, 1898. Cited on p. 160.

[125] S. T. Hedetniemi. *Homomorphisms of Graphs and Automata*. PhD thesis, University of Michigan, 1966. Cited on p. 152.

[126] L. Heffter. Ueber das Problem der Nachbargebiete (On the problem of neighboring areas). *Math. Ann.*, 38(4):477–508, 1891. Cited on pp. 258 and 266.

[127] C. Hierholzer. Ueber die Möglichkeit, Linienzug ohne Wiederholung und ohne Unterbrechung zu umfahren (On the possibility, to travel around a graph without repetition and without cutting). *Math. Ann.*, 6(1):30–32, 1873. Cited on p. 54.

[128] A. J. Hoffman. *Selected Papers of Alan Hoffman with Commentary*. World Sci. Publ., River Edge, New Jersey, 2003. Edited by C. A. Micchelli. Cited on pp. 243 and 245.

[129] A. J. Hoffman and R. R. Singleton. On Moore graphs with diameters 2 and 3. *IBM J. Res. Develop.*, 4:497–504, 1960. Cited on pp. 243 and 244.

[130] I. J. Holyer. The NP-completeness of edge-coloring. *SIAM J. Comput.*, 10(4):718–720, 1981. Cited on p. 147.

[131] J. Hopcroft and R. E. Tarjan. Efficient planarity testing. *J. Assoc. Comput. Mach.*, 21:549–568, 1974. Cited on p. 115.

# I

[132] R. P. Isaacs. Infinite families of nontrivial trivalent graphs which are not Tait colorable. *Amer. Math. Monthly*, 82:221–239, 1975. Cited on p. 165.

[133] H. Izbicki. Reguläre Graphen beliebigen Grades mit vorgegebenen Eigenschaften (Regular graphs of any degree with given properties). *Monatsh. Math.*, 64:15–21, 1960. Cited on p. 299.

# J

[134] F. Jaeger. Flows and generalized coloring theorems in graphs. *J. Combin. Theory Ser. B*, 26(2):205–216, 1979. Cited on p. 184.

[135] M. Jungerman. A characterization of upper-embeddable graphs. *Trans. Amer. Math. Soc.*, 241:401–406, 1978. Cited on p. 275.

# K

[136] J. G. Kalbfleisch. Construction of special edge-chromatic graphs. *Canad. Math. Bull.*, 8:575–584, 1965. Cited on p. 236.

[137] J. G. Kalbfleisch. *Chromatic Graphs and Ramsey's Theorem*. PhD thesis, University of Waterloo, 1966. Cited on pp. 235, 236, and 237.

[138] J. J. Karaganis. On the cube of a graph. *Canad. Math. Bull.*, 11:295–296, 1968. Cited on p. 72.

[139] R. M. Karp. Reducibility among combinatorial problems. In R. E. Miller and J. W. Thatcher, editors, *Complexity of Computer Computations*, IBM Research Symposium Series, pages 85–103. Plenum Press, New York, New York, 1972. Cited on p. 5.

[140] P. J. Kelly. A congruence theorem for trees. *Pacific J. Math.*, 7:961–968, 1957. Cited on p. 301.

[141] A. B. Kempe. On the geographical problem of the four colours. *Amer. J. Math.*, 2(3):193–200, 1879. Cited on p. 157.

[142] A. B. Kempe. How to colour a map with four colours. *Nature*, 21(539):399–400, 1880. Cited on p. 157.

[143] G. Kéry. Ramsey egy tételéről (On a theorem of Ramsey). *Mat. Lapok*, 15(1-3):204–224, 1964. Cited on pp. 232 and 235.

[144] P. A. Kilpatrick. *Tutte's First Colour-Cycle Conjecture*. PhD thesis, University of Cape Town, 1975. Cited on p. 184.

[145] G. R. Kirchhoff. Ueber die auflösung der Gleichungen, auf welche man bei der Untersuchung der linearen Vertheilung galvanischer Ströme geführt wird (On the solution of the equations to which one is led in the investigation of the linear distribution of electric current). *Ann. Phys. Chem.*, 148(12):497–508, 1847. Cited on p. 289.

[146] T. P. Kirkman. On a problem in combinations. *Cambridge and Dublin Math. J.*, 2(10&11):191–204, 1847. Cited on p. 211.

[147] D. J. Kleitman. The crossing number of $K_{5,n}$. *J. Combin. Theory*, 9:315–323, 1970. Cited on p. 126.

[148] W. L. Kocay. On Stockmeyer's nonreconstructible tournaments. *J. Graph Theory*, 9(4):473–476, 1985. Cited on p. 302.

[149] D. Kőnig. Über Graphen und ihre Anwendung auf Determinantentheorie und Mengenlehre (On graphs and their application to determinant theory and set theory). *Math. Ann.*, 77(4):453–465, 1916. Cited on pp. 147 and 203.

[150] D. Kőnig. Graphok és matrixok (Graphs and matrices). *Math. Fiz. Lapok*, 38:116–119, 1931. Cited on pp. 93 and 199.

[151] D. Kőnig. *Theorie der endlichen und unendlichen Graphen* (*Theory of Finite and Infinite Graphs*). Akademische Verlagsgesellschaft, Leipzig, Germany, 1936. Cited on p. 299.

[152] A. V. Kostochka and M. P. Yancey. Ore's conjecture for $k = 4$ and Grötzsch's theorem. *Combinatorica*, 34(3):323–329, 2014. Cited on pp. 150 and 160.

[153] A. V. Kostochka and M. P. Yancey. Ore's conjecture on color-critical graphs is almost true. *J. Combin. Theory Ser. B*, 109:73–101, 2014. Cited on p. 150.

[154] H. V. E. Kronk, R. D. Ringeisen, and A. T. White. On 2-cell imbeddings of complete $n$-partite graphs. *Colloq. Math.*, 36(2):295–304, 1976. Cited on p. 276.

[155] K. Kuratowski. Sur le problème des courbes gauches en Topologie (On the problem of gauche curves in topology). *Fund. Math.*, 15:271–283, 1930. Cited on p. 112.

# L

[156] H. G. Landau. On dominance relations and the structure of animal societies. III. The condition for a score structure. *Bull. Math. Biophys.*, 15:143–148, 1953. Cited on pp. 40 and 46.

[157] S. A. J. L'Huilier. Mémoire sur la polyédrométrie; contenant une démonstration directe du théorème d'Euler sur les polyèdres, et un examen des diverses exceptions auxquelles ce théorème est assujetti (Thesis on polyhedrometry; containing a direct proof of Euler's theorem on polyhedra, and an examination of the various exceptions to this theorem). *Ann. Math. Pures et Appl. (Ann. Gergonne)*, 3:169–189, 1812/13. Cited on p. 253.

[158] L. Lovász. A characterization of perfect graphs. *J. Combin. Theory Ser. B*, 13:95–98, 1972. Cited on p. 154.

[159] L. Lovász. Normal hypergraphs and the perfect graph conjecture. *Discrete Math.*, 2(3):253–267, 1972. Cited on p. 154.

[160] E. M. Luks. Isomorphism of graphs of bounded valence can be tested in polynomial time. *J. Comput. System Sci.*, 25(1):42–65, 1982. Cited on p. 6.

# M

[161] W. Mantel. Problem 28 (Solution by H. Gouwentak, W. Mantel, J. Teixeira de Mattes, F. Schuh and W. A. Wythoff). *Wiskundige Opgaven*, 10:60–61, 1907. Cited on p. 221.

[162] B. Manvel. *On Reconstruction of Graphs*. PhD thesis, University of Michigan, 1970. Cited on p. 304.

[163] K. O. May. The origin of the four-color conjecture. *Isis*, 56(3):346–348, 1965. Cited on p. 156.

[164] W. F. McGee. A minimal cubic graph of girth seven. *Canad. Math. Bull.*, 3:149–152, 1960. Cited on p. 242.

[165] B. D. McKay. Reconstruction of small graphs and digraphs. *Australas. J. Combin.*, 83(3):448–457, 2022. Cited on p. 302.

[166] B. D. McKay and Z. K. Min. The value of the Ramsey number $R(3,8)$. *J. Graph Theory*, 16(1):99–105, 1992. Cited on p. 235.

[167] B. D. McKay, W. Myrvold, and J. Nadon. Fast backtracking principles applied to find new cages. In *Proceedings of the Ninth Annual ACM-SIAM Symposium on*

*Discrete Algorithms (SODA)*, pages 188–191. Association for Computing Machinery, New York, New York, 1998. Cited on p. 242.

[168] B. D. McKay and S. P. Radziszowski. Linear programming in some Ramsey problems. *J. Combin. Theory Ser. B*, 61(1):125–132, 1994. Cited on p. 236.

[169] B. D. McKay and S. P. Radziszowski. $R(4, 5) = 25$. *J. Graph Theory*, 19(3):309–322, 1995. Cited on p. 236.

[170] B. D. McKay and S. P. Radziszowski. Subgraph counting identities and Ramsey numbers. *J. Combin. Theory Ser. B*, 69(2):193–209, 1997. Cited on p. 236.

[171] T. A. McKee. Recharacterizing Eulerian: intimations of new duality. *Discrete Math.*, 51(3):237–242, 1984. Cited on p. 57.

[172] D. J. McQuillan, S. Pan, and R. B. Richter. On the crossing number of $K_{13}$. *J. Combin. Theory Ser. B*, 115:224–235, 2015. Cited on p. 125.

[173] K. Menger. Zur allgemeinen Kurventheorie (On the general theory of curves). *Fund. Math.*, 10:96–115, 1927. Cited on p. 93.

[174] K. Menger. *Kurventheorie (Curve Theory)*. B.G. Teubner, Leipzig, Germany, 1932. Cited on p. 93.

[175] K. Menger. On the origin of the $n$-arc theorem. *J. Graph Theory*, 5(4):341–350, 1981. Cited on p. 93.

[176] H. Meyniel. Une condition suffisante d'existence d'un circuit hamiltonien dans un graphe oriente (A sufficient condition for the existence of a hamiltonian circuit in an oriented graph). *J. Combin. Theory Ser. B*, 14:137–147, 1973. Cited on p. 65.

[177] R. H. Montgomery, A. Pokrovskiy, and B. Sudakov. A proof of Ringel's conjecture. *Geom. Funct. Anal.*, 31(3):663–720, 2021. Cited on p. 213.

[178] J. W. Moon. On subtournaments of a tournament. *Canad. Math. Bull.*, 9:297–301, 1966. Cited on p. 68.

[179] W. Myrvold and J. Woodcock. A large set of torus obstructions and how they were discovered. *Electron. J. Combin.*, 25(1):Paper No. 1.16, 17 pp., 2018. Cited on p. 278.

# N

[180] C. S. J. A. Nash-Williams. Edge-disjoint Hamiltonian circuits in graphs with vertices of large valency. In L. Mirsky, editor, *Studies in Pure Mathematics (Presented to Richard Rado)*, pages 157–183. Academic Press, New York, New York, 1971. Cited on p. 71.

[181] C. S. J. A. Nash-Williams. Hamiltonian arcs and circuits. In M. F. Capobianco, J. B. Frechen, and M. Krolik, editors, *Recent Trends in Graph Theory (Jamaica, NY, USA 1970)*, volume 186 of *Lecture Notes in Math.*, pages 197–210. Springer, Berlin, Germany, 1971. Cited on p. 209.

[182] E. A. Nordhaus, B. M. Stewart, and A. T. White. On the maximum genus of a graph. *J. Combin. Theory Ser. B*, 11:258–267, 1971. Cited on pp. 273 and 276.

[183] Notes. *Nature*, 20(507):275–278, 1879. Cited on p. 157.

# O

[184] Ø. Ore. Note on Hamilton circuits. *Amer. Math. Monthly*, 67:55, 1960. Cited on p. 61.

[185] Ø. Ore. *Theory of Graphs*, volume 38 of *Colloquium Publications*. American Mathematical Society, Providence, Rhode Island, 1962. Cited on p. 201.

[186] Ø. Ore. Hamilton connected graphs. *J. Math. Pures Appl. (9)*, 42:21–27, 1963. Cited on p. 69.

# P

[187] S. Pan and R. B. Richter. The crossing number of $K_{11}$ is 100. *J. Graph Theory*, 56(2):128–134, 2007. Cited on p. 125.

[188] C. Payan. Sur le nombre d'absorption d'un graphe simple (On the absorption number of a simple graph). *Cahiers du Centre d'Études de Recherche Opérationnelle*, 17(2-4):307–317, 1975. Cited on p. 202.

[189] C. Payan and N. H. Xuong. Domination-balanced graphs. *J. Graph Theory*, 6(1):23–32, 1982. Cited on p. 202.

[190] J. P. C. Petersen. Die Theorie der regulären Graphs (The theory of regular graphs). *Acta Math.*, 15(1):193–220, 1891. Cited on pp. 8, 196, and 206.

[191] J. P. C. Petersen. Sur le théorème de Tait (On the theorem of Tait). *Interméd. Math.*, 5:225–227, Oct. 1898. Cited on pp. 165 and 203.

[192] J. Plesník. Critical graphs of given diameter. *Acta Fac. Rerum Natur. Univ. Comenian. Math.*, 30:71–93, 1975. Cited on p. 90.

# R

[193] S. P. Radziszowski. Small Ramsey numbers. *Electron. J. Combin.*, 1:Dynamic Survey 1, 30 pp., 1994. Cited on p. 236.

[194] F. P. Ramsey. On a problem of formal logic. *Proc. London Math. Soc. (2)*, 30(4):264–286, 1929. Cited on pp. 229, 234, and 236.

[195] L. Rédei. Ein kombinatorischer Satz (A combinatorial theorem). *Acta Litt. Sci. Szeged*, 7:39–43, 1934. Cited on pp. 65 and 66.

[196] B. A. Reed. $\omega$, $\Delta$, and $\chi$. *J. Graph Theory*, 27(4):177–212, 1998. Cited on p. 153.

[197] R. D. Ringeisen. Determining all compact orientable 2-manifolds upon which $K_{m,n}$ has 2-cell imbeddings. *J. Combin. Theory Ser. B*, 12:101–104, 1972. Cited on p. 276.

[198] G. Ringel. Über drei kombinatorische Probleme am $n$-dimensionalen Würfel und Würfelgitter (About three combinatorial problems on the $n$-dimensional cube and cube lattice). *Abh. Math. Sem. Univ. Hamburg*, 20:10–19, 1955. Cited on p. 259.

[199] G. Ringel. Problem 25. In *Theory of Graphs and its Applications (Proc. Sympos. Smolenice, 1963)*, page 162. Publ. House Czechoslovak Acad. Sci., Prague, Czechoslovakia, 1964. Cited on p. 213.

[200] G. Ringel. Das Geschlecht des vollständigen paaren Graphen (The genus of the complete bipartite graph). *Abh. Math. Sem. Univ. Hamburg*, 28:139–150, 1965. Cited on p. 259.

[201] G. Ringel and J. W. T. Youngs. Solution of the Heawood map-coloring problem. *Proc. Nat. Acad. Sci. U.S.A.*, 60:438–445, 1968. Cited on pp. 257 and 258.

[202] H. E. Robbins. A theorem on graphs, with an application to a problem in traffic control. *Amer. Math. Monthly*, 46(5):281–283, 1939. Cited on pp. 39 and 91.

[203] N. Robertson, D. P. Sanders, P. D. Seymour, and R. Thomas. The four-colour theorem. *J. Combin. Theory Ser. B*, 70(1):2–44, 1997. Cited on pp. 159 and 166.

[204] N. Robertson and P. D. Seymour. Graph minors. XX. Wagner's conjecture. *J. Combin. Theory Ser. B*, 92(2):325–357, 2004. Cited on p. 277.

[205] N. Robertson, P. D. Seymour, and R. Thomas. Hadwiger's conjecture for $K_6$-free graphs. *Combinatorica*, 13(3):279–361, 1993. Cited on p. 160.

[206] N. Robertson, P. D. Seymour, and R. Thomas. Tutte's edge-colouring conjecture. *J. Combin. Theory Ser. B*, 70(1):166–183, 1997. Cited on p. 166.

[207] N. Robertson, P. D. Seymour, and R. Thomas. Cyclically five-connected cubic graphs. *J. Combin. Theory Ser. B*, 125:132–167, 2017. Cited on p. 166.

[208] N. Robertson, P. D. Seymour, and R. Thomas. Excluded minors in cubic graphs. *J. Combin. Theory Ser. B*, 138:219–285, 2019. Cited on p. 166.

[209] A. Rosa. On certain valuations of the vertices of a graph. In P. Rosenstiehl, editor, *Theory of Graphs (Internat. Sympos., Rome, 1966)*, pages 349–355. Gordon and Breach, New York, New York, 1967. Cited on pp. 211, 212, and 213.

[210] N. E. Rutt. Concerning the cut points of a continuous curve when the arc curve, $AB$, contains exactly $N$ independent arcs. *Amer. J. Math.*, 51(2):217–246, 1929. Cited on p. 93.

# S

[211] D. P. Sanders and Y. Zhao. Planar graphs of maximum degree seven are class I. *J. Combin. Theory Ser. B*, 83(2):201–212, 2001. Cited on p. 165.

[212] N. Sauer. Extremaleigenschaften regulärer Graphen gegebener Taillenweite, I (Extremal properties of regular graphs of given girth, I). *Österreich. Akad. Wiss. Math.-Natur. Kl. S.-B. II*, 176:9–25, 1967. Cited on p. 240.

[213] A. Schrijver. Paths and flows—a historical survey. *CWI Quarterly*, 6(3):169–183, 1993. Cited on p. 93.

[214] A. Schrijver. *Combinatorial Optimization. Polyhedra and Efficiency. Vol. A*, volume 24 of *Algorithms and Combinatorics*. Springer-Verlag, Berlin, Germany, 2003. Cited on p. 146.

[215] A. J. C. Schwenk. Problem 6434. *Amer. Math. Monthly*, 90(6):403, 1983. Cited on p. 288.

[216] A. J. C. Schwenk and O. P. Lossers. Solution to problem 6434. *Amer. Math. Monthly*, 94(9):885–886, 1987. Cited on p. 288.

[217] M. Sekanina. On an ordering of the set of vertices of a connected graph. *Spisy Přírod. Fak. Univ. Brno*, 1960:137–141, 1960. Cited on p. 72.

[218] P. D. Seymour. Sums of circuits. In J. A. Bondy and U. S. R. Murty, editors, *Graph Theory and Related Topics (Proc. Conf., Univ. Waterloo, Waterloo, Ont., 1977)*, pages 341–355. Academic Press, New York, 1979. Cited on p. 57.

[219] P. D. Seymour. Nowhere-zero 6-flows. *J. Combin. Theory Ser. B*, 30(2):130–135, 1981. Cited on p. 184.

[220] Y. Shitov. Counterexamples to Hedetniemi's conjecture. *Ann. of Math. (2)*, 190(2):663–667, 2019. Cited on p. 152.

[221] S. K. Stein. Convex maps. *Proc. Amer. Math. Soc.*, 2:464–466, 1951. Cited on p. 128.

[222] R. Steinberg. The state of the three color problem. In J. Gimbel, J. W. Kennedy, and L. V. Quintas, editors, *Quo Vadis, Graph Theory?*, volume 55 of *Ann. Discrete Math.*, pages 211–248. North-Holland, Amsterdam, The Netherlands, 1993. Cited on pp. 162 and 184.

[223] P. K. Stockmeyer. The falsity of the reconstruction conjecture for tournaments. *J. Graph Theory*, 1(1):19–25, 1977. Cited on p. 302.

[224] P. K. Stockmeyer. Erratum to: "The falsity of the reconstruction conjecture for tournaments". *J. Graph Theory*, 62(2):199–200, 2009. Cited on p. 302.

[225] G. Szekeres. Polyhedral decompositions of cubic graphs. *Bull. Austral. Math. Soc.*, 8:367–387, 1973. Cited on pp. 57 and 165.

[226] G. Szekeres and H. S. Wilf. An inequality for the chromatic number of a graph. *J. Combin. Theory*, 4(1):1–3, 1968. Cited on p. 142.

[227] T. Szele. Kombinatorische Untersuchungen über den gerichteten vollständigen Graphen (Combinatorial studies of the directed complete graph). *Math. Fiz. Lapok*, 50:223–256, 1943. Cited on p. 66.

# T

[228] P. G. Tait. On the colouring of maps. *Proc. R. Soc. Edinburgh*, 10:501–503, 1880. Cited on p. 165.

[229] The On-line Encyclopedia of Integer Sequences (OEIS). Published electronically at https://oeis.org/. Cited on p. 40.

[230] R. Thomas. Recent excluded minor theorems for graphs. In J. D. Lamb and D. A. Preece, editors, *Surveys in Combinatorics, 1999*, volume 267 of *London Math. Soc. Lecture Note Ser.*, pages 201–222. Cambridge University Press, Cambridge, England, 1999. Cited on p. 166.

[231] C. Thomassen. Landau's characterization of tournament score sequences. In G. T. Chartrand, editor, *The Theory and Applications of Graphs*, pages 589–591. Wiley, New York, New York, 1981. Cited on p. 46.

[232] C. Thomassen. The graph genus problem is NP-complete. *J. Algorithms*, 10(4):568–576, 1989. Cited on p. 256.

[233] S. Toida. Properties of a Euler graph. *J. Franklin Inst.*, 295:343–345, 1973. Cited on p. 57.

[234] P. Turán. Eine Extremalaufgabe aus der Graphentheorie (An extremal problem in graph theory). *Math. Fiz. Lapok*, 48:436–452, 1941. Cited on p. 222.

[235] P. Turán. A note of welcome. *J. Graph Theory*, 1(1):7–9, 1977. Cited on p. 126.

[236] W. T. Tutte. On Hamiltonian circuits. *J. London Math. Soc.*, 21:98–101, 1946. Cited on p. 121.

[237] W. T. Tutte. A family of cubical graphs. *Proc. Cambridge Philos. Soc.*, 43:459–474, 1947. Cited on p. 242.

[238] W. T. Tutte. A contribution to the theory of chromatic polynomials. *Canad. J. Math.*, 6:80–91, 1954. Cited on p. 183.

[239] W. T. Tutte. A short proof of the factor theorem for finite graphs. *Canadian J. Math.*, 6:347–352, 1954. Cited on p. 193.

[240] W. T. Tutte. A theorem on planar graphs. *Trans. Amer. Math. Soc.*, 82:99–116, 1956. Cited on p. 121.

[241] W. T. Tutte. On the algebraic theory of graph colorings. *J. Combin. Theory*, 1(1):15–50, 1966. Cited on pp. 165 and 184.

# U

[242] S. M. Ulam. *A Collection of Mathematical Problems*, volume 8 of *Interscience Tracts in Pure and Applied Mathematics*. Interscience Publishers, New York, New York, 1960. Cited on p. 301.

# V

[243] O. Veblen. An application of modular equations in analysis situs. *Ann. of Math. (2)*, 14(1-4):86–94, 1912/13. Cited on p. 56.

[244] V. G. Vizing. On an estimate of the chromatic class of a $p$-graph. *Diskret. Analiz*, 3:25–30, 1964. Cited on p. 145.

[245] V. G. Vizing. Some unsolved problems in graph theory. *Uspehi Mat. Nauk*, 23(6):117–134, 1968. Cited on p. 202.

[246] W. F. A. von Dyck. Beiträge zur Analysis situs (Contributions to topology). *Math. Ann.*, 32(4):457–512, 1888. Cited on p. 266.

[247] S. Říha. A new proof of the theorem by Fleischner. *J. Combin. Theory Ser. B*, 52(1):117–123, 1991. Cited on p. 73.

# W

[248] K. Wagner. Bemerkungen zum Vierfarbenproblem (Comments on the four-color problem). *Jahresber. Dtsch. Math.-Ver.*, 46:26–32, 1936. Cited on p. 128.

[249] K. Wagner. Über eine Eigenschaft der ebenen Komplexe (About a property of the flat complexes). *Math. Ann.*, 114(1):570–590, 1937. Cited on pp. 117 and 159.

[250] K. Walker. Dichromatic graphs and Ramsey numbers. *J. Combin. Theory*, 5:238–243, 1968. Cited on p. 236.

[251] K. Walker. An upper bound for the Ramsey number $M(5,4)$. *J. Combin. Theory Ser. A*, 11:1–10, 1971. Cited on p. 236.

[252] D. J. A. Welsh and M. B. Powell. An upper bound for the chromatic number of a graph and its application to timetabling problems. *Comput. J.*, 10(1):85–86, 1967. Cited on p. 141.

[253] P. A. L. Wernicke. Über den kartographischen Vierfarbensatz (On the four-color map theorem). *Math. Ann.*, 58(3):413–426, 1904. Cited on p. 110.

[254] A. T. White. *Graphs of Groups on Surfaces*, volume 188 of *North-Holland Mathematics Studies*. North-Holland, Amsterdam, The Netherlands, 2001. Cited on p. 127.

[255] H. Whitney. Congruent graphs and the connectivity of graphs. *Amer. J. Math.*, 54(1):150–168, 1932. Cited on pp. 87, 89, and 95.

[256] J. E. Williamson. Panconnected graphs. II. *Period. Math. Hungar.*, 8(2):105–116, 1977. Cited on p. 70.

[257] J. G. Wilson. New light on the origin of the four-color conjecture. *Historia Math.*, 3:329–330, 1976. Cited on p. 156.

[258] D. R. Woodall. Sufficient conditions for circuits in graphs. *Proc. London Math. Soc. (3)*, 24:739–755, 1972. Cited on p. 65.

[259] D. R. Woodall. Cyclic-order graphs and Zarankiewicz's crossing-number conjecture. *J. Graph Theory*, 17(6):657–671, 1993. Cited on p. 126.

# X

[260] N. H. Xuong. How to determine the maximum genus of a graph. *J. Combin. Theory Ser. B*, 26(2):217–225, 1979. Cited on p. 275.

# Y

[261] J. W. T. Youngs. Minimal imbeddings and the genus of a graph. *J. Math. Mech.*, 12:303–315, 1963. Cited on pp. 255 and 266.

# Z

[262] M. Zając. A short proof of Brooks' theorem. arXiv:1805.11176 [math.CO]. Cited on p. 143.

[263] J. Zaks. The maximum genus of Cartesian products of graphs. *Canadian J. Math.*, 26:1025–1035, 1974. Cited on p. 276.

[264] K. Zarankiewicz. On a problem of P. Turán concerning graphs. *Fund. Math.*, 41:137–145, 1954. Cited on p. 126.

[265] A. A. Zykov. On some properties of linear complexes. *Mat. Sbornik N.S.*, 24(66)(2):163–188, 1949. Cited on p. 153.

# Index

Printed in the United States
by Baker & Taylor Publisher Services